T0305775

Statistical Physics of Biomolecules

AN INTRODUCTION

Statistical Physics of Biomolecules

AN INTRODUCTION

Daniel M. Zuckerman

CRC Press
Taylor & Francis Group
Boca Raton London New York

CRC Press is an imprint of the
Taylor & Francis Group, an **informa** business

CRC Press
Taylor & Francis Group
6000 Broken Sound Parkway NW, Suite 300
Boca Raton, FL 33487-2742

© 2010 by Taylor and Francis Group, LLC
CRC Press is an imprint of Taylor & Francis Group, an Informa business

No claim to original U.S. Government works

ISBN 13: 978-1-4200-7378-2 (hbk)

Library of Congress Cataloging-in-Publication Data

Zuckerman, Daniel M.
   Statistical physics of biomolecules : an introduction / Daniel M. Zuckerman.
      p. cm.
   Includes bibliographical references and index.
   ISBN 978-1-4200-7378-2 (alk. paper)
      1. Biophysics. 2. Statistical physics. 3. Biomolecules. I. Title.

QH505.Z83 2010
572--dc22   2009050600

Visit the Taylor & Francis Web site at
http://www.taylorandfrancis.com

and the CRC Press Web site at
http://www.crcpress.com

For my parents,
who let me think for myself.

Contents

Preface

The central goal of this book is to answer "Yes" to the question, "Is there statistical mechanics for the rest of us?" I believe the essentials of statistical physics can be made comprehensible to the new generation of interdisciplinary students of biomolecular behavior. In other words, most of us can understand most of what's important. This "less is more" approach is not an invitation to sloppiness, however. The laws of physics and chemistry do matter, and we should know them well. The goal of this book it to explain, in plain English, the classical statistical mechanics and physical chemistry underlying biomolecular phenomena.

The book is aimed at students with only an indirect background in biophysics. Some undergraduate physics, chemistry, and calculus should be sufficient. Nevertheless, I believe more advanced students can benefit from some of the less traditional, and hopefully more intuitive, presentations of conventional topics.

The heart of the book is the statistical meaning of equilibrium and how it results from dynamical processes. Particular attention is paid to the way averaging of statistical ensembles leads to the free energy and entropy descriptions that are a stumbling block to many students. The book, by choice, is far from comprehensive in its coverage of either statistical mechanics or molecular biophysics. The focus is on the main lines of thought, along with key examples. However, the book does attempt to show how basic statistical ideas are applied in a variety of seemingly complex biophysical "applications" (e.g., allostery and binding)—in addition to showing how an ensemble view of dynamics fits naturally with more familiar topics, such as diffusion.

I have taught most of the first nine chapters of the book in about half a semester to first-year graduate students from a wide range of backgrounds. I have always felt rushed in doing so, however, and believe the book could be used for most of a semester's course. Such a course could be supplemented by material on computational and/or experimental methods. This book addresses simulation methodology only briefly, and is definitely not a "manual."

Acknowledgments

I am grateful to the following students and colleagues who directly offered comments on and corrections to the manuscript: Divesh Bhatt, Lillian Chong, Ying Ding, Steven Lettieri, Edward Lyman, Artem Mamonov, Lidio Meireles, Adrian Roitberg, Jonathan Sachs, Kui Shen, Robert Swendsen, Michael Thorpe, Marty Ytreberg, Bin Zhang, Xin Zhang, and David Zuckerman. Artem Mamonov graciously provided several of the molecular graphics figures, Divesh Bhatt provided radial distribution data and Bin Zhang provided transition path data. Others helped me obtain the understanding embodied herein (i.e., I would have had it wrong without them) including Carlos Camacho, Rob Coalson, David Jasnow, and David Wales. Of course, I have learned most of all from my own mentors: Robijn Bruinsma, Michael Fisher, and Thomas Woolf. Lance Wobus, my editor from Taylor & Francis, was always insightful and helpful. Ivet Bahar encouraged me throughout the project. I deeply regret if I have forgotten to acknowledge someone. The National Science Foundation provided support for this project through a Career award (overseen by Dr. Kamal Shukla), and much of my understanding of the field developed via research supported by the National Institutes of Health.

I would very much appreciate hearing about errors and ways to improve the book.

Daniel M. Zuckerman
Pittsburgh, Pennsylvania

1 Proteins Don't Know Biology

1.1 PROLOGUE: STATISTICAL PHYSICS OF CANDY, DIRT, AND BIOLOGY

By the time you finish this book, hopefully you will look at the world around you in a new way. Beyond biomolecules, you will see that statistical phenomena are at work almost everywhere. Plus, you will be able to wield some impressive jargon and equations.

1.1.1 CANDY

Have you ever eaten trail mix? A classic variety is simply a mix of peanuts, dried fruit, and chocolate candies. If you eat the stuff, you'll notice that you usually get a bit of each ingredient in every handful. That is, unsurprisingly, trail mix tends to be well mixed. No advanced technology is required to achieve this. All you have to do is shake.

To understand what's going on, let's follow a classic physics strategy. We'll simplify to the essence of the problem—the candy. I'm thinking of my favorite discoidally shaped chocolate candy, but you are free to imagine your own. To adopt another physics strategy, we'll perform a thought experiment. Imagine filling a clear plastic bag with two different colors of candies: first blue, then red, creating two layers. Then, we'll imagine holding the bag upright and shaking it (yes, we've sealed it) repeatedly. See Figure 1.1.

What happens? Clearly the two colors will mix, and after a short time, we'll have a fairly uniform mixture of red and blue candies.

If we continue shaking, not much happens—the well-mixed "state" is stable or "equilibrated." But how do the red candies know to move down and the blue to move up? And if the two colors are really moving in different directions, why don't they switch places after a long time?

Well, candy clearly doesn't think about what it is doing. The pieces can only move randomly in response to our shaking. Yet somehow, blind, random (nondirected) motion leads to a net flow of red candy in one direction and blue in the other. This is nothing other than the power of diffusion, which biomolecules also "use" to accomplish the needs of living cells. Biomolecules, such as proteins, are just as dumb as candy—yet they do what they need to do and get where they need to go. Candy mixing is just a simple example of a random process, which must be described statistically like many biomolecular processes.

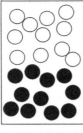 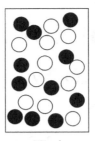

Unmixed   Mixed

FIGURE 1.1 Diffusion at work. If a bag containing black and white candies is shaken, then the two colors get mixed, of course. But it is important to realize the candies don't know where to go in advance. They only move randomly. Further, once mixed, they are very unlikely to unmix spontaneously.

PROBLEM 1.1

Consider the same two-color candy experiment performed twice, each time with a different type of candy: once with smooth, unwrapped candy of two colors, and a second time using candy wrapped with wrinkly plastic. How will the results differ between wrapped and unwrapped candy? Hint: Consider what will happen based on a small number of shakes.

1.1.2 CLEAN YOUR HOUSE, STATISTICALLY

One of the great things about statistical physics is that it is already a part of your life. You just need to open your eyes to it.

Think about dirt. In particular, think about dirt on the floor of your house or apartment—the sandy kind that might stick to your socks a bit. If you put on clean socks and walk around a very dirty room, your socks will absorb some of that dirt—up to the point that they get "saturated." (Not a pleasant image, but conceptually useful!) With every additional step in the dirty room, some dirt may come on to your socks, but an approximately equal amount will come off. This is a kind of equilibrium.

Now walk into the hallway, which has just been swept by your hyper-neat housemate. Your filthy socks are now going to make that hallway dirty—a little. Of course, if you're rude enough to walk back and forth many times between clean and dirty areas, you will help that dirt steadily "diffuse" around the house. You can accelerate this process by hosting a party, and all your friends will transport dirt all over the house (Figure 1.2).

On the other hand, your clean housemate might use the party to his advantage. Assuming no dirt is brought into the house during the party (questionable, of course, but pedagogically useful), your housemate can clean the house without leaving his room! His strategy will be simple—to constantly sweep his own room, while allowing people in and out all the time. People will tend to bring dirt in to the clean room from the rest of the house, but they will leave with cleaner feet, having shed some dirt and picking up little in return. As the party goes on, more and more dirt will come

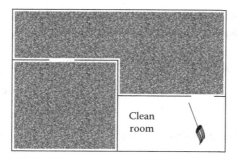

Clean
room

FIGURE 1.2 Clean your house statistically. A sweeper can stay in his room and let the dirt come to him, on the feet of his guests. If he keeps sweeping, dirt from the rest of the house can be removed. Similarly, proteins that only occupy a tiny fraction of the volume of a cell or beaker can "suck up" ligands that randomly reach them by diffusion.

into the clean room from people circulating all over the house, but there will be no compensating outflow of dirt. (Practical tip: very young guests can accelerate this process substantially, if they can be kept inside.)

PROBLEM 1.2

Consider a house with two rooms, one clean and one dirty. Imagine a number of people walk around randomly in this house, without bringing in more dirt. (a) Qualitatively, explain the factors that will govern the rate at which dirt travels from the clean to the dirty room—if no sweeping or removal of dirt occurs. (b) How does sweeping affect the process? (c) What if newly arriving guests bring in dirt from outside?

1.1.3 MORE SERIOUSLY...

In fact, the two preceding examples were very serious. If you understood them well, you can understand statistical biophysics. The candy and dirt examples illustrate fundamental issues of equilibrium, and the dynamical approach to equilibrium.

Everywhere in nature, dynamical processes occur constantly. In fact, it is fair to say that the execution of dynamics governed by forces is nature's only work. For instance, molecules move in space, they fluctuate in shape, they bind to one another, and they unbind. If a system is somewhat isolated, like the rooms of a closed house, an equilibrium can occur. The equilibrium can be changed by changing the "external conditions" like the total amount of dirt in the house or the size of the house. A biomolecular system—whether in a test tube or a living cell—can similarly be changed in quite a number of ways: by altering the overall concentration(s), changing the temperature or pH, or covalently changing one of the molecules in the system.

We won't be able to understand molecular systems completely, but hopefully we can understand the underlying statistical physics and chemistry. The principles described in this book are hardly limited to biology (think of all the candy and dirt in the world!), but the book is geared to an audience with that focus.

1.2 GUIDING PRINCIPLES

To get your mind wiggling and jiggling appropriately, let's discuss some key ideas that will recur throughout the book.

1.2.1 PROTEINS DON'T KNOW BIOLOGY

Biology largely runs on the amazing tricks proteins can perform. But when it comes down to it, proteins are simply molecules obeying the laws of physics and chemistry. We can think of them as machines, but there's no ghost inside. Proteins are completely inanimate objects, whose primary role is to fluctuate in conformation (i.e., in shape or structure). Biology, via evolution, has indeed selected for highly useful structural fluctuations, such as binding, locomotion, and catalysis. However, to understand these highly evolved functions in a given molecule or set of molecules, it is very informative to consider their spontaneous "wigglings and jigglings," to paraphrase the physicist Richard Feynman. To put the idea a slightly different way: Biology at molecular lengthscales is chemistry and physics. Therefore, you can hope to understand the principles of molecular biophysics with a minimum of memorization and a maximum of clarity.

1.2.2 NATURE HAS NEVER HEARD OF EQUILIBRIUM

The fancy word "equilibrium" can mislead you into thinking that some part of biology, chemistry, or physics could be static. Far from it: essentially everything of scientific interest is constantly moving and fluctuating. Any equilibrium is only apparent, the result of a statistical balance of multiple motions (e.g., candy shaken upward vs. down). So an alternative formulation of this principle is "Nature can't do statistical calculations." Rather, there tend to be enough "realizations" of any process that we can average over opposing tendencies to simplify our understanding.

Like proteins, nature is dumb. Nature does not have access to calculators or computers, so it does not know about probabilities, averages, or standard deviations. Nature only understands forces: push on something and it moves. Thus, this book will frequently remind the reader that even when we are using the comfortable ideas of equilibrium, we are actually talking about a balance among dynamical behaviors. It is almost always highly fruitful to visualize the dynamic processes underlying any equilibrium.

1.2.2.1 Mechanistic Promiscuity?

While we're on the subject, it's fair to say that nature holds to no abstract theories at all. Nature is not prejudiced for or against particular "mechanisms" that may inspire controversy among humans. Nature will exploit—indeed, cannot help but exploit—any mechanism that passes the evolutionary test of continuing life. Therefore, this book attempts to steer clear of theorizing that is not grounded in principles of statistical physics.

1.2.3 ENTROPY IS EASY

Entropy may rank as the worst-explained important idea in physics, chemistry, and biology. This book will go beyond the usual explanation of entropy as uncertainty and beyond unhelpful equations, to get to the root meaning in simple terms. We'll also see what the unhelpful equations mean, but the focus will be on the simplest (and, in fact, most correct) explanation of entropy. Along the way, we'll learn that understanding "free" energy is equally easy.

1.2.4 THREE IS THE MAGIC NUMBER FOR VISUALIZING DATA

We can only visualize data concretely in one, two, or three dimensions. Yet large biomolecules "live" in what we call "configuration spaces," which are very high dimensional—thousands of dimensions, literally. This is because, for example, if a protein has 10,000 atoms, we need 30,000 numbers to describe a single configuration (the x, y, and z values of every atom). The net result is that even really clever people are left to study these thousands of dimensions in a very partial way, and we'll be no different. However, we do need to become experts in simplifying our descriptions, and in understanding what information we lose as we do so.

1.2.5 EXPERIMENTS CANNOT BE SEPARATED FROM "THEORY"

The principles we will cover are not just of interest to theorists or computationalists. Rather, because they are actually true, the principles necessarily underpin phenomena explored in experiments. An auxiliary aim of this book, then, is to enable you to better understand many experiments. This connection will be explicit throughout the book.

1.3 ABOUT THIS BOOK

1.3.1 WHAT IS BIOMOLECULAR STATISTICAL PHYSICS?

In this book, we limit ourselves to molecular-level phenomena—that is, to the physics of biomacromolecules and their binding partners, which can be other large molecules or small molecules like many drugs. Thus, our interest will be primarily focused on life's molecular machines, proteins, and their ligands. We will be somewhat less interested in nucleic acids, except that these molecules—especially RNA—are widely recognized to function not only in genetic contexts but also as chemical machines, like proteins. We will not study DNA and RNA in their genetic roles.

This book hopes to impart a very solid understanding of the principles of molecular biophysics, which necessarily underlie both computer simulation and experiments. It is the author's belief that neither biophysical simulations nor experiments can be properly understood without a thorough grounding in the basic principles. To emphasize these connections, frequent reference will be made to common experimental methods and their results. However, this book will not be a manual for either simulations or experiments.

1.3.2 WHAT'S IN THIS BOOK, AND WHAT'S NOT

1.3.2.1 Statistical Mechanics Pertinent to Biomolecules

The basic content of this book is statistical mechanics, but not in the old-fashioned sense that led physics graduate students to label the field "sadistical mechanics." Much of what is taught in graduate-level statistical mechanics courses is simply unnecessary for biophysics. Critical phenomena and the Maxwell relations of thermodynamics are completely omitted. Phase transitions and the Ising model will be discussed only briefly. Rather, the student will build understanding by focusing on probability theory, low-dimensional models, and the simplest molecular systems. Basic dynamics and their relation to equilibrium will be a recurring theme. The connections to real molecules and measurable quantities will be emphasized throughout.

For example, the statistical mechanics of binding will be discussed thoroughly, from both kinetic and equilibrium perspectives. From the principles of binding, the meaning of pH and pK_a will be clear, and we will explore the genuinely statistical origins of the energy "stored" in molecules like ATP. We will also study the basics of allostery and protein folding.

1.3.2.2 Thermodynamics Based on Statistical Principles and Connected to Biology

Thermodynamics will be taught as a natural outcome of statistical mechanics, reinforcing the point of view that individual molecules drive all observed behavior. Alternative thermodynamic potentials/ensembles and nearly-impossible-to-understand relations among derivatives of thermodynamic potentials will only be mentioned briefly. Instead, the meaning of such derivatives in terms of the underlying probability-theoretic (partition-function-based) averages/fluctuations will be emphasized. Basic probability concepts will unify the whole discussion. One highlight is that students will learn when it is correct to say that free energy is minimized and when it is not (usually not for proteins).

1.3.2.3 Rigor and Clarity

A fully rigorous exposition will be given whenever possible: We cannot shy away from mathematics when discussing a technical field. However, every effort will be made to present the material as clearly as possible. We will attempt to avoid the physicist's typical preference for highly compact equations in cases where writing more will add clarity. The intuitive meaning of the equations will be emphasized.

1.3.2.4 This Is Not a Structural Biology Book

Although structural biology is a vast and fascinating subject, it is mainly just context for us. You should be sure to learn about structural biology separately, perhaps from one of the books listed at the end of the chapter. Here, structural biology will be described strictly on an as-needed basis.

1.3.3 BACKGROUND EXPECTED OF THE STUDENT

Students should simply have a grounding in freshman-level physics (mechanics and electrostatics), freshman-level chemistry, and single-variable calculus. The very basics of multi-variable calculus will be important—that is, what a multidimensional integral means. Any familiarity with biology, linear algebra, or differential equations will be helpful, but not necessary.

1.4 MOLECULAR PROLOGUE: A DAY IN THE LIFE OF BUTANE

Butane is an exemplary molecule, one that has something to teach all of us. In fact, if we can fully understand butane (and this will take most of the book!) we can understand most molecular behavior. How can this be? Although butane (*n*-butane to the chemists) is a very simple molecule and dominated by just one degree of freedom, it consists of 14 atoms (C_4H_{10}). Fourteen may not sound like a lot, but when you consider that it takes 42 numbers to describe a single butane configuration and its orientation in space (x, y, and z values for each atom), that starts to seem less simple. We can say that the "configuration space" of butane is 42-dimensional. Over time, a butane molecule's configuration can be described as a curve in this gigantic space.

Butane's configuration is most significantly affected by the value of the central torsion or dihedral angle, ϕ. (A dihedral angle depends on four atoms linked sequentially by covalent bonds and describes rotations about the central bond, as shown in Figure 1.3. More precisely, the dihedral is the angle between two planes—one formed by the first three atoms and the other by second three.) Figure 1.3 shows only the carbon atoms, which schematically represent the sequence of four chemical groups (CH_3, CH_2, CH_2, CH_3).

In panel (a) of Figure 1.3, we see a computer simulation trajectory for butane—specifically, a series of ϕ values at equally spaced increments in time. This is a day

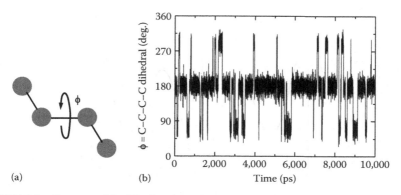

(a) (b) Time (ps)

FIGURE 1.3 Butane and its life. Panel (a) shows a simplified representation of the butane molecule, with only the carbon atoms depicted. Rotations about the central C–C bond (i.e., changes in the ϕ or C–C–C–C dihedral angle) are the primary motions of the molecule. The "trajectory" in panel (b) shows how the value of the dihedral changes in time during a molecular dynamics computer simulation of butane.

in the life of butane. Well, it's really just a tiny fraction of a second, but it tells the whole story. We can see that butane has three states—three essentially discrete ranges of φ values—that it prefers. Further, it tends to stay in one state for a seemingly random amount of time, and then make a rapid jump to another state. In the following chapters, we will study all this in detail: (1) the reason for the quasi-discrete states; (2) the jumps between states; (3) the connection between the jumps and the observed equilibrium populations; and, further, (4) the origin of the noise or fluctuations in the trajectory.

What would we conclude if we could run our computer simulation—essentially a movie made from many frames or snapshots—for a very long time? We could then calculate the average time interval spent in any state before jumping to other states, which is equivalent to knowing the rates for such isomerizations (structure changes). We could also calculate the equilibrium or average fractional occupations of each state. Such fractional occupations are of great importance in biophysics.

A simple, physically based way to think about the trajectory is in terms of an energy profile or landscape. For butane, the landscape has three energy minima or basins that represent the states (see Figure 1.4). As the famous Boltzmann factor ($e^{-E/k_B T}$, detailed later) tells us, probability decreases exponentially with higher energy E. Thus, it is relatively rare to find a molecule at the high points of the energy landscape. When such points are reached, a transition to another state can easily occur, and thus such points are called barriers. Putting all this together, the trajectory tends to fluctuate around in a minimum and occasionally jump to another state. Then it does the same again.

Another technical point is the diversity of timescales that butane exhibits. For instance, the rapid small-scale fluctuations in the trajectory are much faster than the transitions between states. Thinking of butane's structure, there are different types of motions that will occur on different timescales: the fast bond-length and bond-angle fluctuations, as opposed to the slow dihedral-angle fluctuations.

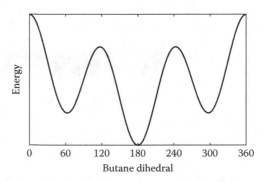

FIGURE 1.4 The "energy" landscape of butane. Butane spends most of its time in one of the three energy minima shown, as you can verify by examining the trajectory in Figure 1.3. Later, we will learn that a landscape like this actually represents a "free energy" because it reflects the relative population of each state—and even of each φ value.

1.4.1 EXEMPLARY BY ITS STUPIDITY

Details aside, butane is exemplary for students of biophysics in the sense that it's just as dumb as a protein—or rather, proteins are no smarter than butane. All any molecule can do is jump between the states that are determined by its structure (and by interactions with nearby molecules). Protein—or RNA or DNA—structures happen to be much more complicated and tuned to permit configurational fluctuations with important consequences that permit us to live!

One of the most famous structural changes occurs in the protein hemoglobin, which alters its shape due to binding oxygen. This "allosteric" conformational change facilitates the transfer of oxygen from the lungs to the body's tissues. We will discuss principles of allosteric conformational changes in Chapter 9.

1.5 WHAT DOES EQUILIBRIUM MEAN TO A PROTEIN?

Proteins don't know biology, and they don't know equilibrium either. A protein "knows" its current state—its molecular configuration, atomic velocities, and the forces being exerted on it by its surroundings. But a protein will hardly be affected by the trillions of other proteins that may co-inhabit a test tube with it. So how can a scientist studying a test tube full of protein talk about equilibrium?

There are two important kinds of equilibrium for us. One is an inter molecular equilibrium that can occur between different molecules, which may bind and unbind from one another. Within any individual molecule, there is also an internal equilibrium among different possible configurations. Both types of equilibriums are worth previewing in detail.

1.5.1 EQUILIBRIUM AMONG MOLECULES

Intermolecular equilibrium occurs among species that can bind with and unbind from one another. It also occurs with enzymes, proteins that catalyze chemical changes in their binding partners.

Imagine, as in Figure 1.5, a beaker filled with many identical copies of a protein and also with many ligand molecules that can bind the protein—and also unbind from it. Some fraction of the protein molecules will be bound to ligands, depending on the affinity or strength of the interaction. If this fraction does not change with time (as will usually be the case, after some initial transient period), we can say that it represents the equilibrium of the binding process. Indeed, the affinity is usually quantified in terms of an equilibrium constant, K_d, which describes the ratio of unbound-to-bound molecules and can be measured in experiments.

But how static is this equilibrium? Let's imagine watching a movie of an individual protein molecule. We would see it wiggling and jiggling around in its aqueous solution. At random intervals, a ligand would diffuse into view. Sometimes, the ligand might just diffuse away without binding, while at other times it will find the protein's binding site and form a complex. Once bound, a ligand will again unbind (after some random interval influenced by the binding affinity) and the process will repeat. So where is the equilibrium in our movie?

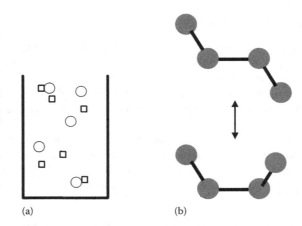

FIGURE 1.5 Two kinds of equilibrium. (a) A beaker contains two types of molecules that can bind to one another. In equilibrium, the rate of complex formation exactly matches that of unbinding. (b) A conformational equilibrium is shown between two (out of the three) states of butane. Transitions will occur back and forth between the configurations, as we have already seen in Figure 1.3.

The equilibrium only exists in a statistical sense, when we average over the behavior of many proteins and ligands. More fundamentally, when we consider all the molecules in the beaker, there will be a certain rate of binding, largely governed by the molecular concentrations—that is, how frequently the molecules diffuse near each other and also by the tendency to bind once protein and ligand are near. Balancing this will be the rate of unbinding (or the typical complex lifetime), which will be determined by the affinity—the strength of the molecular interactions. In equilibrium, the total number of complexes forming per second will equal the number unbinding per second.

Equilibrium, then, is a statistical balance of dynamical events. This is a key idea, a fundamental principle of molecular biophysics. We could write equations for it (and later we will), but the basic idea is given just as well in words.

1.5.2 INTERNAL EQUILIBRIUM

The same ideas can be applied to molecules that can adopt more than one geometric configuration. Butane is one of the simplest examples, and the equilibrium between two of its configurational states is schematized in Figure 1.5. But all large biomolecules, like proteins, DNA, and RNA, can adopt many configurations—as can many much smaller biomolecules, such as drugs.

Again, we can imagine making a movie of a single protein, watching it interconvert among many possible configurations—just as we saw for butane in Figure 1.3. This is what proteins do: they change configuration in response to the forces exerted on them. To be a bit more quantitative, we could imagine categorizing all configurations in our movie as belonging to one of a finite number of states, $i = 1, 2, \ldots$. If our movie was long enough so that each state was visited many times, we could even reliably estimate the average fraction of time spent in each state, p_i (i.e., the fractional population). Since every configuration belongs to a unique state, all the fractions would sum to one, of course.

In a cell, proteins will typically interact with many other proteins and molecules that are large when compared to water. The cell is often called a crowded environment. Nevertheless, one can still imagine studying a protein's fluctuations, perhaps by making a movie of its motions, however complex they may be.

1.5.3 TIME AND POPULATION AVERAGES

Here's an interesting point we will discuss again, later on. Dynamical and equilibrium measurements must agree on populations. To take a slightly different perspective, we can again imagine a large number of identical proteins in a solution. We can also imagine taking a picture ("snapshot") of this entire solution. Our snapshot would show what is called an ensemble of protein configurations. As in the dynamical case, we could categorize all the configurations in terms of fractional populations, now denoted as $\hat{p}_i$.

The two population estimates must agree, that is, we need to have $\hat{p}_i = p_i$. Proteins are dumb in the sense that they can only do what their chemical makeup and configuration allow them to do. Thus, a long movie of any individual protein will have identical statistical properties to a movie of any other chemically identical copy of that protein. A snapshot of a set of proteins will catch each one at a random point in its own movie, forcing the time and ensemble averages to agree.

PROBLEM 1.3

Apply this same argument to the ligand-binding case, showing a connection between a movie and the equilibrium fraction of bound proteins.

1.5.3.1 A Dynamical Description Has More Information Than an Equilibrium Picture

This is a simple but important point. Although dynamical measurements (i.e., from movies) must agree with equilibrium/ensemble measurements, ensemble measurements lack dynamical information. In particular, we cannot learn the rates of binding or of interconversion among configurations based on equilibrium measurements. To give a simple example, an equilibrium description of one person's sleep habits might indicate she sleeps 7/24 of the day on average. But we won't know whether that means 7 h on the trot, or 5 h at night and a 2 h afternoon nap.

1.6 A WORD ON EXPERIMENTS

Experiments can be performed under a variety of conditions that we can understand based on the preceding discussion. Perhaps the most basic class of experiments is the ensemble equilibrium measurement (as opposed to a single-molecule measurement or a nonequilibrium measurement). Examples of ensemble equilibrium measurements are typical NMR (nuclear magnetic resonance) and x-ray structure determination experiments. These are ensemble measurements since many molecules are involved. They are equilibrium measurements since, typically, the molecules have had a long time to respond to their external conditions. Of course, the conditions of an NMR experiment—aqueous solution—are quite different from x-ray crystallography where the protein is in crystalline form. Not surprisingly, larger motions and fluctuations are expected in solution, but some fluctuations are expected whenever the temperature is nonzero, even in a crystal. (Interestingly, protein crystals tend to be fairly wet: they contain a substantial fraction of water, so considerable fluctuations can be possible.) All this is not to say that scientists fully account for these fluctuations when they analyze data and publish protein structures, either based on x-ray or NMR measurements, but you should be aware that these motions must be reflected in the raw (pre-analysis) data of the experiments.

Nonequilibrium measurements refer to studies in which a system is suddenly perturbed, perhaps by a sudden light source, temperature jump, or addition of some chemical agent. As discussed above, if a large ensemble of molecules is in equilibrium, we can measure its average properties at any instant of time and always find the same situation. By a contrasting definition, in nonequilibrium conditions, such averages will change over time. Nonequilibrium experiments are thus the basis for measuring kinetic processes—that is, rates.

Recent technological advances now permit single-molecule experiments of various kinds. These measurements provide information intrinsically unavailable when ensembles of molecules are present. First imagine two probes connected by a tether consisting of many molecules. By tracking the positions of the probes, one can only measure the average force exerted by each molecule. But if only a single molecule were tethering the probes, one could see the full range of forces exerted (even over time) by that individual. Furthermore, single-molecule measurements can emulate ensemble measurements via repeated measurements.

Another interesting aspect of experimental measurements is the implicit time averaging that occurs; that is, physical instruments can only make measurements over finite time intervals and this necessarily involves averaging over any effects occurring during a brief window of time. Think of the shutter speed in a camera: if something is moving fast, it appears as a blur, which reflects the time-averaged image at each pixel.

To sum up, there are two key points about experiments. First, our "theoretical" principles are not just abstract but are 100% pertinent to biophysical behavior measured in experiments. Second, the principles are at the heart of interpreting measured data. Not surprisingly, these principles also apply to the analysis of computer simulation data.

1.7   MAKING MOVIES: BASIC MOLECULAR DYNAMICS SIMULATION

The idea of watching a movie of a molecular system is so fundamental to the purpose of this book that it is worthwhile to sketch the process of creating such a movie using computer simulation. While there are entire books devoted to molecular simulation of various kinds, we will present a largely qualitative picture of molecular dynamics (MD) simulation. Other basic ideas regarding molecular simulation will be described briefly in Chapter 12.

MD simulation is the most literal-minded of simulation techniques, and that is partly why it is the most popular. Quite simply, MD employs Newton's second law ($f = ma$) to follow the motion of every atom in a simulation system. Such a system typically includes solute molecules such as a protein and possibly a ligand of that protein, along with solvent like water and salt. (There are many possibilities, of course, and systems like lipid membranes are regularly simulated by MD.) In every case, the system is composed of atoms that feel forces. In MD, these forces are described classically—that is, without quantum mechanics, so the Schrödinger equation is not involved. Rather, there is a classical "forcefield" (potential energy function) U, which is a function of all coordinates. The force on any atomic coordinate, x_i, y_i, or z_i for atom i, is given by a partial derivative of the potential: for instance, the y component of force on atom i is given by $-\partial U/\partial y_i$.

MD simulation reenacts the simple life of atom: an atom will move at its current speed unless it experiences a force that will accelerate or decelerate it. To see how this works (and avoid annoying subscripts), we'll focus on the single coordinate x, with velocity $v = dx/dt$ and force $f = -dU/dx$. We can use Newton's law ($a = dv/dt = f/m$) to describe the approximate change in speed a short time Δt after the current time t. If we write $dv/dt \simeq \Delta v/\Delta t$, we find that $a\Delta t = v(t + \Delta t) - v(t)$. We can then write an equation for the way velocity changes with time due to a force:

$$v(t + \Delta t) = v(t) + a(t)\Delta t = v(t) + \left[\frac{f(t)}{m}\right]\Delta t. \qquad (1.1)$$

As in freshman physics, we can also integrate this equation under the assumption that speed and acceleration are constant over the small time interval Δt. The integration yields an equation for the position at time $t + \Delta t$, and should also look familiar:

$$x(t + \Delta t) = x(t) + v(t)\Delta t + \left[\frac{f(t)}{2m}\right](\Delta t)^2, \qquad (1.2)$$

where $f(t)/m = a(t)$.

It is straightforward to read such an equation: the x coordinate at time $t + \Delta t$ is the old coordinate plus the change due to velocity, as well as any additional change if that velocity is changing due to a force.

To implement MD on a computer, one needs to calculate forces and to keep track of positions and velocities. When this is done for all atoms, a trajectory (like that shown in Figure 1.3 for butane) can be created. More schematically, take a look at Figure 1.6. To generate a single new trajectory/movie-frame for butane without any solvent, one needs to repeat the calculation of Equation 1.2 42 times—once for each of the x, y, and z components of all 14 atoms! Sounds tedious (it is), but this is a perfect task for a computer. Any time a movie or trajectory of a molecule is discussed, you should imagine this process.

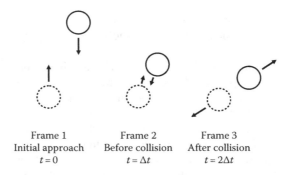

Frame 1   Frame 2   Frame 3
Initial approach   Before collision   After collision
$t = 0$ $t = \Delta t$ $t = 2\Delta t$

FIGURE 1.6 Making movies by molecular dynamics. Two colliding atoms are shown, along with arrows representing their velocities. The atoms exert repulsive forces on one another at close range, but otherwise are attractive. Note that the velocities as shown are not sufficient to determine future positions of the atoms. The forces are also required.

PROBLEM 1.4

Make your own sketch or copy of Figure 1.6. Given the positions and velocities as shown, sketch the directions of the forces acting on each atom at times $t = 0$ and $t = \Delta t$. Can you also figure out the forces at time $t = 2\Delta t$?

1.8   BASIC PROTEIN GEOMETRY

While there's no doubt that proteins are dumb, we have to admit that these polypeptides, these heteropolymers of covalently linked amino acids (or "residues"), are special in several ways. The main noteworthy aspects involve protein structure or geometry, and every student of molecular biophysics must know the basics of protein structure. There are a number of excellent, detailed books on protein structure (e.g., by Branden and Tooze), and everyone should own one of these.

1.8.1   PROTEINS FOLD

Of prime importance here are some basic structural ideas, and then some of the detailed chemical geometry. The most essential idea is that proteins fold or tend to adopt fairly well-defined structures (see Figure 1.7). Thus, if you encounter two chemically identical copies of the same protein, they should look basically the same. But don't forget that proteins cannot be frozen into single configurations, or else they would not be able to perform their functions. You should imagine that, depending on the conditions, a protein will inhabit a set of states permitted by its sequence and basic fold. If conditions change, perhaps when a ligand arrives or departs, the protein's dominant configurations can change dramatically.

1.8.2   THERE IS A HIERARCHY WITHIN PROTEIN STRUCTURE

The hierarchical nature of protein structure is also quite fundamental. The sequence of a protein, the ordered list of amino acids, is sometimes called the primary structure

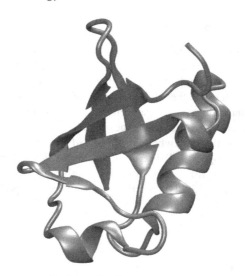

FIGURE 1.7 An experimentally determined structure of the protein ubiquitin. Although the coordinates of every atom have been determined, those are usually omitted from graphical representations. Instead, the "secondary structures" are shown in ribbon form, and other "backbone" atoms are represented by a wire shape.

(for reasons that will soon become clear). On a slightly larger scale, examining groups of amino acids, one tends to see two basic types of structure—the alpha helix and beta sheet, as shown schematically in the ubiquitin structure in Figure 1.7. (The ribbon diagram shown for ubiquitin does not indicate individual amino acids, let alone individual atoms, but is ideal for seeing the basic fold of a structure.) The helix and the sheet are called secondary structures, and result from hydrogen bonds between positive and negative charges. In structural biology, every amino acid in a protein structure is usually categorized according to its secondary structure—as helix, sheet or "loop," with the latter indicating neither helix nor sheet. Moving up in scale, these secondary structures combine to form domains of proteins: again see the ubiquitin structure above. This is the level of tertiary structure. Many small proteins, such as ubiquitin, contain only one domain. Others contain several domains within the same polypeptide chain. Still others are stable combinations of several distinct polypeptide chains, which are described as possessing quaternary structure. Multidomain and multichain proteins are quite common and should not be thought of as rarities.

1.8.3 THE PROTEIN GEOMETRY WE NEED TO KNOW, FOR NOW

The preceding paragraph gives the briefest of introductions to the complexity (and even beauty) of protein structures. However, to appreciate the physics and chemistry principles to be discussed in this book, a slightly more precise geometric description will be useful. Although our description will be inexact, it will be a starting point for quantifying some key biophysical phenomena.

1.8.4 THE AMINO ACID

Amino acids are molecules that covalently bond together to form polypeptides and proteins. With only one exception, amino acids have identical backbones—the chemical groups that bond to one another and form a chain. However, from one of the backbone atoms (the alpha carbon) of every amino acid, the chemically distinct side chains branch off (see Figure 1.8). The specific chemical groups forming the side-chains give the amino acids their individual properties, such as electric charge or lack thereof, which in turn lead to the protein's ability to fold into a fairly specific shape. If all the amino acids in the polypeptide chain were identical, proteins would differ only in chain length and they would be unable to fold into the thousands of highly specific shapes needed to perform their specific functions.

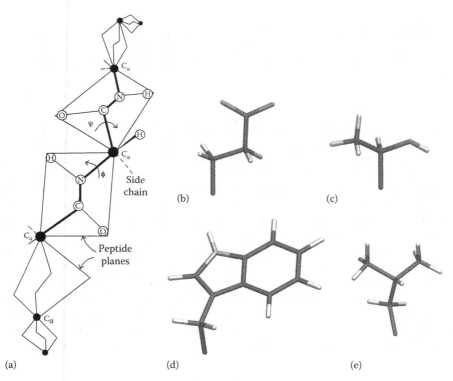

(a)   (d)   (e)

FIGURE 1.8 The protein backbone and several amino acid side chains. (a) The backbone consists of semirigid peptide planes, which are joined at "alpha carbons" (C_α). Each type of amino acid has a distinct side chain, which branches from the alpha carbon. Several examples are shown in (b)–(e). The figure also suggests some of the protein's many degrees of freedom. Rotations about the N–C_α and C_α–C bonds in the backbone (i.e., changes in the ϕ and ψ dihedral angles) are the primary movements possible in the protein backbone. Side chains have additional rotatable angles.

1.8.5 THE PEPTIDE PLANE

Although the biochemical synthesis of proteins relies on amino acids as the basic units, a clearer picture of protein structure emerges when one considers the peptide plane. The peptide plane is a fairly planar group of atoms, which extend from one alpha-carbon to the next, in the protein backbone. Figure 1.8 shows a number of peptide planes, joined at their corners where the alpha carbons are located. Although not rigidly planar, the planes are fairly stiff because of special electron sharing (pi-bonding) among the atoms of the plane. Hence, for the kind of rough sketch of protein structure we want, it is reasonable to assume the C–N dihedral angle in the middle of the plane is fixed in a flat conformation. The only remaining degrees of freedom are two dihedral angles per residue, named ϕ and ψ ("phi" and "psi"). As shown in Figure 1.8, these angles describe rotations about the N–C and C–C bonds.

1.8.6 THE TWO MAIN DIHEDRAL ANGLES ARE NOT INDEPENDENT

If life were simple, each of the ϕ and ψ dihedrals would be able to take on any value, independent of the other angle. Yet the Ramachandran plane of ϕ and ψ coordinates, sketched in Figure 1.9, shows that the shaded ranges of allowed ψ values depend on the particular value of ϕ. To see this, compare two horizontal lines at different ϕ values: for some lines, there are no allowed ψ values, while at others there might be one or two different ranges. (Ramachandran is the scientist who first explained the interdependence, or correlation, among these two key structural coordinates.) This correlation, as in many molecular cases, stems primarily from "steric" hindrances— that is, simple excluded-volume or van der Waals clashes among the atoms—that prevent the simultaneous occupation of certain ϕ and ψ values. It's worthwhile to build or view a molecular model to get a feel for this. In any case, understanding the mathematical consequences of such correlations will be a critical part of this book.

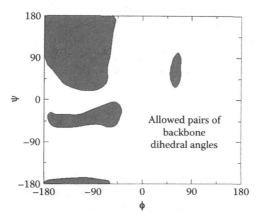

FIGURE 1.9 Ramachandran correlations. Only (ϕ, ψ) pairs in the shaded areas are possible. Other values are prevented by steric clashes among backbone atoms. Thus, the two angles ϕ and ψ are correlated: a given value of one angle determines the possible range(s) of the other.

1.8.7 CORRELATIONS REDUCE CONFIGURATION SPACE, BUT NOT ENOUGH TO MAKE CALCULATIONS EASY

Correlation in the Ramachandran plane is an example of a reduction of the allowed configuration space due to molecular geometry. In other words, there are many fewer configurations than one might guess based on the number of degrees of freedom. For instance, considering one-degree increments, one might have expected 360×360 possible (ϕ, ψ) pairs, but in fact, many fewer are allowed. Yet such correlations make the mathematical analysis of proteins complicated: to a good approximation, everything depends on everything else.

But surely, you might say, computers are perfect for dealing with such complexities! Unfortunately, many protein calculations are beyond the reach of any computer. Consider even a highly simplified scenario that assumes there are only five permitted Ramachandran states per alpha carbon (per amino acid). Since proteins can contain hundreds of residues, this implies an impossibly huge number of configurations. For instance, with 100 residues, we have 5^{100}–10^{70} configurations. Even a computer that could generate a trillion configurations per second (and there is no such computer) would require more than 10^{50} years to enumerate all such possibilities! (You can compare 10^{50} years to the estimated age of the universe.) Further, the five-state model is itself is a gross simplification, ignoring peptide-plane fluctuations and side-chain configurations, each of which can have a dramatic impact on the structure. The bottom line here is that it is impossible to enumerate the configurations of even a small protein! Proteins may be dumb, but they're complicated!

1.8.8 ANOTHER EXEMPLARY MOLECULE: ALANINE DIPEPTIDE

Figure 1.10 shows another molecule we will turn to frequently to illustrate important concepts in the statistical mechanics of molecules. This is the alanine "dipeptide" molecule, which does not consist of two amino acids, but of a single one (alanine) extended to have two full peptide planes on either side of the side chain. Because it has full peptide planes mimicking those of a protein, it can be said to have standard dihedral angles. These two dihedrals—ϕ and ψ—exhibit some of the behavior we expect in real proteins. More importantly, alanine dipeptide serves as a great model of a molecule with two "important" degrees of freedom. The two dihedral angles may be considered nontrivial in that the value of either one is affected by the other, as we will study in some detail below.

1.9 A NOTE ON THE CHAPTERS

Although every chapter of the book discusses fundamental theory, some chapters are more focused on applications. The heart of the statistical and thermodynamic description of molecules is covered in Chapters 2 through 7, which build from one- and two-dimensional examples to complex systems. The underlying dynamical basis of statistical behavior is emphasized with a separate introductory-level chapter (Chapter 4).

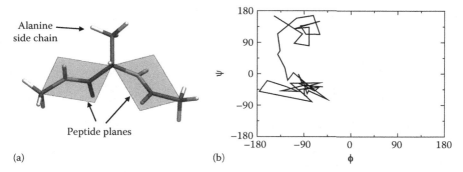

(a)   (b)   φ

FIGURE 1.10 (a) The alanine "dipeptide" molecule. Besides butane, this is the other molecule we will consider most frequently to understand the statistical mechanics of molecules. Alanine dipeptide is constructed to have two full peptide planes, and hence φ and ψ dihedrals just as in a protein (compare with Figure 1.8). These two dihedrals' angles make the molecule intrinsically more complex and interesting than butane. (b) We show a short trajectory from a molecular dynamics simulation of alanine dipeptide: the line traces out the time-ordered sequence of (φ, ψ) values from configurations in the trajectory.

Even a basic understanding of molecular biophysics would be incomplete without Chapter 8 on water and Chapter 9 on binding. More advanced material begins in Chapter 10 with a somewhat detailed discussion of kinetics, along with the essentials of its application to allostery and protein folding. Chapter 11 applies ensemble/distribution ideas to the description of dynamics. Finally, Chapter 12 attempts to sketch key ideas of statistically sound molecular simulation.

FURTHER READING

Branden, C. and Tooze, J., *Introduction to Protein Structure,* 2nd edition, Garland Science, New York, 1999.

Cantor, C.R. and Schimmel, P.R., *Biophysical Chemistry, Part I,* W.H. Freeman, New York, 1980.

Phillips, R., Kondev, J., and Theriot, J., *Physical Biology of the Cell,* Garland Science, New York, 2009.

Rhodes, G., *Crystallography Made Crystal Clear,* 3rd edition, Academic Press, Amsterdam, the Netherlands, 2006.

Rule, G.S. and Hitchens, T.K., *Fundamentals of Protein NMR Spectroscopy,* Springer, Berlin, Germany, 2005.

2 The Heart of It All: Probability Theory

2.1 INTRODUCTION

Probability theory is at the heart of so much of science. You will use it every day, either explicitly or tacitly, and whether you perform experiments or calculations. (The end of this chapter highlights a famous experimental paper published in *Science*, which makes fundamental use of statistical theory.) But beyond being practically useful, probability theory is a way of understanding nearly the whole fields of biophysics and computational biology. It is a fundamental theoretical foundation and perspective, not just a collection of mathematical techniques.

Most directly, probability theory is the essence of statistical mechanics. In fact, I would argue that there is nothing more to statistical mechanics than a solid understanding of probability theory. How can this be, you might say, since statistical mechanics speaks of mysterious quantities like entropy, specific heat, and isobaric ensembles? Well, pay attention, and I hope you will see how.

There are only a few critical ideas one needs to understand in probability theory, but unfortunately, these ideas emerge disguised and transformed in many ways. It can be a real challenge to grasp the diversity of statistical phenomena encountered by the student and researcher. The goal of this chapter is to provide a strong understanding of the basics with sufficient illustrations so that the student can begin to say, "Oh, this is just like situation XYZ, which I've seen before."

2.1.1 THE MONTY HALL PROBLEM

Before we start, it's worth mentioning a nontrivial math problem you can talk about at a party. Plus, understanding the problem's solution will highlight the fundamental meaning of probability.

Monty Hall was the host of a TV show, called *Let's Make a Deal*. I actually watched this show as a child, on days when I was so unfortunate as to stay home from school due to illness. The show featured a game in which a contestant was shown three identical boxes (or curtains or...) to choose from. Inside one was a fabulous prize, like an all-expenses paid vacation to a beautiful beach, while the others contained prizes like inexpensive steak knives. To play the game, the contestant would choose a box, but it was not immediately revealed whether the vacation was selected. Instead, the game's second phase would begin. Monty Hall would indicate one of the two remaining boxes and reveal that it contained steak knives. Of course, at least one of the remaining two had to have steak knives, but the contestant was now told which one, specifically.

At this point, only two of the boxes could contain the vacation prize, and the contestant was given a second choice. She could switch to the other box if she wanted. Thus the question for students of probability theory: Should she switch?

How can one even think about this problem? Well, the real question is whether the contestant has a greater likelihood to win by switching. And what does this mean? Really, there's only one meaning: you have to imagine performing the identical game/procedure/experiment a large number of times, and determining the fractions of the various possible outcomes. This is the fundamental meaning of probability. You always need to imagine a million games, a million molecules, a million monkeys— whatever you are trying to understand in a probabilistic way.

PROBLEM 2.1

Imagine playing the Monty Hall game a million times. If you always make the final switch at the end, what fraction of the time do you expect to win? Explain your answer by describing the expected outcomes before and after the (possible) switch. Start your analysis from the first phase, with the original three choices.

2.2   BASICS OF ONE-DIMENSIONAL DISTRIBUTIONS

The Monty Hall problem is understood by the probabilities for three (and later two) choices. The probability is distributed among the choices according to what is appropriately termed a "distribution." We need to understand distributions very thoroughly.

2.2.1   WHAT IS A DISTRIBUTION?

The most important mathematical object—with which you must be completely comfortable—is a probability distribution, which mathematicians would call a pdf or probability density function. (Later we'll see the reason why the mathematical name is quite sensible.) A distribution simply gives the likelihood or frequency with which each of a set of events or situations can occur (Figure 2.1). Perhaps the simplest case is the uniform distribution for rolls of a fair die (i.e., one of two dice), for which the probability p is distributed according to

$$p(j) = \frac{1}{6}, \quad \text{for } j = 1, 2, \ldots, 6. \tag{2.1}$$

You should feel satisfied that all the probabilities sum to 1, as this indicates that 100% of all possibilities are accounted for. We say that a distribution is properly normalized when the sum of probabilities is 1.

PROBLEM 2.2

Normalization for a nonuniform distribution. Let's consider an unfair die, which has (literally) been weighted so that the roll of 4 is twice as likely any of the other 5 rolls. What is the probability of each roll, given that the total must sum to 1?

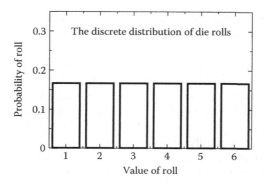

FIGURE 2.1 The discrete probability distribution for all possible rolls of a single fair die. Each roll has the same probability, 1/6.

2.2.1.1 Continuous Variables

More generally, we might be interested in a continuous variable, say x, which can take on real-number values—for instance, if x denotes the height of biophysics students. (Later we will study continuous molecular coordinates.) For now, let us imagine the distribution of student heights follows a well-known functional form, such as the Gaussian shown in Figure 2.2, with mean $\langle x \rangle$ and standard deviation σ (we'll discuss these quantities in detail below), so that

$$p_G(x) \propto \exp\left[-\frac{(x - \langle x \rangle)^2}{2\sigma^2}\right]. \tag{2.2}$$

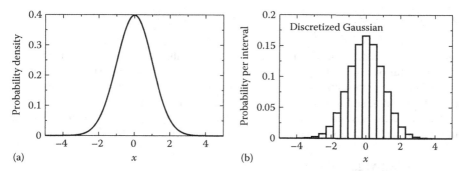

FIGURE 2.2 A Gaussian distribution, also called a normal distribution. We shall see later why Gaussians are indeed "normal" when we discuss the central limit theorem. The pictured distribution has a mean $\langle x \rangle$ of zero and standard deviation $\sigma = 1$. However, you should be able to imagine a distribution with the same shape but different values of the two parameters. The panel (a) shows a standard continuous representation of the Gaussian (i.e., a probability density function), while the panel (b) shows a discretized version. The discrete version can be converted to an approximate density if the probabilities are divided by the constant interval width of 0.04.

In this equation, the symbol "$\propto$" means "proportional to," so that unlike in Equation 2.1, we have not specified the absolute or total probability of a given height.

The idea of proportionality is important. To be mathematically careful, we can imagine there is a number—call it c, which is constant but unknown—multiplying every value on the right-hand side of Equation 2.2. Therefore, Equation 2.2 tells us that what we know exactly are relative probabilities: the probability of a student having average height $\langle x \rangle$ compared to (put in a ratio over) the probability of the height $\langle x \rangle + \sigma$ is $c/c \exp(-1/2) = \exp(+1/2)$. In such a ratio, the unknown constant cancels c out, and we get a fully specified answer.

This issue of relative, as opposed to absolute, probabilities may seem like an unimportant detail, but understanding the difference will later help you to understand concepts like "free energy changes."

The way to think of discrete distributions, such as Equation 2.1, and continuous ones on the same footing is to have in mind clearly the idea of normalization: the probability of all possible occurrences must be 1. To make this concrete, note that the integral of the right-hand side of Equation 2.2 is given by

$$\int_{-\infty}^{+\infty} dx \, \exp\left[-\frac{(x - \langle x \rangle)^2}{2\sigma^2}\right] = \sqrt{2\pi}\sigma. \qquad (2.3)$$

(Don't worry about deriving this result.) In turn, Equation 2.3 means that if we define the probability density

$$\rho_G(x) = \left(\sqrt{2\pi}\sigma\right)^{-1} \exp\left[-\frac{(x - \langle x \rangle)^2}{2\sigma^2}\right] \qquad (2.4)$$

then the integral of this function from minus to plus infinity is exactly 1. In other words, the function ρ_G is normalized.

2.2.1.2 Units Matter!

However, the function ρ_G still does not give absolute probabilities unless we are a bit careful. Indeed, if x has units of length, then ρ_G must have units of inverse length if its integral over all values is truly one. One, after all, is a dimensionless number! Therefore, ρ_G represents the probability per unit length (in the present context of student heights), and it is a density. Recalling the idea of summing rectangles from calculus, we can say that there is a total probability of approximately $\rho_G(x_0)\Delta x$ in an interval of width Δx about the value x_0. This point is illustrated in Figure 2.2.

PROBLEM 2.3

On the meaning of "small." In order for the expression $\rho_G(x_0)\Delta x$ to accurately yield the probability contained in the interval Δx, that interval must be small. By comparison to what must the interval be small? Equivalently, how should $\rho_G(x)$ behave in the interval? Answer without equations.

Another way to think about densities is to consider what they lack. In fact, it is meaningless to speak of the probability of single value of x when there are a continuum of possibilities: rather, you can either use the probability in an interval, or consider a ratio of probabilities of two specific x values.

In general, we shall use the Greek letter "rho" (ρ) to denote a probability density, as in Equation 2.4.

As a final note, perhaps you have noticed that it doesn't make sense to discuss a distribution of student heights—which are always positive—using a distribution that ranges from minus to plus infinity. This is indeed true, and suggests the next problem.

PROBLEM 2.4

Try to think of a sense in which the distribution of student heights (h values) could be approximately Gaussian. What size relation (large/small) should hold between $\langle h \rangle$ and σ?

2.2.2 MAKE SURE IT'S A DENSITY!

Because the issue of density is so important—both conceptually and mathematically—we'll discuss it a bit more. A density is always normalized, as we noted above, so its integral over all possible values is one. However, it's important to emphasize that this is not a restriction on possible distributions or densities. Indeed, almost any functional form can be turned into a density (as can any specification of relative probabilities).

If one starts from only a description of the relative probabilities, denoted $w(x)$, this can be transformed into a proper density. We already saw this with ρ_G above. As another example, we could set $w(x) = 1$ over the interval from 0 to 10. In this case, the corresponding pdf is $\rho(x) = 0.1$, which is properly normalized. More generally, to convert from a relative probability or weighting w to a pdf over the interval from a to b (which could be infinite), one simply divides by the total integral:

$$\rho(x) = \frac{w(x)}{\int_a^b dx\, w(x)} . \tag{2.5}$$

PROBLEM 2.5

Calculate the pdf for each of the following choices for the relative probability w: (a) $w(x) = 5$ in the interval from 0 to 10, (b) $w(x) = x^{-4}$ in the interval from 1 to ∞, and (c) $w(x) = e^{-5x}$ in the interval from 0 to ∞.

2.2.3 THERE MAY BE MORE THAN ONE PEAK: MULTIMODALITY

There is no rule that forces a probability distribution to have only one peak. Imagine the distribution of heights of all members of all families with a child in elementary school. Undoubtedly, this distribution would show one peak for the heights of the parents and another for the heights of the young children.

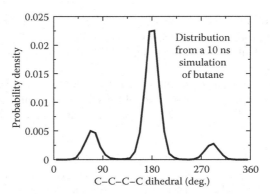

FIGURE 2.3   The distribution of butane's main dihedral. The plotted probability density was calculated from the 10 ns molecular dynamics simulation depicted in Figure 1.3. Although butane is a symmetric molecule, the distribution does not mirror itself about the central value of 180°, due to the finite length of the simulation.

In a scientific context, it is well known that many molecules possess more than one stable conformation, so a distribution of an appropriately chosen molecular coordinate could also be multimodal. Butane provides the perfect example. Look back at the trajectory shown in Figure 1.3, which clearly shows three states. Indeed, if we histogram that 10 ns trajectory, we can construct the three-peaked distribution of Figure 2.3.

2.2.4 CUMULATIVE DISTRIBUTION FUNCTIONS

Here is an example of a new mathematical object in probability theory, which contains absolutely no additional information compared to the distributions (densities) we just discussed. The cumulative distribution function (cdf) is downright redundant! The process of understanding it will, however, reinforce your grasp of distribution functions and also have some practical payoffs.

In words, the definition of the cdf is simple: the amount of probability contained in the pdf (density) starting from the lower limit until a specified point. Thus, the cdf is always less than or equal to 1. The mathematical notation can be a bit intimidating for the novice, but it simply puts in an equation exactly what we just said:

$$\mathrm{cdf}(x) = \int_{-\infty}^{x} dx' \mathrm{pdf}(x') = \int_{-\infty}^{x} dx' \rho(x'). \tag{2.6}$$

Having the function's argument appear in the upper limit of integration takes some getting used to, but it involves no advanced calculus. The definition is simply telling us that the density must be integrated from negative infinity up to the argument of the cdf—up to x in Equation 2.6. See Figure 2.4.

PROBLEM 2.6

Why does the cdf contain exactly the same information as a pdf?

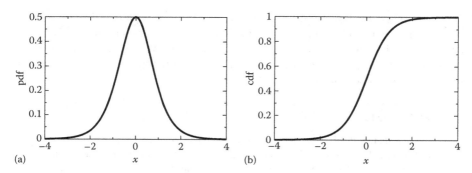

FIGURE 2.4   The cdf of panel (b) is a useful way of visualizing the pdf in (a). The cdf indicates how much probability occurs to the left of its argument in the original pdf. For a smooth distribution with a single peak, the cdf has a characteristic sigmoidal shape. By following particular y-axis values of the cdf down to their corresponding x values, we see that exactly 50% of the probability occurs to the left of zero, and that nearly 90% occurs to the left of 1 ($x < 1$). Correspondingly, nearly 80% of the probability lies between -1 and 1.

Cumulative distribution functions are extremely useful for describing data, and are probably underutilized in the biophysical literature. For instance, if one considers a distribution of measurements or simulation results, it's clear that those in the "tails" (extreme values) of the pdf typically are least likely. But these unlikely events may be important. Yet it's difficult to get a quantitative feel even for where the tails begin just by looking at a pdf. From the cdf, on the other hand, if we're interested only in the middle 90% of values, we can easily determine the upper and lower limits of such measurements by tracing lines across from cdf values of 0.05 and 0.95. Similarly, we can determine precisely where the most extreme 10% of the data is.

PROBLEM 2.7

What is the name of the cdf of a Gaussian? Is an exact functional form (i.e., an equation without an integral) known for it? Hint: The answer is not in this book!

PROBLEM 2.8

How would the cdf for a discrete distribution be constructed? Give a simple example.

2.2.4.1   Generating Nonuniform Random Numbers from a cdf

Thinking ahead to computational applications, the cdf is a powerful tool for generating random numbers distributed according to any distribution. You may already know that there are many computer algorithms available for generating (apparently) random numbers that are uniformly distributed in the interval from 0 to 1. Knowing an analytical or numerical representation of the cdf of the desired distribution permits a trivial conversion to this desired distribution. In visual terms, the method works by

(1) computing a random number between 0 and 1; (2) in a plot of the cdf such as Figure 2.4, drawing a horizontal line at the random value across to the cdf function; and (3) then drawing a vertical line down to the x axis. Values of x obtained from such a procedure (where the vertical lines hit the axis) will be distributed according to the distribution described by the cdf.

PROBLEM 2.9

Make a sketch of this procedure. For a unimodal distribution such as that shown in Figure 2.4, explain qualitatively why this approach for random-number generation works—that is, why it generates more values in the peak of the pdf and fewest in the tails.

PROBLEM 2.10

Sketch a "bimodal" distribution (a pdf with peaks at x_a and x_b), along with its cdf.

2.2.5 AVERAGES

We have already used the notation $\langle x \rangle$ to indicate the mean of the variable x, but we should make a proper definition. We also want to look at the connection to "sampling," since you may find yourself calculating an average on the computer. First, the formal mathematical approach: for a coordinate x distributed according to a distribution ρ, the average of any function f (including the function x itself!) is calculated by

$$\langle f(x) \rangle = \langle f \rangle = \int dx f(x)\, \rho(x)\,, \tag{2.7}$$

where we omitted the argument from the second set of brackets to remind you that the average doesn't depend on the "name" of the variable, but rather on the two functions f and ρ. We also omitted the limits of integration to suggest that the equation applies very generally (x could be a multidimensional vector) and for whatever limits of integration are appropriate. In the simplest formal example of Equation 2.7, the mean is given by $\langle x \rangle = \int dx\, x\, \rho(x)$.

We can consider concrete examples from simple distributions. First, assume a uniform or "rectangular" distribution of x values in the interval zero to one: $0 < x < 1$. For this distribution, $\rho(x) = 1$ in the allowed range. Thus, the average is given by $\langle x \rangle = \int_0^1 dx\, x = 1/2$. This example just reassures us that calculus delivers the same answer as common sense! What if we have the nonuniform distribution $\rho(x) = 2x$ in the same interval? The average of this distribution is $\langle x \rangle = \int_0^1 dx\, 2x^2 = 2/3$, that is, when larger x values are weighted more, the average shifts to the right—again consistent with intuition.

In biophysics, the distributions of interest typically are not normalized, but we can still calculate averages. (The crucial example is the Boltzmann factor where we only know the distribution up to a constant of proportionality: $\rho(x) \propto w(x) = \exp[-U(x)/k_B T]$, where U is the potential energy, k_B is Boltzman's constant and T

is the temperature). As a reminder, we use the symbol w to denote an unnormalized probability distribution—also called a weighting factor. When only a weighting factor is known, the average of f is given by

$$\langle f \rangle = \frac{\int dx f(x)\, w(x)}{\int dx\, w(x)}.$$ (2.8)

PROBLEM 2.11

Show that Equation 2.8 follows from Equation 2.5.

2.2.6 SAMPLING AND SAMPLES

Despite the bland name, "sampling" is one of the most important ideas in computation—whether in biology, physics, or any other field of study. Biophysics is no exception. Understanding what sampling should be like in the ideal case—and by contrast, how it proceeds in practice—will be incredibly valuable to anyone who runs biomolecular simulations or reads about them.

We described previously an approach for generating random numbers from a cumulative distribution function. If you skipped that paragraph, or didn't understand it, go back to it now and make sure you "get" it.

Sampling refers to generating a set of numbers or objects according to a specified distribution. The resulting samples necessarily contain only a finite number of elements, and Figure 2.5 gives simple examples of the resulting histograms. Whether one has sufficient sampling depends very sensitively on what one is trying to calculate. For instance, it takes increasingly larger samples to properly calculate higher "moments" of a distribution (the mean is the first moment, while the second measures fluctuations about the mean; more on this later).

Because we will often be computing averages from a computer simulation (rather than calculating integrals with paper and pencil), we need to consider how to calculate averages using samples. In the same way that the integrals of Equations 2.7 and 2.8 account for more frequent occurrences with heavier weighting, we want our sample-based averages to do the same. Fortunately, this happens automatically! Sampling, after all, is the process of generating coordinates in proportion to the distribution ρ under consideration. Thus it turns out that if one has a properly distributed sample with N values, then what looks like a linear average is actually a weighted average:

$$\langle f \rangle \doteq \frac{1}{N} \sum_{i=1}^{N} f(x_i),$$ (2.9)

where we use the symbol "$\doteq$" to denote the fact that the right-hand side is actually a computational estimate of the left; that is, there is only true equality here in the limit $N \to \infty$, and the estimate may be quite lousy for small N. It is critical in

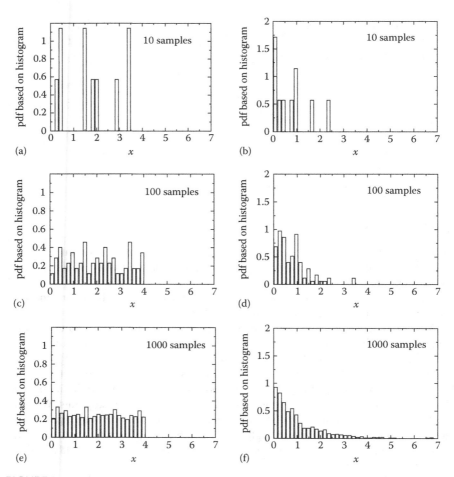

FIGURE 2.5 Histograms produced from several small-sized samples. Panels (a), (c), (e) are based on a rectangular distribution, where $\rho(x) =$ const. whereas panels (b), (d), (f) result from an exponential distribution given by $\rho(x) \propto \exp(-x)$. Larger samples are progressively included (10, 100, 1000 numbers) from top to bottom. A key point is that it is difficult to determine the type of distribution without a large number of samples. One could even mistake the number of peaks in a distribution from a small sample—for instance, from the top or middle panels.

Equation 2.9 that the set of values $\{x_1, x_2, \ldots\}$ be generated according to ρ—or equivalently, according to $w \propto \rho$. It might be more appropriate to use a notation like $\sum^\rho$, but this is rarely done.

PROBLEM 2.12

By considering the ratio of weights at two arbitrary values of x, explain why sampling according to ρ is the same as sampling from w.

PROBLEM 2.13

Write a computer program that can randomly generate integers from 1 to 6, with equal probability, and which can do so N times (for any N). This is your simulated dice roller. Run your program three times for each of $N = 6$, 60, and 600. (a) Plot the resulting nine histograms, and comment on the differences occurring for the same N and among different N. (b) Also calculate the mean of each of the nine data sets and comment on these values.

Thinking ahead to molecular systems, the prospects for good sampling would seem rather dire. As you may already know, the backbone configuration of a protein is largely (though not completely!) specified by two dihedral angles per amino acid. A small protein might contain 100 amino acids, or 200 important dihedrals. To divide configuration space into multidimensional regions or "bins," one could make a rather crude division of the possible values of each dihedral into 10 intervals. This results in 10^{200} bins, which one could never hope to sample. It's too many even for the fastest supercomputer! On the other hand, many or even most bins may be irrelevant because of "steric" clashes between overlapping atoms. Nevertheless, even a tiny fraction of 10^{200} is likely to be prohibitively large—which is indeed the case, as far as we know. These stubborn numbers are worth thinking about.

Not only can a complex system possess more (possibly important) bins than one could hope to sample, but it may also be difficult to generate even properly distributed small samples. In fact, you should assume that wrongly distributed samples are generated from any large molecular simulation (e.g., for a protein or nucleic acid)! Figure 2.6 explains why. Typical simulations generate molecular configurations in a dynamical or evolutionary way (e.g., by "molecular dynamics" simulation): each configuration is necessarily similar to the previous one. This is in sharp contrast to the cdf method above, where coordinate values are generated completely independent of previous data. Even though there is fairly good agreement among scientists on what the mathematical distribution of configurations should be (following the famous Boltzmann factor), almost all current biomolecular simulations are dynamical and of insufficient length to follow the correct distribution. This is another fact to think about for a long time.

If you are interested in this, take a look ahead at sampling problems in the large protein rhodopsin (Figure 12.4). Even supercomputers can't necessarily sample large biomolecules well!

2.2.7 THE DISTRIBUTION OF A SUM OF INCREMENTS: CONVOLUTIONS

The title of this section sounds like it was designed to intimidate students without experience in probability theory. On the contrary, however, think of it as a chance to see that some fancy-sounding concepts can be understood in a relatively simple way.

Fortunately, the section title itself gets you over the first hurdle: A convolution, in fact, is nothing more than the distribution of a sum of "increments"—that is, a sum of more than one variable. The simplest example would be rolling two dice instead of one, so that the value of each roll would be considered an increment summing to the whole.

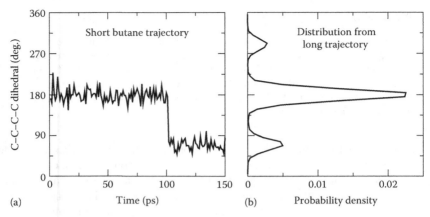

FIGURE 2.6 Correlated sampling in butane. Panel (a) depicts time-correlated or "evolutionary" sampling, which is the typical result of modern computer simulations: every new configuration is not chosen at random from the full distribution shown in panel (b). Rather, new configurations (and hence dihedral angles) evolve from preceding values, and hence the discrete points of the trajectory are appropriately connected with lines. The trajectory at left has not even visited the third state revealed in the pdf based on a much longer simulation (10 ns). In fact, the behavior shown here is typical of any complex system—whether a small molecule or protein. Evolutionary (or dynamical) sampling will only produce the correct distribution after a very long simulation. You can look ahead to Figure 12.4 to see the same effect in a large protein simulation, done on a supercomputer!

PROBLEM 2.14

While the distribution of individual die rolls is uniform, as given in Equation 2.1, the distribution for two dice most certainly is not. (a) With paper and pencil, calculate the probabilities for all possible two-dice rolls and graph the function. (b) Explain qualitatively how a uniform distribution turns into one with a peak.

Convolutions are critical to understanding numerous biological phenomena. Diffusion, for instance, which is responsible for much of biological "transport," is nothing more than a convolution of elementary steps. The distribution of stepping times in motor proteins, moreover, can be analyzed with simple convolution principles to reveal whether there are unobserved substeps (i.e., increments) contributing to the overall motion.

2.2.7.1 The Integral Definition of Convolution

To describe convolutions of continuous variables (i.e., distributions of sums of continuous variables), we need to generalize the simple thinking of the two-dice homework problem. We are talking about any density ρ and two "increments" drawn from (distributed according to) ρ, which we'll call x_1 and x_2. We want to know the distribution ρ_u of the sum of these increments, $u = x_1 + x_2$. Thus, the distribution for u must take into account all possible combinations of the increments, which may be done mathematically by the relation

$$\rho_u(u) = \int_{-\infty}^{\infty} dx_1\, \rho(x_1)\, \rho(u - x_1), \qquad (2.10)$$

where x_2 does not appear explicitly because, by the definition of u, it is always equal to $u - x_1$, which is indeed the argument of the second ρ. Note that ρ_u appropriately has units of a density because two densities and one integration appear on the right-hand side of Equation 2.10. The equation should be "read" in the following way: to obtain the probability of a particular value of u, we need to integrate over all possible combinations x_1 and x_2 that add up to u—weighted according to their associated probabilities, $\rho(x_1)\,\rho(x_2) = \rho(x_1)\,\rho(u - x_1)$.

PROBLEM 2.15

By integrating Equation 2.10 over u, followed by an x_1 integration, show that ρ_u is properly normalized.

PROBLEM 2.16

The convolution of two uniform distributions (say for $0 < x \leq 1$) is a symmetric triangular distribution. Without calculation, sketch this distribution and explain the reason for this shape. You may want to consult your result for Problem 2.14.

The convolution of two exponential distributions with $x > 0$, echoing the situation with the dice, has a very different shape from the original. In fact, a "gamma distribution" results, which is rounded on top.

PROBLEM 2.17

Using paper and pencil, show that the convolution of two identical exponential distributions is indeed a gamma distribution. Use $\rho(x) = ke^{-kx}$ with $x > 0$, and set $k = 1$. Also calculate the convolution of this two-increment distribution with a third exponential, and sketch all three functions on the same graph.

PROBLEM 2.18

To study dice again, modify the program you wrote for Problem 2.13 so that it sums M repeated rolls of a single die. Generate 300 sums for $M = 2, 3, 4, 5$ and plot the histogram for each M. Compare your $M = 2$ histogram with your result above to ensure it is correct, and comment on the evolving shape of the distribution as M increases.

More generally, convolutions can describe an arbitrary *number* of increments distributed according to an arbitrary *set* of distributions.

PROBLEM 2.19

Write the generalization of Equation 2.10 if x_1 is distributed according to ρ_1 and x_2 according to ρ_2.

We note that convolutions may also be studied using Fourier transforms. Students who are "fluent" in the use of Fourier methods will actually find the use of reciprocal space to be mathematically easier than the convolution of Equation 2.10. However, our "real space" convolutions are more physically and statistically informative.

2.2.8 Physical and Mathematical Origins of Some Common Distributions

While we can view distributions as purely mathematical objects without worrying about their origins, any student of the physical or biological sciences can benefit from understanding some key examples in more depth.

2.2.8.1 The Gaussian Distribution Really Is "Normal" because of the Central Limit Theorem

The central limit theorem tells us that the distribution of a sum of many small increments, almost regardless of their initial distributions, will follow a Gaussian distribution. A key point to realize is that the convoluted distribution—that is, the distribution of the sum—will be most Gaussian near its peak, and less so further away. As more increments are included, the distribution becomes more Gaussian (more "normal") further from its peak, and you can think of measuring distance in units of standard deviations. Because so many observable phenomena are complicated "convolutions" of many incremental causes, the Gaussian distribution is indeed "normal." To convince yourself, on the other hand, that a convolution usually won't be Gaussian sufficiently far out in the tails, consider summing increments that individually have a limited range, say $-a < x < b$. For n such increments the convoluted distribution must be exactly zero when $x > nb$, which clearly differs from the function (2.4).

2.2.8.2 The Exponential Distribution and Poisson Processes

Now consider the ubiquitous exponential distribution. As you may recall from a physics or chemistry class, this distribution comes up in radioactive decay. But it actually comes up in a huge number of processes, called Poisson processes, that are straightforward to understand. A Poisson process is one in which there is a constant probability per unit time of an event occurring (e.g., of a nucleus decaying or of a molecule isomerizing). To describe the situation mathematically, let $N(t)$ be the number of systems in the initial state—for example, un-decayed or un-isomerized—at time t, and assume that all "events" decrease this population. If we imagine a large number, $N(t = 0)$, of identical systems prepared in the initial state, then the mathematical statement of the event probability being constant over time is

$$\frac{dN(t)}{dt} = -k \times N(t), \tag{2.11}$$

where k is the constant probability, which evidently has units of inverse time. In other words, for all the $N(t)$ systems remaining in the initial state, a constant fraction of them can be expected to transform via the Poisson process in a small interval of time. The function that satisfies Equation 2.11 is just the exponential, $N(t) = \text{const.} \times e^{-kt}$. The constant must be set to $N(0)$ in the problem we have set up.

So where is the distribution in all this? Well, that depends on how you look at the problem. Just above, we called $N(t)$ the number of systems in the initial state, but we might have said, equivalently, that $N(t)/N(0)$ is the fraction of "surviving" systems. Thus the distribution (density) ρ_s of survival times is proportional to $N(t)/N(0) = e^{-kt}$. If we normalize this density, it is given exactly by

$$\rho_s(t) = ke^{-kt}. \tag{2.12}$$

This is physically sensible since k has units of inverse time, so that Equation 2.12 is a density of a time variable.

2.2.8.3 The Dirac Delta Function

In some ways, the "delta function" is the easiest distribution of all. Quite simply, the distribution $\rho(x) = \delta(x - x_0)$ means that x_0 is the only value of x that occurs. That is the most important thing to know about it.

Yet because the δ function is referred to often, it's worth understanding it a bit better. In a sense, $\delta(x)$ is a way to write a discrete distribution in terms of a continuous function. Delta functions can be used in combination when the distribution consists of a number of points, such as if $\rho(x) = \sum_i c_i \delta(x - x_i)$, where the c_i are constants which must sum to 1 for the proper normalization of ρ.

There are three key properties to know about the delta function. First, $\delta(x) = 0$ for any $x \neq 0$—that is, whenever the argument of the function is not zero. Second, the delta function is normalized in the usual way, so that

$$\int\limits_{-\infty}^{\infty} dx\, \delta(x) = \int\limits_{-\infty}^{\infty} dx\, \delta(x - x_0) = 1. \tag{2.13}$$

Third, the delta function can "single out" a certain argument for an integrand, in the following way:

$$\int dx f(x)\, \delta(x - x_0) = f(x_0). \tag{2.14}$$

Note that writing the form $\delta(x)$ is equivalent to setting $x_0 = 0$.

The delta function, then, is just an infinite spike at the point where its argument is zero. However, such a sudden divergence is mathematically problematical. To be careful about it, one can write the delta function as the limit of a series of increasingly spiky functions and, surprisingly, many different functional forms could be used.

A common form to consider is the Gaussian in the limit of infinitely small standard deviation ($\sigma \to 0$). If you won't be considering derivatives, you can make use of a simple series of increasingly tall and narrow rectangles of unit area.

PROBLEM 2.20

(a) Show that a distribution that is the limit of increasingly narrow rectangular functions of unit area exhibits the property (2.14). (b) Explain in words why you expect that an infinitely narrow Gaussian distribution will exhibit the same property.

2.2.9 Change of Variables

Inevitably in research, one decides that coordinates that used to seem important are no longer best. For example, instead of Cartesian coordinates, you might prefer "internal" coordinates for a molecule, which consist of a combination of bond lengths and angles. Or, you might want to change from velocity to kinetic energy.

Changing the argument of a probability distribution is fairly straightforward when the functional relationship between the two variables is known. Say you want to convert a pdf $\rho(x)$ to a function of some other variable y, and assume further that you know the functional relation and its inverses: $y = y(x)$ and $x = x(y)$. The main danger is that you cannot simply make a variable substitution, because $\rho(y) \neq \rho(x(y))$. Rather, you need to recall changing variables in calculus for integration, in order to preserve normalization and the meaning of a pdf. Then, imagining two integrands forced to match so that $\rho(x)\,dx = \rho(y)\,dy$, one has the relation

$$\rho(y) = \rho(x(y))\,\frac{dx}{dy}. \tag{2.15}$$

Notice that we have been mathematically sloppy because the two ρ symbols in Equation 2.15 evidently are not the same mathematical functions. However, such sloppiness is common in the literature, so look out for it.

The derivative in Equation 2.15 is often called a Jacobian, which we will encounter again when we study molecular coordinates.

PROBLEM 2.21

What distribution of the variable $u = x^2$ corresponds to a Gaussian distribution for x? Explain whether any information is lost if we use only the variable u instead of x.

2.3 FLUCTUATIONS AND ERROR

One of the basic lessons of statistical mechanics in general—and molecular biophysics, in particular—is the critical importance of fluctuations (spontaneous random motions). In biology, most proteins can be thought of as little machines. Without fluctuations or spontaneous motions, these proteins would be static objects and life would cease. (The very interesting processes by which random fluctuations are converted to purposeful or directed motion are discussed briefly in Chapter 10.)

Another way of underscoring the importance of fluctuations is to imagine that all you knew about every distribution was its average. The example of dogs comes to mind, inspired by an article in *The New Yorker* magazine. Imagine you are planning to get a dog and all you know about various breeds is their average behavior—especially their average tendency to violence. It seems that, surprisingly, the average poodle is more likely to be violent than the average pit bull. Unfortunately, when pit bulls do become violent, they have a much higher chance to kill someone. In other words, fluctuations—the tails of distributions—do matter.

2.3.1 VARIANCE AND HIGHER "MOMENTS"

For most purposes, fluctuations can be described statistically/probabilistically by a single quantity: the variance, σ^2. Most of you are undoubtedly familiar with the square root of the variance, called the standard deviation. The common notion of the standard deviation as the "width" of a distribution is perfectly correct—see Figure 2.7—although in this section we will formalize other important aspects of the variance.

We can define the variance σ^2 as an average, following Section 2.2.5. It is the average squared deviation from the mean,

$$\sigma^2 = \text{var}(x) \equiv \langle (x - \langle x \rangle)^2 \rangle, \tag{2.16}$$

where the angled brackets, as usual, indicate an average over x values distributed according to the distribution under consideration. Therefore, the standard deviation

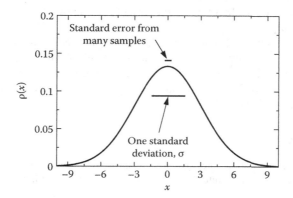

FIGURE 2.7   The natural scale vs. the scale of statistical uncertainty. Every distribution has a natural scale, which is typically quantified by the standard deviation, σ. Here, $\sigma = 3$. This is just the commonsense "width" of the distribution and has nothing to do with how many samples or measurements have been generated from the distribution. By contrast, the statistical uncertainty in a series of measurements from a distribution depends on how many independent samples have been obtained, and is quantified by the standard error (Equation 2.22). With a large number of measurements (independent samples), the statistical uncertainty can become much smaller than the standard deviation.

σ, which has the same units as the variable under consideration x, indeed measures the "width" of a distribution. In a Gaussian distribution (2.4), the parameter σ that appears is exactly the standard deviation.

There is an important alternative way of writing the variance, as noted in the following homework problem.

PROBLEM 2.22

Starting from Equation 2.16, show that $\sigma^2 = \langle x^2 \rangle - \langle x \rangle^2$, explaining your mathematical steps in words.

A key point is that the variance is a characteristic of the distribution at hand (and therefore of the physical or mathematical process that produced it). The variance is often very different from error or uncertainty or precision. Imagine you are interested in an inherently variable physical process: you may be making perfect measurements or simulations that are characterized by large variance—such as the lengths of whales—but this does not mean you are making a large error! We will return to this very important point.

It is worth mentioning that the variance is also known as the second "central moment." "Moments" of a distribution simply refer to the averages of powers of the variable, so that the nth moment is $\langle x^n \rangle$. Central moments, denoted μ_n, are moments of deviations from the mean,

$$\mu_n \equiv \langle (x - \langle x \rangle)^n \rangle . \tag{2.17}$$

Central moments, such as the variance, are important because the mean may not be zero, in which case the ordinary second moment would not correspond to the width of the distribution.

Surprisingly, not every moment can be calculated, even in principle. It is possible you may encounter a "long-tailed" distribution where the variance, in the words of the mathematicians, "does not exist." That is, the average to be computed in Equation 2.16 could diverge to infinity—for example, if $\rho(x) \propto x^{-5/2}$ and $1 < x < \infty$—even though the mean might be finite.

PROBLEM 2.23

By using the appropriate integrals, show that the variance of $\rho(x > 1) \propto x^{-5/2}$ diverges but its mean is finite. Explain the reason for this result in words.

2.3.2 THE STANDARD DEVIATION GIVES THE SCALE OF A UNIMODAL DISTRIBUTION

One of our overall goals is to develop physical intuition for examining many different situations. A question that often arises in science is whether a quantity is big or small, significant or not. A physicist will always want to know the answer to a more precise question, "Is it large (or small) compared to the important scale(s)?"

Statistics can help us set the scale when we have a set of data. In particular, the standard deviation, σ, gives an estimate for the overall scale ("width") of a distribution (see Figure 2.7). The fundamental example of this, of course, is the Gaussian, where including two-standard deviations from the mean captures 95% of the probability. It's not important whether 100% of the probability is included or not, but rather whether a small number (much less than 10) of standard deviations includes the bulk of the probability.

2.3.3   THE VARIANCE OF A SUM (CONVOLUTION)

One of the most important statistical results concerns the variance of a sum (i.e., of a convoluted distribution). The result is the essence of diffusion, which underpins so many biological and other scientific processes.

If we sum variables drawn from an arbitrary set of distributions, what is the variance of the sum of these variables? A simple example would be determining the variance in plant growth over 24 h periods, based on knowing, separately, the variances for day and night growth. Perhaps, more pertinently, you might want the variance of a process constructed from many incremental distributions, as in diffusion.

PROBLEM 2.24

Describe in words the procedure you would use to estimate the variance of a sum of N variables, based on repeated draws from the individual distributions $\rho_1, \rho_2, \ldots, \rho_N$. In this problem, do not write down the integrals, but rather describe how to estimate the variance based on hypothetical data.

Our basic task will be to calculate, formally, the variance of the sum

$$S = X_1 + X_2 + \cdots + X_N. \tag{2.18}$$

Here, we follow the mathematicians and use a capital letter for the variables because we imagine we have not yet assigned any specific value for each (thus, S and the X_j are called "random variables" in contrast to the x_i appearing in Equation 2.9, which are specific values already drawn from the distribution).

To calculate the variance of the sum S, it is useful to define deviations from the means of variables as $\Delta S = S - \langle S \rangle$ and $\Delta X_i = X_i - \langle X_i \rangle$. We then have

$$
\begin{aligned}
\sigma_S^2 &= \langle \Delta S^2 \rangle \\
&= \langle [(X_1 + X_2 + \cdots + X_N) - \langle X_1 + X_2 + \cdots + X_N \rangle]^2 \rangle \\
&= \langle (\Delta X_1 + \Delta X_2 + \cdots + \Delta X_N)^2 \rangle \\
&= \langle \Delta X_1^2 \rangle + \langle \Delta X_2^2 \rangle + \cdots + \langle \Delta X_N^2 \rangle \\
&\quad + 2\langle \Delta X_1 \Delta X_2 \rangle + 2\langle \Delta X_1 \Delta X_3 \rangle + \cdots \\
&= \langle \Delta X_1^2 \rangle + \langle \Delta X_2^2 \rangle + \cdots + \langle \Delta X_N^2 \rangle \equiv \sum_{i=1}^{N} \sigma_i^2,
\end{aligned}
\tag{2.19}
$$

where σ_i^2 is the variance of the distribution of X_i, by definition. Most of the equalities result from simple rearrangements, but not the last line. For the last line, we use the fact that the value of $\Delta X_1 = X_1 - \langle X_1 \rangle$ is completely independent of ΔX_2, since all variables are sampled separately. It follows that, for a given ΔX_1 value, averaging over all ΔX_2 will yield exactly zero by the definition of the mean, and this is true for any ΔX_1. Hence, all the terms in the form $\langle \Delta X_i \Delta X_j \rangle$ with $i \neq j$ exactly vanish. Note that our mathematics here did not ever use the fact that all the X variables were to be drawn from the same distribution, so Equation 2.19 applies to any set of distributions.

Equation 2.19 is a simple but remarkable result: variances can simply be added up. Note that this does not apply to standard deviations, only to their squares.

2.3.4 A Note on Diffusion

Have you ever seen a drop of ink placed into some water? Although at first localized to the spot where it landed, the color gradually spreads out. Such spreading is an example of diffusion, which is omnipresent in biology. Small molecules diffuse in the cell and some bacteria perform diffusive random walks as described in the book by Howard Berg. (See 'Further Reading' at chapter's end.)

The main mathematical fact about simple diffusion is that the mean squared distance grows linearly with time. That is, if the process starts at time $t = 0$ with $x = 0$, then such growth is described by

$$\langle [x(t)]^2 \rangle = \text{const.} \times t. \tag{2.20}$$

This is basically a variance growing with time, if we imagine averaging over the individual trajectories $x(t)$ of many ink particles. All the while, the mean is unchanged: $\langle x \rangle = 0$.

Equation 2.20 should look strange at first. From the perspective of simple mechanics (Newton's laws), we expect particles to move roughly at constant speed, so that the distance—and not its square—would grow linearly with time ($x = vt$). Thus, diffusive behavior is qualitatively different from deterministic mechanics.

Yet, the behavior of the mean squared distance is a trivial example of the rule for summing variances, which we just learned about—recall Equation 2.19. When we think of diffusion probabilistically, it's almost obvious. Imagine starting a million ink "particles" at some time, and measure all their distance changes after some short time Δt. We expect each to move in some random direction, so that the simple average (mean) distance will be zero, but the variance will have some finite value, reflecting the random spreading of the particles. What will happen in the next time interval Δt? Well, we expect the same random motions, independent of what occurred before—that is, a statistically independent process. This exactly fulfills the criteria set out for summing independent variances (for variances of convolutions of independent variables). After a little bit of thought, then, it's obvious that diffusion must obey the famous rule of Equation 2.20, almost regardless of what forces are driving the local trajectories.

We will revisit diffusion in Chapters 4 and 10.

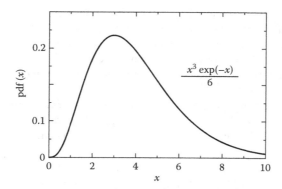

FIGURE 2.8 An asymmetric distribution. Specifically, a gamma distribution is shown. In contrast to the Gaussian distribution (Figure 2.2), values to the left and right of the mean are not equally probable.

2.3.5 BEYOND VARIANCE: SKEWED DISTRIBUTIONS AND HIGHER MOMENTS

While the second central moment (i.e., the variance) describes the width of a distribution, if you think of a distribution such as in Figure 2.8, it is clear that such a function is characterized by more than the mean and variance. (To put it more carefully, knowing only the mean and variance does not permit one to distinguish among distributions of *a priori* unknown functional forms.) Accordingly, statisticians frequently discuss higher moments. The most important are the skew, defined as μ_3/σ^3, and the strange-sounding kurtosis, $(\mu_4/\sigma^4 - 3)$, which measure how much the tails of the distribution bulge in comparison to the peak.

PROBLEM 2.25

Re-sketch Figure 2.8 for yourself and explain, based only on the shape of the distribution, why the mean of this distribution differs from its most probable value (the peak)? Should the mean be bigger or smaller than the peak value?

PROBLEM 2.26

What mathematical condition defines a symmetric distribution, bearing in mind that the mean need not be zero? Also, show that $\mu_3 = 0$ for symmetric distributions.

2.3.6 ERROR (NOT VARIANCE)

Perhaps you've heard that error decreases as the square root of the number of measurements, which is true for experimental as well as numerical data. Of course, it makes sense that error decreases with more measurement, but why is one functional form ubiquitous? We can now answer this question, and hopefully appreciate the crucial difference between variance and error.

Our strategy—as in Section 2.3.3—will be to consider the variance in the distribution of a sum of many "increments." Each increment can be thought of as a measurement, and considering the sum is a simple step away from the average, which we will examine too. In the spirit of imagining repetition of the same experiment (or computation) many times, let us consider the sum of N such measurements, which we will again denote as $S = \sum_{j=1}^{N} X_j$.

But if we think now of the variance of the average measurement—that is, the variance of S/N—we find something new and quite important. For our case, where all the X_i variables of Equation 2.19 are identically distributed, you can show that

$$\operatorname{var}\left(\frac{S}{N}\right) = \frac{\operatorname{var}(X)}{N}. \tag{2.21}$$

PROBLEM 2.27

Derive Equation 2.21 from Equation 2.19 by substituting $S \to S/N = (X_1 + \cdots + X_N)/N$. Do not yet set all $X_i = X$, for this will mean you are studying a single variable, rather than a sum. Rather, determine the appropriate factors of N that now occur in all the lines of Equation 2.19, and only set $X_i = X$ at the very end.

In words, our new result means that *the standard deviation of the average of N measurements decreases as $\sqrt{N}$ regardless of the variability in the individual measurements. Further, it is indeed the variability in the average of many measurements that determines error. Notice the difference from the variability (variance) of individual measurements.*

PROBLEM 2.28

Consider a distribution of length measurements, perhaps of whales, that have a standard deviation of 1 m. The preceding discussion makes clear that with enough measurements, we can reduce the error to any level we want—say, to 1 cm. To the uncertainty in what quantity does this "error" apply? Explain in words how it can be so much smaller than the standard deviation.

PROBLEM 2.29

In many computer simulations, measurements will not be fully independent but correlated due to the dynamical nature of common simulation techniques, as shown in Figure 2.6. Suppose that of N measurements, only $N/100$ can be considered independent. Does the error still decrease in proportion to $1/\sqrt{N}$? Explain.

It is natural, given the preceding discussion, to define the "standard error of the mean" (std-err), which sets the scale of uncertainty for an estimate of the mean.

Specifically, we define

$$\text{std-err} = \sqrt{\frac{\text{var}(X)}{N}} = \frac{\sigma}{\sqrt{N}}, \tag{2.22}$$

given N independent measurements of X. Unlike the standard deviation, the standard error most definitely depends on the number of independent samples or measurements obtained for the quantity of interest. See Figure 2.7 for a simple illustration of this point.

I will end this section with an overheard conversation. When I was a postdoc at the Johns Hopkins School of Medicine, I usually parked on the roof of my assigned parking garage. At the end of every day, I trudged up the many flights of stairs. One day, there were two men discussing their data—scientists, apparently—who remained out of sight as they kept ahead of me by a flight of stairs. One of them said, "In the old days, we always used the standard deviation, but now they're using the standard error. I don't get it."

Make sure you "get it." The difference is subtle but crucial.

2.3.7 CONFIDENCE INTERVALS

If you've understood error, you'll have no trouble understanding confidence intervals. These intervals indicate the likelihood that a given function of the coordinates (or the coordinates themselves) will fall within a given "interval," which is simply a range of values specified by a maximum and a minimum. Thus, a made-up example is, "There is a 90% chance that the mean age of graduate students is between 25 and 26." In other words, based on the finite amount of data we have, we are only able to estimate the average age within a full year, but our data indicates this is 90% likely.

The standard error $\sigma/\sqrt{N}$ sets the scale of confidence intervals for the mean. This is just another way of saying that the standard error gives a measure of the error itself: the size of confidence interval is a measure of uncertainty. More advanced statistical analyses show that roughly a 90% confidence interval for the mean of a unimodal distribution is given by the mean plus or minus twice the standard error.

2.4 TWO+ DIMENSIONS: PROJECTION AND CORRELATION

If you recall that protein configuration space has thousands of dimensions (x, y, z values for every atom), you may worry that this will be quite a long chapter because we are only now up to two dimensions! However, probability theory in two dimensions will serve as a model for any number of dimensions. Considering more dimensions will not greatly complicate the ideas at play.

As we will see below, two new and absolutely critical things arise in two and higher dimensions: projection and correlation. Correlation is intertwined with conditional probabilities, which are also useful.

Two-dimensional distributions or densities must obey the same basic rules of probability theory that apply in one dimension. As always, all the probability must sum—or integrate—to one. Thus, the normalization rule for two-dimensional densities is simply

$$\int dx\,dy\,\rho(x,y) = 1,$$ (2.23)

where I have purposely been sloppy in a mathematical sense: the function ρ now has two arguments, so it is clearly a different function from any ρ we have previously considered. Nevertheless, you will be expected to stay alert to such changes, which happen frequently in the biophysical literature, and make sure you always understand exactly what a function specifies. Its arguments will provide important clues.

The new two-dimensional density ρ also has different units (in a physics sense) from a one-dimensional ρ. This is clear from Equation 2.23 since ρ is multiplied by the infinitesimals $dxdy$ that have combined units of length^2 if x and y are lengths. This indicates that ρ is now a true density in a two-dimensional sense—for example, the number of leaves on the ground in a park, per square meter.

PROBLEM 2.30

(a) Write down a two-dimensional Gaussian distribution that depends only on $x^2 + y^2$. Include correct units. (b) Show that it is properly normalized.

2.4.1 PROJECTION/MARGINALIZATION

Projection is a critically important idea: it is a procedure that occurs in a good fraction of papers written about molecules of any kind, or indeed whenever a high-dimensional system is studied. The reason is simple. Projection is the means by which we plot (on two-dimensional paper) data we have for complex systems. In fact, we are already guilty of projecting data in the case of butane. Figure 2.3 shows only 1 of 36 molecular coordinates, after all.

Projection is even more critical in the study of large biomolecules. Imagine a protein containing thousands of atoms, in which the distance between two particular atoms is chemically important. In such a case, one might choose to study the distribution of values of that key distance. In other words, from a sample of fully detailed protein structures, each containing thousands of coordinates, we single out one coordinate and study its distribution. We are "projecting" a high-dimensional distribution onto a single coordinate.

When we project, we are integrating over all coordinates except the one onto which we project. Thus, when we project a two-dimensional distribution onto the first coordinate, say x, we integrate over the second variable,

$$\rho(x) = \int dy\,\rho(x,y).$$ (2.24)

PROBLEM 2.31

Show that the projected distribution is properly normalized if the original distribution was.

More generally, one can project onto a subset of variables $\{x_1, x_2, \ldots, x_n\}$ by integrating over all other coordinates $\{x_{n+1}, x_{n+2}, \ldots\}$.

$$\rho(x_1, x_2, \ldots, x_n) = \int dx_{n+1} dx_{n+2} \ldots \rho(x_1, x_2, \ldots, x_n, x_{n+1}, x_{n+2}, \ldots). \quad (2.25)$$

Perhaps you can see why two dimensions are ideal for understanding projection. In a typical high-dimensional case, one has no choice but to project onto one or two dimensions in order to visualize one's data. However, a two-dimensional distribution can be visualized directly—along with projections onto one dimension—enabling us literally to see what is going on.

Figure 2.9 illustrates projection and its pitfalls. One can easily underestimate the number of peaks in a distribution. On the other hand, some projections may give a perfectly adequate and easier-to-understand picture of a complex distribution. Proceed carefully!

Figure 2.9 also suggests the reason why statisticians call a projected distribution a "marginal" distribution. The joint distribution is projected to the margins (edges) of the graph.

2.4.2 CORRELATIONS, IN A SENTENCE

If two variables (or functions) are correlated then one is affected when the other is varied. This rather obvious statement has important scientific and mathematical implications. We begin to explore these in the next sections.

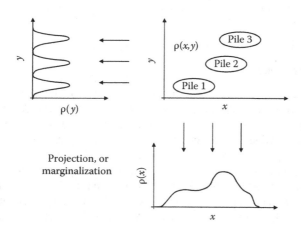

FIGURE 2.9 The projection procedure and its pitfalls. A two-dimensional probability distribution $\rho(x, y)$ is projected separately onto both x and y axes. The original ρ may be thought of as representing piles of leaves on the ground whose height is not constant but "tails off" at the edges. Evidently, the projection onto y correctly captures the three distinct piles, though the projection onto x does not. For a complex system with many variables, such as a biomolecule, it may be difficult or impossible to determine a low-dimensional set of coordinates onto which an informative projection can be made.

2.4.3 STATISTICAL INDEPENDENCE

We will first consider the absence of correlation, which is often called statistical independence (or simply "independence") and is quite important in itself. An example is that the number of pennies in the pocket of a professor in Pittsburgh should be statistically independent of the price of gold in London on any given day. In mathematical terms, the distribution of a set of statistically independent variables factorizes in a simple way, namely, if x and y are independent, then

$$\rho(x, y) = \rho_x(x)\,\rho_y(y). \tag{2.26}$$

Here I have included subscripts on the single-variable distributions to emphasize that they are different functions.

PROBLEM 2.32

Show whether or not the two-dimensional Gaussian of Problem 2.30 can be written in the form of the right-hand side of Equation 2.26.

PROBLEM 2.33

(a) Given a distribution $\rho(x, y)$ as in Equation 2.26, write down the integrals defining ρ_x and ρ_y. (b) Note the connection to projection, but explain why a projected distribution might or might not describe an independent variable.

In general, you should assume variables are *not* independent unless you have a good reason to believe they are!

2.4.4 LINEAR CORRELATION

While there are many subtle types of correlation, as will be sketched below, the simplest linear correlations are easy to understand and quantify. A typical example that commonly appears in the literature is the comparison between results for the same quantities obtained by different means. For instance, you may numerically estimate a set of quantities $A(j)$ for $j = 1, 2, \ldots$ (denoted $A_{\text{calc}}(j)$) that also can be measured experimentally—$A_{\text{meas}}(j)$. Each j value could represent a different ligand molecule and A could be its binding affinity to a given protein. As sketched in Figure 2.10a, for every j one can plot a point whose location is defined by the coordinates $(x, y) = (A_{\text{calc}}(j), A_{\text{meas}}(j))$. This is called a "scatter plot." Perfect agreement would mean exact equality between the measured and calculated values, in which case all points would fall on a diagonal line passing through the origin. It is clear that one can get a visual feel for the level of agreement or correlation using a scatter plot.

A "correlation coefficient" quantifies the degree of linear correlation. (Let us be clear that linear correlation is not the whole story, as depicted in Figure 2.10.) The mathematical basis for the correlation coefficient can be understood as a contrast to the case of statistical independence, given above in Equation 2.26. In particular,

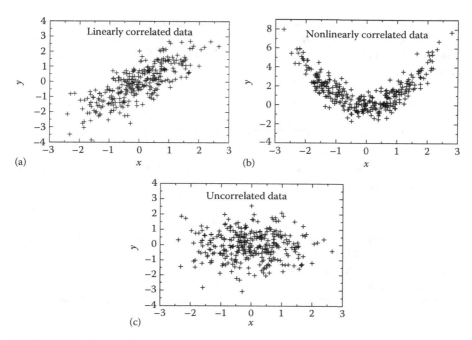

FIGURE 2.10 Examples of linearly correlated data (a), nonlinearly correlated data (b), and uncorrelated data (c). Note that the data set in (b), which is clearly correlated, would yield a Pearson coefficient of zero by Equation 2.29.

imagine computing the average of the product of two independent or uncorrelated variables. We would be able to factorize this product as follows:

$$\text{Uncorrelated: } \langle x\,y \rangle = \int dx\,dy\,x\,y\,\rho(x,y) = \int dx\,x\,\rho_x(x) \int dy\,y\,\rho_y(y)$$
$$= \langle x \rangle\,\langle y \rangle. \tag{2.27}$$

The factorizability for independent variables seen in Equation 2.27 suggests a simple route for quantifying correlation. In particular, if the equality in Equation 2.27 does not hold, then the two variables are correlated. Noting the identity $\langle xy \rangle - \langle x \rangle\,\langle y \rangle = \langle (x - \langle x \rangle)\,(y - \langle y \rangle) \rangle$, we can define linear correlation as occurring whenever

$$\langle (x - \langle x \rangle)\,(y - \langle y \rangle) \rangle \neq 0. \tag{2.28}$$

Note that the deviations from the respective means of x and y arise naturally here.

Yet as they are written, Equations 2.26 and 2.28 are useless for practical purposes. There is always statistical and other error in our calculations (based on finite data), so one could never convincingly demonstrate statistical independence using these relations: What would a "small" deviation from zero mean? In fact, what is small anyway?!?

What we are missing is what physicists would call a *scale* for the problem. You would not call a basketball player short if she were 0.002 km tall! For statistics, it is the width of a distribution which provides the natural scale—that is, the standard deviation σ.

PROBLEM 2.34

Why is the standard deviation appropriate here and not the standard error?

The Pearson correlation coefficient, r, is therefore defined using a scaled version of Equation 2.28,

$$r^2 = \frac{\langle (x - \langle x \rangle)\,(y - \langle y \rangle) \rangle}{\sigma_x\,\sigma_y}. \tag{2.29}$$

Because the natural scales for each variable have been built into the denominator, the correlation coefficient can be compared to 1. That is, $|r| \ll 1$ implies no (or minimal) linear correlation, whereas r close to positive or negative one implies strong correlation.

Linear regression provides a highly related analysis that can be found in any statistics textbook, and will not be discussed here.

2.4.5 MORE COMPLEX CORRELATION

As you were warned above, correlations are far from a simple business. While everything looks pretty for linearly correlated data or for uncorrelated data, it is clear from Figure 2.10b that more complex situations can arise. Note that the parabolic shape of the nonlinearly correlated data guarantees that the Pearson correlation coefficient of Equation 2.29 will be close to zero: for every value of y (and hence for every value of $\Delta y = y - \langle y \rangle$), there is a nearly equal likelihood of finding a positive or negative value of $x \simeq x - \langle x \rangle$ (since $\langle x \rangle \simeq 0$) of equal magnitude.

Things only get worse with increasingly complex systems, such as molecules that may contain hundreds or thousands of coordinates. In general, coordinates characterizing nearby atoms are correlated, and furthermore, there may be multi-dimensional correlations. In other words, any given coordinate may affect and be affected by many different coordinates and combinations thereof.

2.4.5.1 The Simplest and Most General Way of Understanding Correlations

Let's return to what we first said about correlations: they are present between two variables if "one is affected when the other is varied." What does this really mean in terms of distributions? For one thing, it means that distributions of correlated variables cannot be written in the statistically independent form of Equation 2.26. But there is a simple visual test that checks this exact criterion.

Correlations are one issue where visual inspection is often better than a simplified mathematical measure like the Pearson (linear) coefficient. That is, it is easy to tell

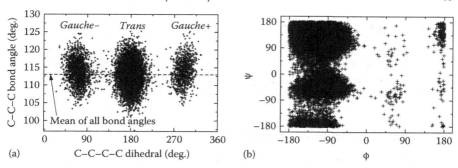

FIGURE 2.11 Molecular correlations, weak and strong. Scatter plots are shown in which each point represents a pair of coordinate values based on a single configuration from a long computer simulation. (a) For butane, subtle correlations are visible in two regards: the mean C–C–C bond angle of the central *trans* state is slightly lower than that of either of the *gauche* states; also, the *gauche* distributions are slightly diagonal. Correlations in alanine dipeptide (b) are stronger. This can be seen by comparing ψ distributions at different fixed ϕ values.

visually that the distribution of Figure 2.10b is correlated. But we can be more careful about this: what is it exactly about the parabolic shape that indicates correlation? The question is answered with a simple test: First, choose any x value and visually determine the distribution of y values for a small range around this particular x by covering all other x values, perhaps with two paper scraps. Now, shift your attention to another x value and determine this second y distribution. If the two distributions of y values are different—that is, if they depend on the x value—then x and y are correlated by definition. Further, if you can find any two x values for which the y distributions differ, then there is correlation. The same applies for using y values as references.

Molecular examples of correlated data are shown in Figure 2.11.

PROBLEM 2.35

Using paper and pencil, illustrate the procedure just described by making two sketches based on the distribution of Figure 2.10b, one for each of two x values. Repeat this exercise for two y values.

2.4.5.2 Mutual Information

The "mutual information" between two variables quantifies this general notion of correlation. It is defined by

$$I_{x,y} = \int dx\, dy\, \rho(x,y)\, \ln \frac{\rho(x,y)}{\rho_x(x)\, \rho_y(y)} \geq 0 \,, \qquad (2.30)$$

where ρ_x and ρ_y are the usual projections. Equality holds in the case of statistically independent variables. The definition seems very nonintuitive at first, but the problem below will help to clarify the meaning.

PROBLEM 2.36

(a) Explain why the definition (2.30) of mutual information captures the general concept of correlations, by first examining the limiting case of statistically independent variables and then the general case. (b) Could a function besides the logarithm be used to capture the degree of correlation?

2.4.6 PHYSICAL ORIGINS OF CORRELATIONS

Correlations can seem so abstract, until one considers simple physical examples. Imagine your heart as the center of a coordinate system and consider the distribution of locations of your right hand (assuming you stay standing in one position/orientation). Clearly your hand can only occupy certain locations, and the horizontal x and y values, for instance, are dependent on the height of your hand (z value). This is correlation. Think now of the location of your right elbow: clearly, this is highly correlated with the location of your right hand—and even correlated with your left foot, although more weakly.

PROBLEM 2.37

Explain in words why the locations of your foot and hand are correlated.

The schematic Ramachandran plot from Figure 1.9 is an important molecular example of correlations. Using the logic of Section 2.4.5, it is easy to see that the permitted ϕ values certainly depend on the value of ψ—and *vice versa*. What is the origin of this correlation? Every combination of ϕ and ψ indicates the local geometry of two neighboring peptide planes, and all their atoms. Quite simply, some of these configurations entail steric clashes, the physically prohibited overlapping of atoms. Indeed, steric considerations are the dominant source of correlations in molecular configurations. A large fraction of configuration space is simply ruled out by the impossibility of atoms overlapping. At the same time, attractions (e.g., van der Waals or electrostatic) can also create correlations in the sense of promoting a tendency for two atoms to be near one another. Note that the Pearson coefficient might give an ambiguous assessment of whether the Ramachandran distribution is correlated, but the direct method of Section 2.4.5 immediately indicates the strong correlations.

Other examples of molecular correlations are shown in Figure 2.11. Even a molecule as simple as butane exhibits correlations among coordinates, although they are weak. In particular, the figure shows that a bond angle and the main dihedral of butane are correlated. Another example comes from the "alanine dipeptide" molecule (Section 1.8.8), which consists of two peptide planes and an alanine (CH_3) side chain. Because alanine dipeptide has complete peptide planes on either side of its alpha carbon, the ϕ and ψ angles are defined as in an ordinary polypeptide or protein. Figure 2.11 makes clear not only that ϕ and ψ are correlated, but that the distribution is quite similar to the schematic Ramachandran plot of Figure 1.9. It is due to this similarity that tiny alanine dipeptides are often considered the simplest peptide-like systems.

2.4.7 Joint Probability and Conditional Probability

The phrases "joint probability" and "conditional probability" are sure to make the beginning student a little nervous. Fortunately, we have already studied joint probability and, furthermore, the two intimidating-sounding objects are closely related.

First of all, a joint probability distribution is simply one that gives the probability for more than one variable at a time—such as $\rho(x, y)$. Another example would be the distribution of all coordinates of all atoms in a molecule. These joint distributions already contain in them all of the (non-time-dependent) information we might need. For instance, they contain all the correlation information—whether linear or not. They also contain all conditional probability information.

"Conditional probability" refers to a probability distribution restricted by a certain "condition." For example, given a condition on one of two coordinates (such as the value of x), what is the distribution of the other coordinate y? The answer is written as $\rho(y|C)$ and can be read as "the distribution for y given condition C," or "the distribution for y conditioned on C." To make this still more concrete, picture a two-dimensional distribution (perhaps one of the panels from Figure 2.10) and ask the following: Given that $1 < x < 2$, what is the distribution of y? Graphically, this distribution could be determined by shading a vertical region between the lines $x = 1$ and $x = 2$ and examining the distribution of points within the region, as suggested by Figure 2.12. We would write it as $\rho(y|1 < x < 2)$.

The graphical construction in Figure 2.12 should provide a strong hint that there is a fairly simple way to construct the equation for a conditional probability distribution given the full (joint) distribution. Indeed there is, as we will see below. Unfortunately, the rule for doing so would look complicated if we wrote it down for a case like that described above, so we will first consider the discrete version. Consider a set of discrete events $\{(a_1, b_1), (a_1, b_2), \ldots, (a_2, b_1), \ldots\}$ each of which occurs with a probability $p(a_i, b_j)$. Perhaps a is the color of a car's exterior and b the color of its

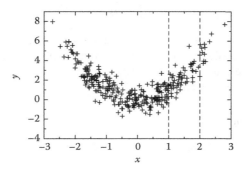

FIGURE 2.12 Constructing a conditional probability. Given the condition that $1 < x < 2$, which means only the points between the dashed vertical lines are considered, one could construct a corresponding distribution for y by Equation 2.32. In this case, without normalization, we have $\rho(y|1 < x < 2) \propto \int_1^2 dx\, \rho(x, y)$.

seats, so that p gives the probability of the two occurring together. Then we can "cut up" this distribution self-consistently, via Bayes' rule

$$p(a_i|C) = \frac{\sum_{b_j \in C} p(a_i, b_j)}{p(C)},$$ (2.31)

where we have used the condition C instead of a particular b value because the condition could include more than one b value. For instance, we could use Bayes' rule to answer the question: "What fraction of cars with blue or green exteriors have white seats?" The denominator on the right-hand side is the probability of all events consistent with the condition: $p(C) = \sum_{b_j \in C} \sum_i p(a_i, b_j)$. This denominator simply ensures proper normalization: perhaps blue and green cars make up only 10% of all cars, so if the conditional probability distribution for white interiors is to be properly normalized, all probabilities must be divided by 0.1 (multiplied by 10). Note that the rule (2.31) would be equally valid if we had considered some b_j in place of a_i on the left-hand side.

PROBLEM 2.38

Make up a 3×2 table of nonzero probabilities for any two correlated discrete variables. Then determine some conditional probability of your choice using Equation 2.31.

Let's return now to the example of our vertical stripe $1 < x < 2$ in a continuous distribution $\rho(x, y)$, as depicted in Figure 2.12. Intuitively, determining the conditional probability $\rho(y \mid 1 < x < 2)$ is fairly simple. We want to consider only the distribution within the stripe, we want to combine probability for equal y values, and we want to normalize the resulting distribution by dividing by a suitable constant. The analog of Equation 2.31 is therefore

$$\rho(y|C) = \frac{\int_C dx\, \rho(x, y)}{\int \int_C dx\, dy\, \rho(x, y)}.$$ (2.32)

The notation $\int_C$ means to integrate over all values consistent with the condition C. While Equation 2.32 looks messy, the numerator is simply a projection onto y of the region selected by the condition and the denominator's integral is simply the constant necessary to ensure normalization.

PROBLEM 2.39

Show that $\rho(y|C)$ in Equation 2.32 is properly normalized.

2.4.8 Correlations in Time

We have mostly been concerned with correlations among spatial variables, but the logic of correlations applies equally to time dependence. Following the reasoning of

Section 2.4.5, one can determine whether time correlations are present by asking, "Is the value of interest at a certain time affected by values at previous times?" Thus, for instance, the location of a soccer ball during a game is correlated with previous locations—at short-enough times. In molecular biophysics, time correlations are particularly important in understanding dynamic phenomena and simulation trajectories. For instance, Figure 2.6 shows a butane trajectory that, for short-enough intervals between different dihedral values, is clearly correlated. Another important notion, to which we will return in a subsequent chapter, is of a correlation time—beyond which values of interest are not correlated.

PROBLEM 2.40

Imagine a test tube containing two proteins that are far from one another and do not interact. Assuming you can measure whatever you want about these two molecules, give an example of a quantity that should be correlated in time and give another example of something that should not be correlated.

2.4.8.1 Time-Dependent Conditional Probability: A Quick Preview of Langevin (Stochastic) Dynamics

Despite its daunting name, you probably already think about time-dependent conditional probabilities. For instance, "if it's cloudy now, it's likely to rain soon," is just such a statement because the future probability (for rain) depends on a current condition (clouds). The probability for rain, in other words, is not independent of the condition.

Some useful descriptions of molecules function the same way. Although the dynamics of a molecule, in principle, are fully deterministic, this is not true if we omit some information. Imagine our old friend from the introductory chapter, the butane molecule—but now immersed in some solvent. The motions of the butane molecule will be determined by the intramolecular forces (within itself), as well as by the intermolecular forces between butane and the solvent. But if we watch only the motions of butane, ignoring solvent, some of the movement will appear random (stochastic). In particular, based on knowing the present conditions of only the butane molecule (locations and velocities of all butane atoms plus the intramolecular forces) at one point in time, we cannot exactly predict the future configurations. Rather, we can only do so in a probabilistic sense. That is, we can give a time-dependent conditional probability: given the present values of butane's variables, we can expect that repeated experiments or simulations will yield a distribution of future values. The need for such a description arises from the failure to account for solvent effects explicitly—you might call it apparent randomness.

PROBLEM 2.41

Make a time-dependent conditional probability statement (words only) regarding the location of a soccer ball during a game—or regarding any other activity.

2.5 SIMPLE STATISTICS HELP REVEAL A MOTOR PROTEIN'S MECHANISM

Almost every good experimental or computational paper includes an analysis of statistical uncertainty, but this can seem like an unimportant detail many times. It's easy to ignore. On the other hand, many papers have statistical analysis at their heart, and one such experimental paper is now described.

Molecular motors have attracted intense interest in recent years, particularly with the advent of experimental techniques capable of studying single molecules. One famous example from the group of Paul Selvin (see the paper by Yildiz et al.) is the experimental determination that the motor protein myosin V walks "hand-over-hand." A key development reported in the paper was the ability to determine the location of a fluorescent dye within a resolution of about a nanometer. If you read the paper, you will see that what is actually measured is a distribution (i.e., histogram) of the locations of a series of photons. In fact, as the paper says, "the fundamental goal is to determine the center or mean value of the distribution... and its uncertainty, the standard error of the mean." Not surprisingly, then, as more photons are collected, the statistical uncertainty can be made much smaller than the width of the distribution, which is reported to be about 300 nm (recall Figure 2.7). Armed with 10,000 or more photon counts, the researchers were thus able to determine the "walking" mechanism of the motor and publish a paper in *Science*!

2.6 ADDITIONAL PROBLEMS: TRAJECTORY ANALYSIS

The following problems are computational analyses of a molecular trajectory—that is, a series of configurational snapshots show the time evolution of the molecule, as shown in Figure 1.10. While many coordinates may be present, the problems here assume that the values of four ("internal") coordinates—θ_1, θ_2, ϕ, and ψ—are available for 10,000 time points. Further, ϕ and ψ are presumed to correspond to peptide dihedral angles, such as in a protein or in the alanine dipeptide, while the θ_i are simple bond angles.

A trajectory from a simulation of any protein or peptide could be analyzed the same way—by focusing on a single residue. The simplest choice, however, is alanine dipeptide (Figure 1.10).

PROBLEM 2.42

Distributions and normalization. From your trajectory, using 40 bins, plot (a) the histogram of ϕ values, (b) the corresponding numerically normalized pdf, and (c) the corresponding cdf. Explain how you calculated (b) and (c). (d) Re-plot the histogram as in (a) except with 400 and then 4000 bins, and explain the importance of the size of the bin.

PROBLEM 2.43

(a) Repeat Problem 2.42(a) using only the first half of the trajectory. Comment on the issue of dynamical simulation based on comparing the two plots. (b) To illustrate your points, also plot ϕ vs. time.

PROBLEM 2.44

Scale in a multimodal distribution. Normally, the variance can be used to estimate an overall scale for a distribution. Explain why this is not the case for the ϕ distribution. Estimate the scales involved using simple visual inspection ("by eye").

PROBLEM 2.45

Distributions, variance, error. For θ_1 (a) plot the histogram of the distribution, (b) compute the standard deviation, and (c) compute the standard error. (d) Explain whether the standard deviation corresponds to the scale of the distribution in this case. (e) What does the standard error give the scale for?

PROBLEM 2.46

The Ramachandran plane, correlations, and projection. (a) Make a "Ramachandran" scatter plot using the ϕ and ψ angles. (b) Explain, in words, why these dihedrals are correlated, using the logic of Section 2.4.5.1. (c) Using 40 bins, plot the projections onto ϕ and ψ. (d) What information has been lost in the process of projecting?

PROBLEM 2.47

Time correlations. (a) Plot each of the four coordinates against time (on separate plots). (b) By inspection, estimate how long it takes for each coordinate to become de-correlated (to lose "memory" of its earlier values) and comment on your estimates.

PROBLEM 2.48

Conditional probability. (a) Plot the ϕ distribution based only on configurations where $-30 < \psi < 0$. (b) Explain the connection to Equation 2.32.

FURTHER READING

Berg, H.C., *Random Walks in Biology*, Princeton University Press, Princeton, NJ, 1993.
Spiegel, M.R., Schiller, J.J., and Srinivasan, R.A., *Schaum's Outline of Probability and Statistics*, McGraw-Hill, New York, 2000.
Yildiz, A. et al., *Science*, 300:2061–2065, 2003.

3 Big Lessons from Simple Systems: Equilibrium Statistical Mechanics in One Dimension

3.1 INTRODUCTION

Statistical mechanics is a branch of physics that explores the connection between the microscopic and the observable. It is intimately related to thermodynamics, and indeed involves the intimidating-sounding concepts of entropy, fluctuations, and specific heat. However, from the microscopic viewpoint of statistical mechanics, these concepts are relatively simple to understand. And having grasped the ideas in a statistical mechanics context, we can readily understand their thermodynamic meaning.

In no other setting is statistical mechanics as simple as in one dimension. Of course, we sacrifice a deep understanding of some complex systems by looking at a single dimension, but by and large, the key concepts of statistical mechanics and thermodynamics can be understood in one dimension. Most of the understanding we will gain here extends immediately to complex, high-dimensional systems.

Our "secret weapon" in studying statistical mechanics will be our grasp of elementary probability theory. Perhaps our key task in this chapter is translation: we must learn to see that statistical mechanics is little more than probability theory set in a physical context. The connection comes from the famous Boltzmann factor, which explains the fundamental relationship between energy and (relative) probability.

3.1.1 LOOKING AHEAD

Of course, no molecular system is one dimensional, so are we wasting our time here? I believe we are absolutely making the best use of our efforts to study one-dimensional systems first, and for two reasons. First, the intuition we can get in one dimension is directly applicable to molecular systems, as we will see at the end of the chapter. It is not the whole story, but it is much of the story. Think back for a minute to butane (Figures 1.3 and 1.4), where the most important behavior occurs in the single dimension of the C–C–C–C dihedral angle.

Equally importantly, it's not easy to understand general statistical mechanics starting from complex systems. We could study the most general formula first, and show that it can be reduced to some simple cases. Unfortunately, general formulas are almost impossible to understand, even after a few special cases. In my opinion, most people learn best by first understanding simple examples, with generalization done afterward. Accordingly, we will generalize to multidimensional, multiparticle systems—such as molecular systems—in subsequent chapters.

3.2 ENERGY LANDSCAPES ARE PROBABILITY DISTRIBUTIONS

If you have even a superficial acquaintance with biophysics, physics, or chemistry, you probably have seen an energy landscape such as that presented in Figure 3.1. This "landscape" is just a visual representation of the potential energy function, which in one dimension is $U(x)$. (The function U should not be confused with the internal energy you may have learned about in earlier study of thermodynamics. Thermodynamics deals strictly with average quantities.) The function U is just an ordinary "real-valued" function of its argument x, so that for any x, it yields the energy value.

Along with the temperature T, the potential energy function essentially determines everything we will want to know about our systems. (This includes much of the dynamical—time-dependent—behavior that we shall not study in this chapter.) Knowledge of the potential function is equivalent to knowledge of the probability distribution, and probability is what we want. After all, *probability is observed behavior, described statistically.*

The connection between probability and energy is made explicit by the famous Boltzmann factor. The Boltzmann factor tells us that the probability of a "microstate" or configuration—that is, the x value in one dimension—is given by a simple exponential relation

$$\text{pdf}(x) \equiv \rho(x) \propto \exp\left[\frac{-U(x)}{k_B T}\right], \tag{3.1}$$

where k_B is the universal Boltzmann constant and T is the absolute temperature.

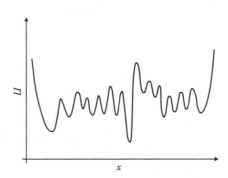

FIGURE 3.1 A one-dimensional energy landscape exhibiting several barriers and local minima. Such a figure is often used to represent, schematically, a "rough" energy landscape in a complex molecule. In fact, however, this sketch vastly under-represents the number of minima expected in a large biomolecule.

The appropriate units for k_B are determined by the units of the potential U (or *vice versa*, depending on which you start from). In MKS units, $k_B = 1.4 \times 10^{-23} \text{J K}^{-1}$.

There are four important features of the Boltzmann relation: (1) The relative probability of a microstate decreases as the energy increases. In other words, lower-energy microstates are more likely (note that a microstate is a single x value, not a basin as in Figure 3.1). Basins will be addressed later. (2) The Boltzmann factor gives relative probabilities only, since the normalizing prefactor is absent (a point familiar to us from continuous distributions discussed earlier). In principle, any continuous distribution can be normalized, but in most statistical mechanics problems of interest—typically high dimensional—the normalization is unknown and difficult to calculate even when U is easy to evaluate. (3) The relative probabilities of all microstates become more equal as temperature increases. This is easy to see by taking the ratio of Boltzmann factors for two different x values and letting T become large. The key physical consequence is that states that are unlikely at one temperature become more likely as T increases. (4) Adding a constant to the overall energy function U—that is, the same constant for all x values—is irrelevant because the relative probabilities remain unchanged; that is, the ratio of the probabilities for any two configurations will not change when a constant is added to U.

PROBLEM 3.1

Starting from MKS units, convert $k_B T$ at $T = 300\,\text{K}$ to units of kcal/mol, which are the conventional units of chemistry and biology.

PROBLEM 3.2

Consider the ratio of probabilities for two x values, say, x_1 and x_2. Show that this ratio does not change when the potential energy function is shifted by any constant $U(x) \to U(x) + U_c$.

One caveat in our discussion of the Boltzmann factor: for simplicity, we are excluding from consideration the kinetic energy, $\text{KE} = (m/2)v^2$, where m is the mass and v the velocity of our one-dimensional particle. In fact, the total energy $E = U(x) + \text{KE}(v)$ should be used in the Boltzmann factor. Nevertheless, we make no error to exclude the kinetic energy, as will be detailed in Section 3.8. Indeed, we can quickly see why this is the case, by noting that the full Boltzmann factor is factorizable:

$$e^{-E/k_B T} = e^{-U(x)/k_B T} \cdot e^{-(m/2)v^2/k_B T}. \tag{3.2}$$

The first factor depends only on the variable x and the second only on v, which means that the variables are statistically independent (Section 2.4.3). In turn, this means that the distribution of velocities does not affect the distribution of positions, and Equation 3.1 is indeed correct.

Derivations of the Boltzmann factor rely on the system of interest being in contact with a much larger system (a "bath") at the temperature T. The two systems can freely exchange energy, which happens physically via molecular collisions. Thus, if the system of interest gains energy, the bath loses it. In essence, it is the bath's loss of energy that is probabilistically unfavorable for the combined system and bath, because decreased energy restricts possibilities for the larger bath. Readers interested in mathematical derivations can consult the books by Reif, by Dill and Bromberg, and by Phillips et al.

3.2.1 TRANSLATING PROBABILITY CONCEPTS INTO THE LANGUAGE OF STATISTICAL MECHANICS

What we need to learn by the end of this chapter is that equilibrium statistical mechanics is nothing more than elementary probability theory (at least at a fixed temperature). This has to be true, since all equilibrium statistical mechanics can be derived from the Boltzmann factor and suitable variants—all of which specify nothing more than probability densities. Let's begin the process of translating these ideas.

First of all, if we want the exact probability density (rather than just a weighting factor proportional to it), we need only normalize the Boltzmann factor:

$$\text{pdf}(x) \equiv \rho(x) = \frac{\exp\left[-U(x)/k_B T\right]}{\int_V dx \,\exp\left[-U(x)/k_B T\right]}. \tag{3.3}$$

Here, I have specified the "volume" V over which we are integrating. In one dimension, of course, the volume is just a linear region or a set of linear regions.

Similarly, if we want to calculate the average of any function, say $g(x)$, we write our usual weighted average as

$$\langle g \rangle = \frac{\int_V dx \, g(x) \,\exp\left[-U(x)/k_B T\right]}{\int_V dx \,\exp\left[-U(x)/k_B T\right]}. \tag{3.4}$$

The integral appearing in the denominators of both Equations 3.3 and 3.4 is just a normalizing constant. On the other hand, it has physical meaning because it represents a summation of all probability weight in the region of interest. For this and other reasons, the normalizing integral gets a special name, the partition function. (It is sometimes called a configuration integral to emphasize the fact that kinetic energy has been omitted.) The symbol Z or Q is usually given to the partition function. We will use $\hat{Z}$ for the normalizing configuration integral of Equations 3.3 and 3.4, with the symbol "^" to remind us of the absence of the kinetic energy. The full partition function differs from $\hat{Z}$ by a simple factor denoted by λ. We therefore have

$$\text{Partition function} = \hat{Z} \equiv \int_V dx \,\exp\left[\frac{-U(x)}{k_B T}\right] \equiv \lambda Z. \tag{3.5}$$

Below we will see that, because λ has units of length, the full partition function Z is dimensionless.

It's worth noting that, in our probabilistic context, the partition function is not actually a function in the mathematical sense. It's just an integral, which is evaluated to be a plain old number once the potential function U, temperature T, and volume V have been specified. "Plain old numbers" are just plain old numbers like 2 or 3.167 or 5×10^{27}. On the other hand, one can imagine leaving the temperature as a variable T, in which case $Z(T)$ would be a function, yielding a different value for every T. Alternatively, the volume V could be varied, suggesting the function $Z(T, V)$. These and other generalizations will be considered when we study thermodynamics in Chapter 7.

3.2.2 PHYSICAL ENSEMBLES AND THE CONNECTION WITH DYNAMICS

We've probably said more than 10 times that we will explain statistical mechanics in terms of simple probability concepts. But this begs the question, Where is the physics? It turns out the physics is embedded in the connection between dynamics (driven by plain-old mechanics) and ensembles. This connection means that time averages and statistical averages must be the same, as we suggested in Section 1.5.3.

Once again, imagine a large number (on the order of Avogadro's number!) of identical macromolecules in a solution in a test tube, but assume the solution is dilute enough so that the molecules don't interact. You can imagine measuring average properties of the system by taking a "snapshot" at a single moment in time, encompassing all the configurations of all the molecules. In this snapshot, molecular configurations would be statistically distributed, that is, they would form an ensemble or statistical sample. You could also imagine measuring average properties by following a single molecule over a long period of time. In this case, the molecule would undergo its natural fluctuations (dynamics) and occupy different configurations for different fractions of the time. These fractional occupancies (populations) also constitute an ensemble. Since all the macromolecules in the solution are identical, they must experience the same populations over time—except in different, random orders. Hence, a snapshot of the whole solution must catch the molecules with exactly the same distribution as each experiences individually, over time.

Experiments, it is interesting to note, combine both types of averaging. Except for special single-molecule studies, most experiments study large numbers of molecules (e.g., in a solution) and hence average over many molecules simultaneously. On the other hand, no piece of equipment is capable of gathering data instantaneously, and therefore any measurement—even "at a single time point"—in fact averages over some window of time determined by the speed of the instrument.

It is worth mentioning that all the dynamical fluctuations experienced by a molecule necessarily are caused by mechanical forces—typically collisions with solvent molecules. The velocities of collision, in turn, are regulated by temperature (average kinetic energy of molecules in the "environment"). Thus, the chain of logic is complete: mechanics leads to dynamics, regulated by temperature, and in turn leads to configurational fluctuations—and these last constitute the ensemble itself.

3.2.3 SIMPLE STATES AND THE HARMONIC APPROXIMATION

While typical energy landscapes are complex and possess many barrier-separated basins, as depicted in Figure 3.1, important lessons can be learned from the simplest type of potential containing a single basin (see Figure 3.2). It is reasonable to assume

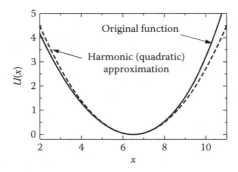

FIGURE 3.2 Single-well one-dimensional potential. We can approximate the potential energy (solid line) near the minimum by a simple "harmonic"—that is, quadratic—potential (dashed line). A single-well potential can be thought of as a simple approximation to one of the states of butane shown in Figure 1.4.

that our single-basin energy function $U(x)$ is smooth, and therefore can be represented as a Taylor series about the value of its minimum x_0. Thus, we write

$$U(x) = U(x_0) + (x - x_0)U'(x_0) + (1/2)(x - x_0)^2 U''(x_0) + \cdots , \qquad (3.6)$$

where you should recognize that the linear term vanishes because the slope at the minimum must vanish: $U'(x_0) = 0$. Because our energy basin is roughly parabolic or "harmonic" in shape, we will approximate $U(x)$ by the constant and quadratic terms in Equation 3.6, as illustrated in Figure 3.2. (In fact, energy basins will almost always be quadratic near enough to their minima; the validity of the harmonic approximation for statistical mechanics requires the approximate energy and the energy to agree reasonably as the energy increases by a few $k_B T$ from the minimum.)

We now focus on the harmonic part of Equation 3.6,

$$U_{\text{harm}}(x_0) \equiv U_0 + \left(\frac{\kappa}{2}\right)(x - x_0)^2, \qquad (3.7)$$

where we have set the energy minimum to be $U_0 = U(x_0)$ and the "spring constant" as $\kappa = U''(x_0)$. With this simple harmonic potential, we can perform a number of interesting statistical mechanics calculations exactly. Indeed, the Boltzmann factor associated with U_{harm} is simply a Gaussian. One is often interested in the average energy, which in our case is given by

$$\langle U_{\text{harm}} \rangle = \frac{\int dx\, U_{\text{harm}}(x)\, \exp[-U_{\text{harm}}(x)/k_B T]}{\int dx\, \exp[-U_{\text{harm}}(x)/k_B T]}. \qquad (3.8)$$

PROBLEM 3.3

Explain in words why this expression is correct, using probability theory concepts.

After substituting the explicit form for U_{harm} into Equation 3.8, this integral can be performed exactly, and one finds

$$\langle U_{\text{harm}} \rangle = U_0 + \frac{k_B T}{2}. \tag{3.9}$$

PROBLEM 3.4

(a) Derive Equation 3.9 by integrating the harmonic potential from negative to positive infinity. Use a table of integrals if necessary. (b) Why is only a minimal error made (mathematically) when the integration limits are (unphysically) extended to infinity, if the harmonic approximation is reasonable for the temperature of interest? This question requires a little bit of careful thought.

Although Equation 3.9 does not look special, it is quite an interesting result. In fact, it is most interesting for what it is *not* in it, namely, the effective spring constant κ. In other words, the average energy of a harmonic well does not depend on how broad or narrow it is. This rather intriguing result is known as the "equipartition theorem" of statistical mechanics because it applies to any system in which a variable appears only quadratically in the energy. It is a useful rule of thumb: energy will tend to partition itself equally among possible "modes" (roughly, degrees of freedom) with $k_B T/2$ alloted to each. In this light, it is obvious why the kinetic energy, which is harmonic in velocity, averages to $k_B T/2$ per degree of freedom. More on this later.

Despite the simple behavior of the average energy, we know there's something else which is important—entropy. However, because we are experts in probability theory, we can actually derive the entropy in a very simple way. This will be pursued below, when we consider the relative populations of "states."

3.2.4 A HINT OF FLUCTUATIONS: AVERAGE DOES NOT MEAN MOST PROBABLE

One of the main points of this entire book is that structural fluctuations in proteins and other molecules are important. In other words, a single structure provides a highly incomplete picture. Statistical mechanics turns these statements into mathematics, since every structure has some probability (proportional to its Boltzmann factor) of occurring. The observed behavior embodies all structures and their probabilities. Here, we examine this idea a bit more concretely, and see that even the most important—most probable—configuration surely falls short of fully describing a system at finite temperature.

The one-dimensional Lennard-Jones potential is an excellent example. It is a nonsymmetric potential, and therefore makes apparent the importance of configurations (x values) other than the most probable (lowest energy) one. This potential is used in typical molecular mechanics potential energy functions ("forcefields") to represent the interaction between two atoms that are not covalently linked. It has the shape shown in Figure 3.3 and the mathematical form

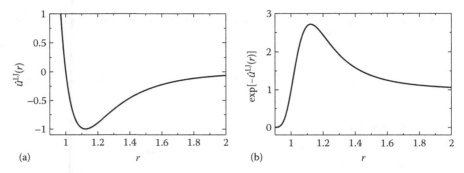

FIGURE 3.3   The shape of the Lennard-Jones potential (a) from Equation 3.10, which models interactions between atoms and an illustrative Boltzmann factor (b) for a fictitious temperature where $k_B T$ is the same as the well depth. Both are plotted as a function of the center-to-center distance between two atoms.

$$U(r) \propto 4\left[\left(\frac{r_0}{r}\right)^{12} - \left(\frac{r_0}{r}\right)^{6}\right] \equiv \hat{u}^{\mathrm{LJ}}(r), \tag{3.10}$$

where r denotes the distance between atomic centers and r_0 is a constant that sets the lengthscale over which the potential varies.

The lower-case symbol "$\hat{u}^{\mathrm{LJ}}$" is defined to be a dimensionless function. The overall factor of four simply makes the minimum of $\hat{u}^{\mathrm{LJ}}$ exactly equal to -1.

The shape of the potential Equation 3.10 is highly asymmetric. In fact, the average value of r can be significantly different from the most probable, as the following exercises will show.

PROBLEM 3.5

In the following, treat the potential of Equation 3.10 as one dimensional (even though it really refers to distance in three dimensions, which brings in some complications). (a) The most probable "configuration" is the one with the lowest energy. Calculate the most probable value of r with paper and pencil. Explain why it is independent of temperature. (b) Assume r can vary in the range $0.01 r_0 < r < 10\, r_0$ and set $r_0 = 1$. Write a simple computer program to calculate the Boltzmann-weighted average r value, $\langle r \rangle$, considering the discrete states occurring at increments of 0.01 in r. Use the dimensionless potential $\hat{u}^{\mathrm{LJ}}$ and the dimensionless temperatures $k_B T = 1, 2, 5, 10$. (c) Explain in words why the average and most probable structures typically differ. (d) Explain in words the variation of the average with temperature. (e) Without calculation, what values do you expect for $\langle r \rangle$ in the limits $k_B T \to 0$ and $k_B T \to \infty$?

It is important to realize that the average configuration is also not a complete description. It, too, is just a single configuration. In fact, the average configuration may be completely meaningless, as the following exercise shows.

PROBLEM 3.6

Sketch a bimodal distribution (two peaks) that is perfectly symmetric about $x = 0$ and that has very low probability at the center. Explain why the average configuration poorly describes the ensemble of possibilities.

The bottom line is that neither the most probable nor the average configuration of a "statistical" system is a good description of it. Never forget this!

3.2.4.1 Aside on the Physical Origin of the Lennard-Jones Function

Interestingly, the r^{-6} term in $\hat{u}^{LJ}$ has a precise physical origin while the r^{-12} term does not. The r^{-6} term quantifies the attractions calculated from quantum mechanical fluctuations of charge density that occur in all neutral atoms. On the other hand, the functional form r^{-12} is an invented term that crudely approximates the (genuine) repulsions among all atoms at close distances. Note also that in the physical case of three dimensions, the distribution based on $\hat{u}^{LJ}$ would require a Jacobian as described in Chapter 5.

3.3 STATES, NOT CONFIGURATIONS

We have already been using the word "state," but what does it really mean? The word state is one of the more slippery concepts in biophysics, mostly because it can be used in so many different ways. Here we will try to use it in a fairly consistent way, to mean a group of configurations belonging to the same energy basin. There may be times when the word is used to group together a number of similar energy basins—for instance, in Figure 3.1, one could say that there are two basic states. But the main point is that a state will be taken to mean a group of configurations or microstates.

Operationally, the main consequence of our definition is that it will not be possible to speak of the potential energy of state. After all, the potential energy is a mathematical function in the traditional sense—it "maps" a set of real numbers (the coordinates) to a single real number (the energy). But a state consists of multiple configurations, each with its own specific set of coordinates and hence potential energy. One cannot say a state has a particular potential energy, although one can speak of the average potential energy as we did above when we considered the harmonic approximation to a single-well (single-basin) landscape.

As we hinted above, the energy is not the whole story—we will also need to consider the entropy.

3.3.1 RELATIVE POPULATIONS

It can be argued that probability, unlike energy, is the whole story. In fact, we will now see that if we consider relative probabilities of states, the average energy and entropy emerge simply and naturally. On a concrete level, if you imagine trying to describe a real situation in a test tube or a cell, who cares about the energy if it can't predict the observed behavior? However, by definition, the observed behavior of a large number

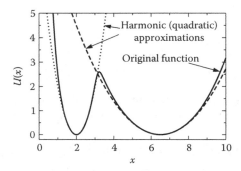

FIGURE 3.4 Double-well one-dimensional potential. We can use the harmonic approximation to estimate the relative populations (probabilities) of the two states (basins). The two wells
can be thought of as a crude approximation to two of the states of a molecule such as butane
(Figure 1.4).

of molecules under identical conditions will be determined by probabilities. In other
words, probabilities are typically observed—not energies.

Our strategy for understanding states will be simple. We will again employ the
harmonic approximation, but now to each of the two wells shown in Figure 3.4. We
will therefore be able to obtain both the relative probabilities and average energies of
the states. Based on these results, we will see that the average energy is not sufficient
to determine probability, and hence we will be led to consider what does. It will turn
out that the widths of the wells are also critical to determining probability, but this
width gets a fancy name—entropy!

We want to consider the relative probabilities of two smooth wells, such as
those depicted in Figure 3.4: how much of the population will be in the left state
compared to the right? Since the probability of any configuration (microstate or
simply the x value, in one dimension) is proportional to the Boltzmann factor, we
need to integrate over all Boltzmann factors in each state (basin). You may say that
we don't know the proportionality constant for the probability—but we don't need
to. Since the two states belong to the same system—that is, are governed by the same
potential energy function—this constant is the same for both. In other words, we can
compare the probability of an x value from the left well to that of a value on the
right (or we could consider two in one basin). Thus, the problem at hand amounts to
summing up these probabilities for each state—that is, integrating over the Boltzmann
factor.

We will use the harmonic approximation to do this, and make sure it's justified.

3.4 FREE ENERGY: IT'S JUST COMMON SENSE. . . IF YOU BELIEVE IN PROBABILITY

In the next few paragraphs you will learn nearly everything you need to know
about free energy. That seems crazy, but it's true. You will see that obscure
thermodynamic/statistical-mechanics jargon can be understood in simple terms.

3.4.1 Getting Ready: Relative Populations

We face a straightforward problem. What are the relative probabilities of the two states shown in Figure 3.4? That is, if the left state is state A and the right one state B, what is the ratio of the two probabilities p_A/p_B? To answer this, all we need to do is "add up" (i.e., integrate) all the probability densities in each state. Really, that's it, and it's easy.

In terms of the equations, it's easy to write down what we need. For instance, the probability for state A is simply

$$p_A = \int_{V_A} dx\, \rho(x) = (\hat{Z})^{-1} \int_{V_A} dx\, \exp\left[-U(x)/k_B T\right], \qquad (3.11)$$

where V_A is the region encompassing the whole left basin, and the factor of $1/\hat{Z}$ ensures that we are using the true probability density. Of course, the probability for state B is identical in form, but the integral is performed over V_B. The presence of the partition function might be cause for worry, since these are usually impossible to calculate (in systems with many dimensions). However, it is critical on a conceptual level to recognize that a partition function is proportional to a sum (integral) of probability—and is therefore the right normalization.

PROBLEM 3.7

If states A and B together comprise all of space, show that Equation 3.11 implies that $p_A + p_B = 1$.

An important simplification occurs, however, when we consider *relative* populations. In particular, the ratio of probabilities, for states A and B, is given by a ratio of configuration integrals:

$$\frac{p_A}{p_B} = \frac{\int_{V_A} dx\, \exp\left[-U(x)/k_B T\right]}{\int_{V_B} dx\, \exp\left[-U(x)/k_B T\right]} \equiv \frac{\hat{Z}_A}{\hat{Z}_B}. \qquad (3.12)$$

Note that overall normalization factors of $\hat{Z}$ have canceled each other, and we have defined the state-specific configuration integrals $\hat{Z}_i$, with $i =$ A or B.

To proceed, we will make use of the harmonic approximation embodied in Equation 3.7. Here, we will need different κ values, κ_a and κ_b, to describe the "spring constants" for each of the two minima (x_A and x_B); note that both minima are set to be zero, exactly, in this special case we are studying—see Figure 3.4. Thus, near x_i for $i =$ A or B, we set $U(x) \simeq (\kappa_i/2)(x - x_i)^2$. While it is true that the potential of Figure 3.4 has been designed to be well approximated by harmonic wells, the basic principles discussed here are true for all reasonably shaped potentials.

We're now ready to integrate the Boltzmann factors needed in Equation 3.12 for the ratio of probabilities. Recalling Problem 3.4, in the harmonic approximation, we have Gaussian integrals, and we extend the limits of integration to $\pm\infty$ since this leads to only a tiny error in the integration for a single basin; that is, using state B as an example, we have

$$\int_{V_B} dx \, \exp\left[\frac{-U(x)}{k_B T}\right]$$

$$\simeq \int_{-\infty}^{\infty} dx \, \exp\left[\frac{-\kappa_b(x - x_B)^2}{2k_B T}\right] = \sqrt{\frac{2\pi k_B T}{\kappa_b}}. \qquad (3.13)$$

The key mathematical point here is that although $U(x)$ describes both states A and B, the quadratic expansion about x_B only describes state B. Thus, extending the integration limits to infinity does not somehow include state A. Rather, all the x values distant from x_B essentially do not weigh in to the integral. If we finally combine the results for both states, we obtain the ratio of probabilities in the harmonic approximation as

$$\frac{p_A}{p_B} \simeq \sqrt{\frac{\kappa_b}{\kappa_a}}. \qquad (3.14)$$

Equation 3.14 indeed makes sense, as the following exercise shows.

PROBLEM 3.8

(a) Is κ larger for narrow or wide wells? Explain. (b) Why then does the result (Equation 3.14) make sense, given the potential as sketched in Figure 3.4?

OK, we've done the math, where's the free energy? For that matter, where's the energy at all? Well, first of all, the spring constants κ do tell us about the shape of the potential energy function as illustrated in the preceding exercise. So somehow the information about U enters Equation 3.14. Yet, on the other hand, it's interesting and critical to note that the average energy cannot distinguish probabilities of narrow vs. broad states. To see this, recall from Equation 3.9 that the average energy does not depend at all on the spring constant κ of a harmonic well. If we take the ratio of the Boltzmann factors of the average energies, we simply get one—regardless of the κ values! That is, $\exp[-\langle U \rangle_A]/\exp[-\langle U \rangle_B] = 1$. (If you think about it, this insensitivity to κ suggests the result holds approximately for potentials with different shapes.) Something clearly is missing from the average-energy description.

Of course, what's missing from a statistical mechanics point of view is entropy and free energy. On the other hand, from the probability point of view, nothing at all is missing! We already know the answer to the question we set out to ask—that is, we know the ratio of probabilities, which is given in Equation 3.14.

3.4.2 FINALLY, THE FREE ENERGY

We therefore take the perspective that we will define free energy and entropy in terms of probabilities. After all, the probabilities are calculated from basic mathematical principles, so you should think of them as completely reliable. Yet because physicists insist on describing systems in terms of energy rather than probability, they must consider a quantity that has come to be called the free energy. Historically, this stems from the development of thermodynamics preceding statistical mechanics.

Here we will define the free energy of a state to be *that energy whose Boltzmann factor gives the correct relative probability of the state*. So, just the way the Boltzmann factor for the energy of a single configuration tells you its relative probability (compared to another configuration), the Boltzmann factor of the free energy of a state indicates its relative probability compared to other states. In other words, the free energy lets us stick with Boltzmann factors. The equation that restates our definition of the free energies (F_A and F_B) for states A and B—or any other pair of states—must therefore be

$$\frac{\exp\left(-F_A/k_B T\right)}{\exp\left(-F_B/k_B T\right)} \equiv \frac{p_A}{p_B}. \tag{3.15}$$

This is the fundamental meaning of free energy: probability. On the other hand, by taking logarithms and comparing with the preceding integrals (3.12), we have the more traditional (and nonintuitive) statistical mechanics relation,

$$F_i = -k_B T \ln\left(\frac{\hat{Z}_i}{\lambda}\right) = -k_B T \ln\left[\lambda^{-1}\left(\int_{V_i} dx \, \exp\left[\frac{-U(x)}{k_B T}\right]\right)\right]. \tag{3.16}$$

(Note that we have been careful to divide $\hat{Z}_i$ by λ to make it dimensionless, which is required when taking a logarithm. The lengthscale λ is discussed in Section 3.8.)

Thus, a word of advice: whenever you see a logarithm, think of exponentiating it and trying to understand the underlying probabilities.

3.4.3 More General Harmonic Wells

In the preceding discussion, we assumed that our harmonic wells always had their minima at exactly the zero of energy. This is not always the case, of course. Thus, in the general harmonic approximation, one assumes that state i is characterized by the potential

$$U(x) \simeq U_i + \left(\frac{\kappa_i}{2}\right)(x - x_i)^2, \tag{3.17}$$

where x_i is the basin minimum and $U_i = U(x_i)$ is the minimum energy—that is, a constant.

It is then easy to show that instead of Equation 3.13, one finds that the configuration integral for a general one-dimensional harmonic potential is

$$\int_{V_i} dx \, \exp\left[\frac{-U(x)}{k_B T}\right] \simeq \int_{-\infty}^{\infty} dx \, \exp\left[\frac{-U_i}{k_B T} - \frac{\kappa_i(x - x_i)^2}{2k_B T}\right]$$

$$= \exp\left(\frac{-U_i}{k_B T}\right)\sqrt{\frac{2\pi k_B T}{\kappa_i}}. \tag{3.18}$$

PROBLEM 3.9

Explain why Equation 3.18 can be derived in one or two lines of math from
Equation 3.13.

The free energy of a harmonic well follows trivially from Equation 3.18 and
amounts to

$$F_i^{\text{harm}} = U_i - k_{\text{B}}T \ln \left(\lambda^{-1} \sqrt{\frac{2\pi k_{\text{B}}T}{\kappa_i}} \right) = U_i + \left(\frac{k_{\text{B}}T}{2} \right) \ln \left(\frac{\kappa_i \lambda^2}{2\pi k_{\text{B}}T} \right). \quad (3.19)$$

Notice that the constant U_i in the energy also enters the free energy as a simple
constant. This can be thought of as a mathematically trivial overall shift, but if one
is comparing two harmonic wells, it is clear that the heights (energy values) of the
minima are highly relevant!

3.5 ENTROPY: IT'S JUST A NAME

What about entropy? Although I've argued that our probabilistic discussion is complete, sooner or later we'll need to communicate with scientists who perversely insist
on discussing entropy. There's only one thing to do about this: stick with what we
know.

We will define the entropy S simply in terms of the free energy and average
energy, both of which have a solid basis in probability concepts. Quite simply, we
will take the entropy to be proportional to the difference between $\langle E \rangle = \langle U \rangle + \langle \text{KE} \rangle$
and F. Our proportionality constant will be the temperature based on thermodynamics
we haven't yet discussed, but it's an arbitrary choice from our current perspective.
Thus, we will define the entropy to be

$$S \equiv \frac{\langle E \rangle - F}{T}. \quad (3.20)$$

(Experienced thermodynamicists will note equivalence to the definition of the
Helmholtz free energy as $F = \langle E \rangle - TS$.) From our current point of view, entropy is
just a kind of "fudge term"—a correction with no obvious basis. However, we will
soon see that it has a fundamental physical meaning. We will also see, in Section 3.8,
that the average kinetic energy occurring as part of Equation 3.20 is just a constant
that does not affect the probabilities of interest.

Our definition (3.20) is mathematically equivalent to

$$e^{-F/k_{\text{B}}T} = e^{+S/k_{\text{B}}} \, e^{-\langle E \rangle/k_{\text{B}}T}. \quad (3.21)$$

Since the Boltzmann factor of the free energy gives the probability of a state, $e^{+S/k_{\text{B}}}$
would appear to indicate how many times the Boltzmann factor of the average energy
needs to be counted in order to correct for the full probability—that is, for the extent
of the state. Below, we will see explicitly that this is true. *Equation 3.21 is one of the
most important in this book* because of the clear picture it provides.

We can make a further simplification to the entropy, since we know (and will demonstrate in Section 3.8) that the average kinetic energy $\langle KE \rangle$ is proportional to temperature. Specifically, $\langle KE \rangle = k_B T/2$ in one dimension—which is not unexpected since $mv^2/2$ is harmonic in v. Thus $\langle KE \rangle/T$, which is part of the entropy definition (3.20), is the simple constant $k_B/2$, and we find that

$$\text{One dimension:} \quad S = \frac{k_B}{2} + \frac{\langle U \rangle - F}{T}. \tag{3.22}$$

One way to think about Equation 3.22 is that it accounts explicitly for the "hidden" second dimension of the velocity, which necessarily accompanies a configurational degree of freedom in a physical system.

3.5.1 ENTROPY AS (THE LOG OF) WIDTH: DOUBLE SQUARE WELLS

To understand the physical meaning of entropy, we will be best served by studying a potential energy even simpler than the harmonic case—the "square"-well system of Figure 3.5. To my mind, this is the most important system in statistical mechanics because it is the only one you can understand fully (including the math!) and it still captures the fundamental physics. The double-square-well system illustrates the connection between probability, energy, and entropy especially well.

We will be interested in the usual ratio of the probabilities of the two states. State A extends from $x = a$ to $x = b$ with constant energy U_A. Similarly, state B extends from $x = c$ to $x = d$ with constant energy U_B.

As before—and by the definition of the probabilities—the desired ratio is given in terms of the state partition functions, so that $p_A/p_B = Z_A/Z_B$. Now, however, everything is trivial to calculate. Since the energy within each state is constant, so too is the Boltzmann factor. Thus, the partition function is an integral over a constant; in other words, it is the area of a simple rectangle. That's about as easy as an integral can be, and you can calculate that $Z_A = \lambda^{-1}(b - a) \exp(-U_A/k_B T)$ and $Z_B = \lambda^{-1}(d - c) \exp(-U_B/k_B T)$. By using the definition of F given in Equation 3.16,

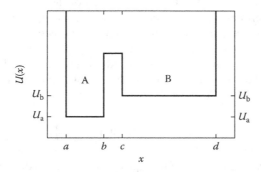

FIGURE 3.5 A double-square-well potential. Everything about this system can be solved with elementary calculations, and the results are highly revealing about entropy.

we therefore find

$$F_A = U_A - k_B T \ln \left(\frac{b-a}{\lambda} \right) \quad \text{and} \quad F_B = U_B - k_B T \ln \left(\frac{d-c}{\lambda} \right). \quad (3.23)$$

PROBLEM 3.10

Starting from the definition of the double-square-well potential, derive Equation 3.23.

Not only can we calculate the free energies exactly, but we can also calculate the exact average energies. In fact, we can just "read them off" from the definition of the square-well U. After all, the average of a constant function must be a constant, even in a confusing business like statistical mechanics. Thus, we have $\langle U \rangle_A = U_A$ and $\langle U \rangle_B = U_B$. (And if you don't believe it, you can find the same result by substituting into Equation 3.4.)

You've probably guessed the next move: use the exact free and average energies to calculate the exact entropy. This is precisely the right thing to do. Using our definition of entropy (3.20), we notice that each $\langle U \rangle_i$ cancels out from F, and we're left with the very simple results

$$S_A = +k_B \left[\ln \left(\frac{b-a}{\lambda} \right) + \frac{1}{2} \right] \quad \text{and} \quad S_B = +k_B \left[\ln \left(\frac{d-c}{\lambda} \right) + \frac{1}{2} \right]. \quad (3.24)$$

This is the big payoff. *Entropy is just the logarithm of the width of a state* (plus an irrelevant constant). That's it. That's all it ever is. You can call entropy "uncertainty" or whatever you like, but it fundamentally measures the extent of a state. This "extent," of course, quantifies the number of ways in which the state can be occupied—see again Equation 3.21—which is sometimes termed degeneracy. A state with larger entropy can be occupied in more different ways (i.e., it has more microstates/configurations and higher degeneracy). The x values in the ranges from a to b and from c to d all represent different possible system "configurations." The λ in the denominator simply sets the scale or units of the lengths (and cancels out when physically pertinent entropy differences are considered). Further, because λ turns out to be a tiny quantum mechanical lengthscale (Section 3.8), the $k_B/2$ terms in the entropy expressions are inconsequential compared to the log terms.

The meaning of entropy as (the log of) width is even clearer if we consider the entropy difference between the two states. The simple subtraction of the results in Equation 3.24 immediately leads to

$$\Delta S = S_B - S_A = k_B \ln \left(\frac{d-c}{b-a} \right), \quad (3.25)$$

which says the entropy difference of the square-well system is determined exactly by the log of the ratio of the widths. This is a general result, so long as the meaning of the "width" of a state is appropriately defined—as we will see in the harmonic case.

3.5.2 Entropy as Width in Harmonic Wells

The same physical picture we just derived also applies to harmonic systems and, by extension, to most potentials of chemical and biological interest. It's a bit less trivial to see this, but the effort is worthwhile since it gives us additional confidence in our intuitive understanding of entropy as basin width.

We can proceed in precisely the same way as before. We will compute entropy via the difference between the exact free and average energies. Thus, by combining Equations 3.9, 3.19, and 3.22, we find

$$S_i = k_B + k_B \ln \left(\frac{\sqrt{2\pi k_B T / \kappa_i}}{\lambda} \right), \tag{3.26}$$

where κ_i is the "spring constant" of the harmonic well. The width of the harmonic well is less obvious in this expression, but it is indeed there, since $s_i \equiv \sqrt{k_B T / \kappa_i}$ has units of length. More specifically, in the probability distribution of x values defined by the Boltzmann factor of a harmonic potential, s_i is just the standard deviation. Since the standard deviation measures the scale of typical fluctuations, it is indeed the effective width of a harmonic well at temperature T.

The factor of $\sqrt{2\pi} \simeq 2.5$ can be thought of as a geometric factor correcting for the shape of the harmonic state, which indeed extends beyond first standard deviation. For the thermodynamics experts, it is also worth noting that our whole classical discussion of statistical mechanics ceases to be valid when the temperature is so low that $s_i \sim \lambda$. In other words, Equation 3.26 is not valid when $T \to 0$.

PROBLEM 3.11

Show that $s_i = \sqrt{k_B T / \kappa_i}$ is the standard deviation of the distribution associated with the general harmonic potential (3.17), regardless of the constant U_i.

The physical and probabilistic meaning of a harmonic well's entropy is most clear when one recalls that the true point of the free energy and its associated quantities is to provide the relative probability of a state when compared to other states. This is the meaning of the fundamental equation (3.15). Therefore, the only quantity with physical and probabilistic meaning is the free energy difference, $F_i - F_j = U_i - U_j - T(S_i - S_j)$. In other words, only entropy differences really matter, and one finds

$$\text{harmonic wells:} \quad S_i - S_j = k_B \left[\ln \left(\frac{s_i}{\lambda} \right) - \ln \left(\frac{s_j}{\lambda} \right) \right] = k_B \ln \left(\frac{s_i}{s_j} \right), \tag{3.27}$$

where the effective width s was defined earlier in this section.

The "take-home" message from the square wells example still stands: entropy is just proportional to the logarithm of the width of the basin. From the harmonic example, we have learned that this width must be interpreted as the range of typical fluctuations for the pertinent temperature. This range makes sense: after all, there is really no other range to consider. A harmonic well goes on forever, with ever-increasing energy, and the thermal energy tells us where to stop.

PROBLEM 3.12

We can consider an alternate definition of the range or width of a harmonic well, and see that it leads to the same result. Do this by defining the range as that value of the deviation from the minimum x_i at which the energy is exactly $k_B T$ above the minimum, and show that the same result (3.27) for the entropy difference is obtained. Show this also holds when the range is defined by an arbitrary multiple of $k_B T$.

We note that from a conventional thermodynamic perspective, Equation 3.26 is problematic, because as the temperature reduces, the entropy becomes large and negative. Really, the entropy should vanish as $T \to 0$, though this is a quantum mechanical result. While the details of this issue are beyond the scope of this chapter, note that λ itself is really a quantum mechanical lengthscale—it contains Planck's constant as we'll see at the end of the chapter. Thus, when s_i becomes smaller than λ, the whole classical picture in which we are working breaks down. In other words, our equations are physically valid for the range of (relatively high) temperatures of interest. Furthermore, no problem arises when one sticks to the strict probability picture without considering the entropy on its own.

PROBLEM 3.13

Consider a particle in a one-dimensional linear potential $U = kx$, which is restricted to values $x > 0$. Calculate (a) the free energy and (b) the entropy. (c) Explain why the entropy makes sense in terms of the physical scale(s) of the system. (d) Does the entropy behave as expected when the temperature decreases?

3.5.3   That Awful $\sum p \ln p$ Formula

This section is for people who have seen entropy expressed in terms of the sum $\sum_j p_j \ln p_j$, and who have been rather baffled by it. I count myself among this group, and I'm not sure I know too many physicists who have an intuitive feel for this formula. It is indeed a critical relation in information theory, computer science, and related fields. Due to its prevalence in the literature, it's worthwhile to understand whether (and how) the formula might be correct.

We should, as always, start from what we believe. Here that will mean the definition of entropy (3.20) in terms of the difference between free energy and average energy, namely, $S = (\langle E \rangle - F)/T$. Given that the $p \ln p$ formula has subscripts that may refer to states, we will divide our continuous variable (x) into discrete regions. We will assume these regions are small enough so that the energy is constant within each one (see Figure 3.6). (This can be accomplished to any degree of approximation desired, since the regions can be made as small as desired.)

Within our discrete model, everything can be calculated exactly, or, at least we can write down exact formulas. First, for any constant-energy "state" j, with energy

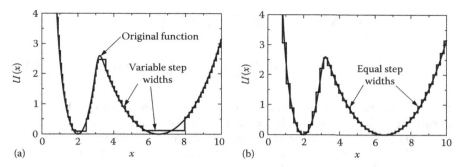

FIGURE 3.6 Stepwise discretization of a continuous one-dimensional potential. (a) The potential is approximated using variable-width intervals, based on the slope of the potential. (b) The plot shows equi-sized intervals. These constructions are useful in understanding the $\sum p \ln p$ entropy formula.

u_j, we can assign the exact probability by integrating over the probability density. This yields

$$p_j = (\lambda Z)^{-1} \int_{V_j} dx \, \exp\left[\frac{-U(x)}{k_B T}\right] = \left(\frac{l_j}{\lambda Z}\right) \exp[-u_j/k_B T], \qquad (3.28)$$

where l_j is the length of the constant-energy state and $Z = \hat{Z}/\lambda$ is the full partition function for the whole system.

The length l_j could vary for different j since the slope of the energy varies: see Figure 3.6a. Note that our construction guarantees that the probability is suitably normalized, so that $\sum_j p_j = 1$.

PROBLEM 3.14

Show that the discrete probabilities p_j are normalized.

With the probabilities in hand, we can now calculate whatever we need. Let's start by "turning around" Equation 3.28 to write the potential energy in terms of the probability, yielding

$$u_j = -k_B T \ln\left(\frac{p_j \lambda Z}{l_j}\right). \qquad (3.29)$$

Using this result, the average potential energy takes on a very interesting appearance, namely,

$$\langle U \rangle \equiv \sum_j p_j u_j = -k_{\mathrm{B}} T \sum_j p_j \ln \left(\frac{p_j \lambda Z}{l_j} \right)$$

$$= -k_{\mathrm{B}} T \sum_j p_j \ln \left(\frac{p_j \lambda}{l_j} \right) - k_{\mathrm{B}} T \ln (Z), \qquad (3.30)$$

where we have used properties of the logarithm and the normalization of the p_j in the last equality. This is rather remarkable: the average potential energy itself looks like the entropy formula, aside from an overall constant term and the lengths l_j! In fact, recalling Equation 3.16, we see that the constant term is just the free energy of the whole system. We therefore find the entropy has *almost* the desired form

$$S \equiv \frac{(\langle E \rangle - F)}{T} = k_{\mathrm{B}} \left[\frac{1}{2} - \sum_j p_j \ln \left(\frac{p_j \lambda}{l_j} \right) \right], \qquad (3.31)$$

which is an exact result for our discrete model.

But wait, don't go away yet! A couple of key points should be made. First, the $\sum p \ln p$ entropy formula is nothing more than what we started with—the difference between the average and free energies. Second, and most important, the l_j factors are critical and indicate that the relation must be applied with care. These factors, in fact, are absent from most presentations of the formula. The significance is that, given our reasonable use of constant-energy ministates, a sort of "local entropy" term comes into play (proportional to $\ln l_j$). It is possible to derive a formula lacking the l_j factors, but then one must assume they are all equal—that is, one's model must take the ministates to be of equal length (or volume in higher dimensions). We can then write

$$\text{Equal volumes only:} \quad S = -k_{\mathrm{B}} \sum_j p_j \ln (p_j) + \text{constant}, \qquad (3.32)$$

where it is fair to say the constant is irrelevant since it is the same for different states (of the same system at some fixed temperature—λ depends on T).

My own opinion is that these entropy relations are of limited intuitive value in biophysics, especially since they could equally well be considered fundamental to the average energy (recall Equation 3.30). However, it is useful to appreciate that there should be no mysteries in biophysics. There have also been some recent interesting computational efforts to apply entropy estimates and bounds based on Equation 3.32.

3.6 SUMMING UP

3.6.1 STATES GET THE FANCY NAMES BECAUSE THEY'RE MOST IMPORTANT

Life is easy when we think in terms of individual configurations x. The Boltzmann factors give relative probabilities of these "microstates" and there's nothing more to it.

It's only when we try to group configurations together into "states" that the trouble starts. (Ironically, although we need the state picture because it is easier to understand, by summing over individual configurations we are actually discarding information about fine details of our system.) Mathematically, when we integrate over configurations/microstates, and then also insist on using energetic terminology instead of probabilistic ideas, we are led (almost) naturally to the free energy and entropy. However, we did see that entropy has a concrete physical meaning in terms of the width or "degeneracy" of a state. Such a physical picture is useful in understanding a number of phenomena (as we will see later). On a pragmatic level, we also need to communicate with other physically oriented scientists, and therefore we must understand the language of free energy and entropy.

3.6.2 It's the Differences That Matter

For configurations or microstates, we want to know the relative populations—that is, the ratio. (The absolute populations require knowing the total population of the system.) Since single configurations have no entropy, by definition, the ratio of populations for configurations i and j depends solely on the energy difference and temperature: $p_i/p_j = \exp\left[-(U_i - U_j)/k_B T\right]$. This is another way of saying that the zero of energy doesn't matter.

For states, similarly, the zero of energy does not matter—but now the energy that matters is the free energy. Hence the free energy difference governs population ratios, and we've seen that $p_a/p_b = \exp\left[-(F_a - F_b)/k_B T\right]$ for states a and b. It's worth noting that, in general, the free energy of a state is itself a function of temperature—for example, $F_a = F_a(T)$—so there is additional temperature dependence beyond that given explicitly in the ratio of Boltzmann factors. (The T-dependence of F comes from its definition as an integral over T-dependent Boltzmann factors. Similarly, both the entropy and average energy of state depend on temperature.)

Table 3.1 highlights key differences between configurations and states. Note that the Boltzmann factor of $-TS$ [i.e., $\exp(+S/k_B)$] effectively gives the degeneracy or width of the state.

TABLE 3.1

Differences between Configurations and States

	Microstate = Single Configuration	State = Set of Configuration
Average potential energy	U of single configuration	$\langle U \rangle$
Entropy	n.a.	$S \sim k_B \ln(\text{width})$
Relative probability	$e^{-U/k_B T}$	$e^{-F(T)/k_B T}$
		$\propto e^{+S/k_B}\, e^{-\langle U \rangle/k_B T}$

Note: The average kinetic energy is a constant and therefore does not affect the relative probability of a state.

3.7 MOLECULAR INTUITION FROM SIMPLE SYSTEMS

How much of what we have learned in one dimension really applies to molecules? Actually, everything. Whether you're in one dimension or three or thousands, probability is always a sum of Boltzmann factors. Hence a "state"—however you want to define that—gains probability in two ways: (1) by consisting of individual configurations that are low in energy, and (2) by including many configurations (i.e., entropically). Never forget that probability is the fundamental thing, because it leads to the observed populations of different states and configurations. Of course, sometimes energy will be more important and sometimes entropy.

To give a simple molecular example, consider the folded state of a protein and the unfolded state (all configurations not folded). In the simplest description, omitting solvent effects, folded states are stabilized because they are relatively low in average energy (or enthalpy, which is basically the same thing and will be considered in Chapter 7). For instance, there might be many energetically favorable hydrogen bonds within a protein. The unfolded "state," on the other hand, consists of so many configurations that it can be even more probable than the folded state under certain conditions, such as high temperature, when low energy becomes less important. As another example, a flexible "loop" of a protein is one that possesses enough entropy compared to the energy of any particular fixed geometry.

On the other hand, there are more subtle effects involving entropy and energy. For instance, hydrophobic "attractions" among nonpolar groups, as we'll see in Chapter 8, are due primarily to entropic effects involving water. On a related note, in molecular systems generally, there is often a trade-off between energy and entropy: almost by definition, an energetically favored state is geometrically well defined (certain atoms must be near others), and so low in entropy. Conversely, a high entropy state is likely to be unable to maintain low-energy interactions because of its flexibility. This is called "entropy–energy compensation."

You see, you already know almost everything!

3.7.1 TEMPERATURE DEPENDENCE: A ONE-DIMENSIONAL MODEL OF PROTEIN FOLDING

The idea of modeling a protein by a one-dimensional system may seem crazy, but in fact, it's quite informative. Fortunately, the model is exactly solvable. And any model that can be solved with paper and pencil will help you to understand the underlying physics much better. We will consider again the double-square-well potential of Figure 3.5, but now the focus will be on what happens as temperature is changed. (The approach is essentially the same as what has been presented in the books by Finkelstein and Ptitsyn and by van Holde et al.)

First of all, which state is which? The basic picture we have of the equilibrium between folded and unfolded protein populations is that the folded state, although comprising many configurations, occupies a much smaller region of configuration space than the unfolded state. Many attractive forces—embodied in hydrogen bonding, the hydrophobic effect, and electrostatics—keep the fold stable. On the other hand, there are many more ways for a protein to be unfolded than folded, so the former has substantially higher entropy. Thus, in Figure 3.5, the lower-energy basin A models the folded state, and the larger (higher entropy) basin B is the unfolded state.

Our basic strategy will be to consider the probabilities of the states as functions of temperature. To this end, we use the notation p_f and p_u for the probabilities of the folded and unfolded state, respectively. Similarly, we will denote the energies and widths by U_i and l_i where $i = $ f or u. Thus, in reference to Figure 3.5, $l_f = b - a$ and $l_u = d - c$.

We can readily write the equilibrium probabilities in this simple model. Because it is always energy and entropy differences that matter, we define $r \equiv l_f/l_u < 1$ and $\Delta U \equiv U_u - U_f > 0$, where the inequalities follow from our assumptions about the states. (Note that $r = \exp(-\Delta S/k_B)$, from Equation 3.24.) Using our knowledge of the meaning of free energy and some simple algebra, we find that

$$p_f(T) = \frac{r}{r + e^{-\Delta U/k_B T}} \quad \text{and} \quad p_u(T) = \frac{e^{-\Delta U/k_B T}}{r + e^{-\Delta U/k_B T}}. \quad (3.33)$$

Predictions of this model for $r = 1/25$ and $\Delta U = 5$ (in units where $k_B = 1$) are shown in Figure 3.7.

The relative sharpness of the transition means that both above and below the transition, only one of the states dominates (i.e., possesses almost all the population). This is sometimes called an "all or none" transition. Further, as you will see from the homework problems, it is the large entropy difference that leads to the sharpness of the transition. We will return to this model, and these issues, when we study thermodynamics in Chapter 7.

PROBLEM 3.15

(a) Derive the state probabilities given in Equation 3.33, including the normalization. (b) Derive the four limiting probabilities $p_i(T = 0)$ and $p_i(T \to \infty)$ for

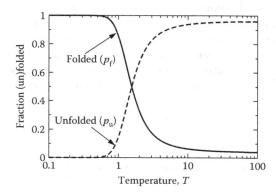

FIGURE 3.7 Predictions of the one-dimensional model of protein (un)folding, based on Figure 3.5 and embodied in Equation 3.33. The steepness of the transition results from the large ratio of state sizes, reflected in the choice $r = 1/25$. The plot uses $\Delta U = 5$ in units where $k_B = 1$ and T is dimensionless.

$i = f, u$. (c) From Equation 3.33, show that the (un) folding transition temperature is given by $T_f = \Delta U / [k_B \ln (r^{-1})]$, where T_f is defined to be the point where both states are equally occupied.

PROBLEM 3.16

The sharpness of the transition—numerically. (a) Make your own plot reproducing Figure 3.7—that is, using the same parameters. (b) Use the analytical result for T_f from the previous problem to redo the plot for three values of r ($r = 0.004$, 0.04, and 0.4) while keeping the value of T_f fixed. In other words, you will need to compute ΔU values so that T_f remains the same in all three plots. (c) Explain the effect of the entropy difference between the states on the sharpness of the transition.

PROBLEM 3.17

The sharpness of the transition—analytically. The sharpness of the transition can be quantified analytically via the slope of the probability plots at the transition. (a) Show why this "sharpness" can be computed from either p_f or p_u. (b) Calculate an appropriate derivative and evaluate it at T_f to show that the derivative can be expressed as $\ln (r^{-1})/4T_f$. (c) Rewrite the result in terms of ΔS and explain its significance.

3.7.2 DISCRETE MODELS

While discrete models are not one dimensional, they are sometimes as simple. Discrete models can also be useful in understanding molecular systems. For instance, the protonation state of an amino acid's side chain should be described discretely: at any point in time, a particular site is either protonated or not (which is reasonably valid despite quantum mechanics). This is also true for the "phosphorylation" state of an amino acid—either a phosphate group is attached or it isn't. More fundamentally, quantum mechanics teaches us that energy levels of molecules are quantized/discretized (see, for instance, the book by McQuarrie and Simon).

The good thing about systems with discrete configurations (or discrete quantum mechanical states) is that partition functions and averages take an even simpler form. For the partition function, if configurations are denoted by x_i, we have

$$Z = \sum_i^{\text{all configs}} e^{-E(x_i)/k_B T} \tag{3.34}$$

and correspondingly, the average of an arbitrary function $f(x)$ is given by

$$\langle f \rangle = Z^{-1} \sum_i^{\text{all configs}} f(x_i)\, e^{-E(x_i)/k_B T}. \tag{3.35}$$

Thus, Equation 3.35 is simply a discrete weighted average.

In the following problems, we shall take an elementary look at one of the most important model systems in statistical physics: the Ising model for magnetic systems.

PROBLEM 3.18

An Ising model with two spins (i.e., two little magnets). Magnetic systems are important in physics generally, but they are fundamental to understanding NMR experiments. The Ising model, which is particularly nice for learning, consists of discrete, possibly interacting, spins—usually in an external magnetic field. Consider two noninteracting (independent) spins that interact with a constant external field of magnitude $H > 0$ pointed in the "up" direction. Assume that each spin can take one of the values $m = \pm 1$ (up or down) in any configuration, and that the magnetic energy of each spin is $-H \cdot m$. (a) Sketch a picture of the system. (b) Make a table of the four possible states of this two-spin system, giving the energy and probability of each at temperature T. The probabilities should sum to one. (c) For this system, the partition function is a sum of Boltzmann factors. Does it play the same role as in a continuous system?

PROBLEM 3.19

Further study of the Ising model. (a) Explain why the form of the energy tends to align the spins and the field. (b) Make a graph showing roughly the probability of each state as a function of temperature. Give specific mathematical comparisons (inequalities) that effectively determine high and low temperature in this problem. (c) Comment on the number of states that are primarily occupied at very high and very low temperatures. (d) What changes in the system if the two spins are coupled so that the energy is decreased when they are parallel? (e) What if they are coupled so that the energy is lower if they are antiparallel?

3.7.3 A NOTE ON 1D MULTI-PARTICLE SYSTEMS

If you have a background in studying gases or liquids, you might be curious about one-dimensional systems of interacting particles. For instance, one can imagine studying a "fluid" of beads confined to a wire, with the interactions between the beads specified. Indeed, this can be done, but not so simply (see the book by Ben-Naim for details). Starting next chapter however, we will examine multiparticle systems in some simple cases.

3.8 LOOSE ENDS: PROPER DIMENSIONS, KINETIC ENERGY

Where is the kinetic energy and what is the meaning of λ? These questions will now be answered.

In statistical physics, the "real" partition function Z is an integral over the Boltzmann factor of the total energy (potential and kinetic), divided by the fundamental scale of Planck's constant, h. Also, real partition functions should describe real particles, so that a mass m must be specified when one writes down the kinetic energy $(1/2)mv^2$. We therefore have, for a single classical particle restricted to one dimensional motion,

$$Z = \frac{m}{h} \int\limits_{-\infty}^{\infty} dx \int\limits_{-\infty}^{\infty} dv \, \exp\left\{ -\frac{\left[U(x) + (1/2)mv^2 \right]}{k_B T} \right\}, \qquad (3.36)$$

which you can check is dimensionless. The factor of Planck's constant h is part of the classical approximation to quantum statistical mechanics, while the overall factor of m results from the fact that fundamental integration is really over momentum (mv), not velocity.

What really matters here? What do we gain from considering kinetic energy, in addition to potential? The answer is, a little bit of understanding, but nothing critical. To see how it works, note that the velocity part of the integral can be factored out (using the definition of the exponential function) and can even be integrated exactly. We find

$$Z = \frac{m}{h} \int\limits_{-\infty}^{\infty} dv \, \exp\left\{ -\frac{\left[(1/2)mv^2 \right]}{k_B T} \right\} \int\limits_{-\infty}^{\infty} dx \, \exp\left\{ -\frac{U(x)}{k_B T} \right\}$$

$$= \left(\frac{m}{h} \right) \sqrt{\frac{2\pi k_B T}{m}} \int\limits_{-\infty}^{\infty} dx \, \exp\left\{ -\frac{U(x)}{k_B T} \right\}$$

$$= \left(\frac{1}{\lambda} \right) \hat{Z}, \qquad (3.37)$$

where $\lambda(T) = h/\sqrt{2\pi m k_B T}$ has resulted from the Gaussian velocity integration and now sets the length scale for the configurational partition function $\hat{Z}$. Because λ sets the scale for fluctuations driven by the temperature, it is sometimes called the thermal wavelength. (It's also called the de Broglie wavelength.)

What happens if you forget to include λ in the full partition function Z? For studying behavior at a single fixed temperature, not much will happen beyond dimensionality problems in your logarithms. Relative populations/probabilities among configurational states, which are the most fundamental thing (since they govern observable behavior), remain unchanged whether or not you remember to include λ. Of course, if your concern is states differing in velocity or kinetic energy, then you should not integrate out v in the first place.

Finally, let's check explicitly that the average kinetic energy is indeed $k_B T/2$ in one dimension, as we claimed earlier. We can use the usual weighted averaging, but now it will turn out that the potential energy integrals cancel out and become irrelevant:

$$\left\langle \frac{mv^2}{2} \right\rangle = Z^{-1} \left(\frac{m}{h} \right) \int dx \int dv \left(\frac{mv^2}{2} \right) e^{-[U(x)+mv^2/2]/k_B T}$$

$$= \frac{\int dv \left(mv^2/2 \right) e^{-(mv^2/2)/k_B T}}{\int dv \, e^{-(mv^2/2)/k_B T}} = \left(\frac{m}{2} \right) \langle v^2 \rangle$$

$$= \left(\frac{m}{2} \right) \left(\frac{k_B T}{m} \right) = \frac{k_B T}{2}, \qquad (3.38)$$

where we used the fact that v is a Gaussian distributed variable with variance $k_B T/m$. The integral in Equation 3.38 embodies the variance itself, aside from a factor of $m/2$, since the mean of v is zero. Of course, this is just another example of a harmonic degree of freedom, and we knew to expect $k_B T/2$.

FURTHER READING

Ben-Naim, A.Y., *Statistical Thermodynamics for Chemists and Biochemists*, Springer, Berlin, Germany, 1992.

Dill, K.A. and Bromberg, S., *Molecular Driving Forces*, Garland Science, New York, 2003.

Finkelstein, A.V. and Ptitsyn, O.B., *Protein Physics*, Academic Press, Amsterdam, the Netherlands, 2002.

McQuarrie, D.A. and Simon, J.D., *Physical Chemistry*, University Science Books, Sausalito, CA, 1997.

Phillips, R., Kondev, J., and Theriot, J., *Physical Biology of the Cell*, Garland Science, New York, 2009.

Reif, F., *Fundamentals of Statistical and Thermal Physics*, McGraw-Hill, New York, 1965.

van Holde, K.E., Johnson, W.C., and Ho, P.S., *Principles of Physical Biochemistry*, Pearson, Upper Saddle River, NJ, 2006.

4 Nature Doesn't Calculate Partition Functions: Elementary Dynamics and Equilibrium

4.1 INTRODUCTION

From the beginning of this book, we have emphasized the intimate connection between equilibrium and dynamics. The philosophy here is to keep that connection explicit in order to reinforce both topics—rather than, as is usual, first treating equilibrium topics in full. The more complete one's physical and statistical intuition, the better. Dynamical topics and their connection to equilibrium therefore will be revisited and reinforced throughout the book.

While some biological and chemical topics can be understood fully from an equilibrium picture, the issues of dynamics and time dependence are absolutely necessary in some instances. As a simple case, every enzyme has an intrinsic rate, to which the machinery of the cell has been "tuned" (and/or *vice versa*). A significant disruption to cellular function could result from arbitrary changes to certain rates (e.g., from mutations). Similarly, the rates for protein folding must be fast enough to permit cell function. Some aggregation phenomena, such as plaque formation in Alzheimer's and related diseases, occur much more slowly than other physiological processes—yet not slowly enough for the victims of such pathologies.

In a more abstract and general sense, one can say that a full dynamical description of a system contains more information than a complete equilibrium description. Recall the example from Chapter 1: to say that someone sleeps 8 h a day is an equilibrium description: $p(\text{sleep}) = 1/3$. Yet this fraction by itself doesn't tell us whether the person stays up late or perhaps takes a separate nap in the afternoon. In a molecular example, if we know all the transition rates among the states of butane (Figure 4.1), we can also compute the equilibrium populations (see below). However, the equilibrium populations do not tell us the rates.

To start to understand dynamics in a more concrete way, we must consider an energy landscape, such as that in Figure 4.1 or Figure 4.4. In the simplest terms, our molecule of interest, perhaps a protein, will tend to stay in either in one state or the other with occasional transitions over the barrier. Recalling from the Boltzmann factor that the high-energy barrier top will rarely be occupied, it is clear that molecular dynamics depend critically on the process of barrier crossing. Part of the goal of this chapter is to gain an initial, semiquantitative understanding of barrier crossing.

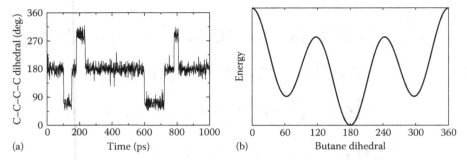

FIGURE 4.1 A butane trajectory and the corresponding schematic (free) energy profile. The trajectory contains substantial dynamical information lacking in the energy profile, such as transition rates between states and intrastate "vibrational" frequencies. The trajectory, analyzed with a histogram as in Figure 2.3, also contains the equilibrium populations implicit in the energy profile. Note that rates can be estimated from a trajectory based on the inverse of the average waiting time for the event of interest.

Further, although details of dynamic processes are governed by details of the energy landscape (its basins and barriers), there are equally important general principles governing the connection between dynamics and equilibrium that are independent of the landscape. This chapter will therefore also start to build a picture of the dynamics/equilibrium connection.

4.1.1 EQUIVALENCE OF TIME AND CONFIGURATIONAL AVERAGES

Dynamical and equilibrium phenomena are not really distinct, because equilibrium is generated by dynamics. Nature has no other tool to work with besides dynamics.

Look at Figure 4.2. As we have discussed already, time and configurational (ensemble) averages of the same quantity will yield the same result. The example of proteins in a test tube observed over a long time is sketched in Figure 4.2. The average energy (or average distance between two atoms or...) measured by considering all proteins at one point in time will be identical to that measured by time averaging for one protein for a long period of time. The reason, again, is that the ensemble observed at any fixed point in time, in fact, is generated by dynamics: each protein exhibits statistically identical dynamic behavior. The ensemble comes from dynamics! Thus, the snapshots in a movie of one molecule will be (statistically) identical to the snapshots seen in the fixed-time ensemble. This argument holds true whether the molecules are interacting or not.

4.1.2 AN ASIDE: DOES EQUILIBRIUM EXIST?

This question was famously (and correctly) answered in the negative in Shang-Keng Ma's book. He argued that no system could be isolated in the time invariant way necessary for true equilibrium. If we prepare a supposedly equilibrium system in a glass chamber, the chamber itself must ultimately change and even decompose over time, as will the table it sits on! The contents of the chamber may also undergo

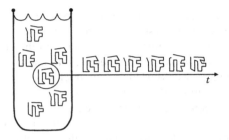

FIGURE 4.2 Why time and ensemble averages have to be the same. On the left is a schematic test tube containing an ensemble of molecules in different configurations, while the time evolution of one of the molecules is shown schematically on the right. Nature is only capable of producing forces and dynamics, so each individual molecule evolves according to the basic laws of physics. If we take a snapshot of all molecules in a test tube, because their dynamics are essentially uncorrelated, we will find the equilibrium ensemble. In other words, the equilibrium ensemble exactly represents the fraction of time a molecule spends in different configurations. Configurations appear more commonly in the equilibrium ensemble because they take up more time in a "movie" of any single molecule.

unexpected spontaneous chemical reactions. From this perspective, equilibrium is relative—defined only in terms of the timescale of observation. Clearly, biological systems and glass chambers have very different timescales for their approximate equilibria.

From a practical point of view, we can say that equilibrium exists to a very good approximation whenever the relaxation processes of interest (e.g., diffusion and conformational motions of a molecule) occur rapidly by comparison to the observation time. Further, the external conditions, such as temperature, should remain approximately constant over the observation time.

Despite Ma's "philosophical" objections, equilibrium is theoretically fundamental. First of all, equilibrium does exist exactly on paper. And it defines the fundamental reference point for many statistical descriptions of system. That is, we can more easily understand nonequilibrium systems by comparison to equilibrium, as we will see below.

4.2 NEWTONIAN DYNAMICS: DETERMINISTIC BUT NOT PREDICTABLE

Newton's laws of motion are fundamental to physics—and indeed to thinking physically—but we cannot always use our intuition derived from Newtonian mechanics to understand biophysical behavior.

Consider a single protein surrounded by water in a test tube, and assume we are interested in studying the dynamics of the protein. Assume further that all atoms and molecules are purely classical, so that there is no shifting electron density, and thus interatomic force laws are constant in time. In this situation, the dynamics are Newtonian and deterministic. Once motion gets started, it is completely determined by the initial conditions (i.e., initial positions and velocities). At first glance, it might

seem that we could determine—in principle, perhaps with a super-duper-computer—
the exact future dynamics of the system. Yet, we cannot.

Every system (e.g., our test tube) is in contact with the environment, which in
turn, is in contact with the whole world. Therefore, to know the exact deterministic
behavior of our protein, we would need to keep track of every atom in the universe!
This is something I call impossible, even in principle.

Let's take a more concrete outlook. Imagine again that our system is completely
classical, but that we are somehow able to make a movie (not a simulation, but a
real movie) of all atoms in our test tube—but not the test tube itself (see Figure 4.3).
We would indeed see all atoms interacting classically. Yet the test tube is in contact
with the air at temperature T, and therefore "test-tube atoms" have the corresponding
kinetic energy. If we had included these atoms in our movie, we would see them
colliding with the water in the tube. But since we don't, the water molecules at
the boundaries of our system appear to be experiencing random collisions—in just
exactly the right way to maintain the temperature as constant. This is the random or
"stochastic" element necessary for almost any realistic description of a constant-and-
finite-temperature system.

The logic just described will tell you that if you look at a protein without solvent,
or with just a small amount, at least the system's boundaries must be described
stochastically. That is, random collisions must somehow be simulated to maintain the
temperature. This is a common approach in many simulations.

We'll see more reasons in Section 4.9 why exact trajectories for molecular systems
cannot be computed—that is, simulated on a computer—even in principle.

As a technical aside, note that in general, kinetic and potential energy can be
exchanged, even in a Newtonian system where the total energy is maintained as
constant. In a large system, then, the kinetic energy of a system may remain relatively
constant even if it is isolated from a "thermal bath" of random collisions, but in a small

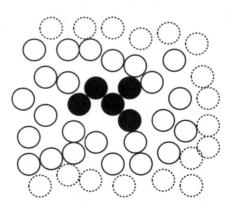

FIGURE 4.3   There is no such thing as a deterministic finite system because no system can
truly be isolated. The cartoon indicates a biomolecule (filled circles) in a solvent (open circles)
contained in a test tube made of atoms (dashed circles). The biomolecule interacts with the
solvent, which interacts with the test tube atoms, which interact with the air.... In a biological
cell, there are even more interactions.

system, the exchange between kinetic and potential energies will result in very large fluctuations in kinetic energy. Even though kinetic energy fluctuations should indeed occur, the absence of a thermal bath could lead to anomalously large fluctuations in a small system.

4.2.1 The Probabilistic ("Stochastic") Picture of Dynamics

Because the thermalizing collisions we just discussed are an intrinsic part of life (and experiments!) at constant temperature, it is often useful to think of dynamics in probabilistic terms. Thus, rather than saying Newton's laws will force our protein to go from its current configuration x_1 exactly to x_2, it is better to speak in terms of a distribution. We can only speak of a conditional probability distribution of future configurations, which is formulated (in one dimension) via

$$\rho(x_2, t_2 | x_1, t_1). \tag{4.1}$$

Thus, given any initial configuration x_1 at time t_1, a protein will have a distribution of possible configurations x_2 at a later time t_2. If we could really specify the configuration of the whole universe, this would not be necessary... but I hope you can see that it's not really an option to think deterministically.

Examples of stochastic trajectories were shown for butane (Figure 1.3) and for alanine dipeptide (Figure 1.10).

This "stochastic" picture is the one that will be used throughout the chapter, and is indeed the basic worldview of biophysics. No outcome for an individual protein or other molecule is certain. However, even in a dynamical context, the average properties can be well described. For instance, given a collection of a large number of molecules, we can say that a certain fraction are likely to make a certain conformational transition in a specified amount of time. In other words, this is the standard logic of probability theory—the goal is to predict the outcomes of a large number of "identical" experiments.

4.3 BARRIER CROSSING—ACTIVATED PROCESSES

Almost every biophysically oriented paper on proteins refers to the presence of energy (or free energy) barriers among conformations. To change conformation, then, these barriers must be surmounted, and doing so is called an activated process. In any protein or large molecule, there will be a large number of minima and barriers—in fact, probably a huge number: millions or billions or more! However, our goal is to understand the basics of any molecular isomerization (conformation change), which can be appreciated even in one dimension.

4.3.1 A Quick Preview of Barrier Crossing

We are already in a position to understand the basics of barrier crossing—and to appreciate that there is more to learn. Let's look at Figure 4.4, which is a typical schematic drawing for a chemical reaction or isomerization, or for a protein conformational transition, including some folding scenarios. The barrier top is often called a "transition state" by chemists.

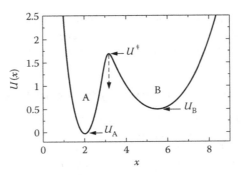

FIGURE 4.4 A one-dimensional model useful in understanding "activated," or barrier-crossing, processes. The dashed arrow schematically illustrates the way a catalyst (e.g., an enzyme) lowers the activation energy $U^{\ddagger}$ but not the "reactant" or "product" energies, U_A and U_B.

We can understand some key features of transitions just by using equilibrium reasoning based on our initial understanding of statistical mechanics. The main point comes from considering relative probabilities via Boltzmann factors. Based on Figure 4.4, if we imagine a system starting in state A (somewhere near the minimum), then the relative probability of the transition sate (barrier top) is given by the ratio $\exp\left(-U^{\ddagger}/k_B T\right)/\exp\left(-U_A/k_B T\right) = \exp\left(-\Delta U/k_B T\right)$, where $U^{\ddagger}$ is the barrier-top energy, U_A is the energy near the state A minimum, and $\Delta U = U^{\ddagger} - U_A$ is the energy difference also called the activation energy. Of course, the activation energy based on starting from state B would be $\Delta U = U^{\ddagger} - U_B$.

You may recognize $\exp\left(-\Delta U/k_B T\right)$ as the Arrhenius factor, which tells us something not only about relative probabilities in equilibrium, but also about dynamics. In fact, we can reframe our equilibrium thinking by asking the following question: If we start our system near the bottom of state A, then after a short amount of time, what are the relative probabilities of the different final configurations (x values)? The simplest way to answer is by saying that we expect x values to occur in proportion to their Boltzmann factors. But if we also recognize that the only way a transition can occur is if the barrier top is reached, then it is clear that the Arrhenius factor must enter the dynamical description. Indeed the rate k of transitions over the barrier is proportional to the Arrhenius factor,

$$k \propto e^{-\Delta U/k_B T}, \tag{4.2}$$

where k and ΔU must both refer to the same transition (e.g., B → A). Qualitatively, the Arrhenius factor tells us that (1) the larger the barrier height (activation energy, ΔU), the slower the transition rate; and (2) for a given barrier, we expect the rate to increase with increasing temperature.

We are not ready to understand transition rates quantitatively, but we can see what is missing. The Arrhenius factor is dimensionless, while a rate is defined to be the number of events per unit time. Somehow, we will eventually need to associate a timescale with the Arrhenius probability. We will take the first steps toward doing so in this chapter.

To understand the time dependence well, we will need to focus in more detail on dynamics. We can see even more concretely why Newtonian dynamics will prove to be less than illuminating—partly because they conserve energy in a deterministic way. Imagine starting a Newtonian trajectory from state A in Figure 4.4; we would need to specify the initial position and velocity. But based on these initial conditions, either the trajectory will cross the barrier or not—it will not if it lacks sufficient energy. This contrasts with the definition of the rate as a probability per unit time.

4.3.2 CATALYSTS ACCELERATE RATES BY LOWERING BARRIERS

Catalysts—whether biological enzymes or inorganic materials—can greatly increase rates of chemical reactions. Reactions, like conformational transitions, can be visualized using the double-well of Figure 4.4.

The key point is that catalysts can only lower the barrier height $U^{\ddagger}$, but they do not affect the intrinsic stability of A or B. After all, the catalyst is typically in contact with the reacting system only for a brief period during the reaction; before and after, the system is unaffected by the catalyst. Thus, the values U_A and U_B, which determine the relative probabilities of A and B (along with the associated entropies), remain unchanged. By the same token, catalysts will necessarily affect the rates in both directions.

4.3.3 A DECOMPOSITION OF THE RATE

Perhaps the most important logical step in understanding rates (which are probabilities per unit time) is to decompose the overall rate k into two factors:

$$k = \nu \times p_{\text{act}}. \tag{4.3}$$

Here ν is the "attempt frequency," which is the number of times per second that the "edge" of the state, near to the barrier, is reached. The other factor is the activation or success probability, p_{act}, which is a number between zero and one describing the likelihood of crossing the barrier once the edge is reached. Such a decomposition can be applied to any system, and the success probability will often be the Arrhenius factor we considered at the outset. However, the physical meaning of the two factors is ambiguous in almost any case besides when there is a simple barrier to be crossed.

Attempt frequencies are new to us. Your intuition should tell you that the broader a state (i.e., the greater the entropy—recall Chapter 3), the lower the attempt frequency. That is, the edge of a broad state will be reached less often than would be the edge of a narrow state. Figure 4.4 can help you to visualize this: the left state, A, clearly would have a higher attempt frequency. The state widths/entropies governing attempt frequencies can vary significantly among molecular systems, particularly for conformational transitions.

For chemical reactions, it is interesting to note, there is a fundamental "thermal timescale," given by $h/k_B T \sim 100$ fs. To see how this scale enters quantitative estimates of chemical reaction rates, see for instance the book by Dill and Bromberg.

4.3.4   MORE ON ARRHENIUS FACTORS AND THEIR LIMITATIONS

While we are not yet experts on multidimensional energy landscapes, we can try to imagine that the curve in Figure 4.4 represents a free energy profile, rather than a simple energy. As we learned in Chapter 3, the free energy simply represents the logarithm of the probability, so our basic discussion about Arrhenius factors and success probability still holds. You could think of the one-dimensional curve as resulting from a projection. The only difference from our pervious discussion is that we would write the Arrhenius factor as $\exp(-\Delta F/k_B T)$. So far, so good.

However, we already have some clues about the dangers of projection, and those fears can be realized here. Specifically, we saw in Figure 2.9 that projection can eliminate distinctions between states and it's easy to believe that projections onto different coordinates could lead to different effective barrier heights, ΔF. So there is an intrinsic worry about attempting to understand kinetics via effective Arrhenius factors in more than one dimension. We will return to this point in Chapter 6.

Yet there are conceptual problems with the Arrhenius picture even in one dimension—though the situation is bound to be worse in higher-dimensional systems. What if the transition region is not a simple single barrier top, as in Figure 4.4, but a region containing multiple minima and barriers? How do we compute the activation energy, even in principle? The answer is: only with difficulty, as will be seen in Chapter 10.

The bottom line is that Arrhenius descriptions must be handled with care!

Nevertheless, the more general attempt-frequency/success-probability formulation of Equation 4.3 is valid even if the Arrhenius picture is not. As we have stressed throughout this book, the safest description is the one that rests on probabilistic considerations. In a dynamical context, rates are indeed just the probabilistic objects we need. The attempt-frequency picture is simply a device to connect the probabilistic and physical (dynamical) pictures, without resorting to oversimplification.

4.4   FLUX BALANCE: THE DEFINITION OF EQUILIBRIUM

Putting physical details aside for now, the dynamic basis of equilibrium can be explained in a very general way that not only is easy to understand, but is also very useful throughout statistical mechanics. The basis is the flow or "flux" of probability.

The general flux balance picture is readily constructed from the definition of equilibrium, namely, a situation in which the probability distribution over the space of interest does not change with time—and where there is also no net flow of probability density anywhere in the system. (It is possible to have net flow with unchanging density: think of stirring a cup of coffee. This situation is called a steady state.) The "space of interest" may be a range of values of a one-dimensional coordinate, or perhaps the set of coordinates describing the positions of all molecules in a fluid, or even the set of all coordinates that describe a protein's configuration.

The flux balance idea is straightforward. Consider an equilibrium system: for example, a fluid in a test tube. If we focus on a particular part of the test tube, perhaps a cubic millimeter, then the condition of equilibrium implies that the number of fluid molecules in that sub-volume does not change over time on average. However, since

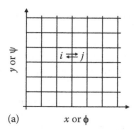
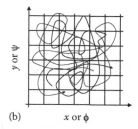

(a) x or ϕ (b) x or ϕ

FIGURE 4.5 Flux balance in equilibrium. Whether in real space or configuration space, the average overall flow of probability fluxes must be balanced in equilibrium. This applies to any pair of "states" or any pair of regions in the space of interest, and is the definition of equilibrium. The two pairs of coordinates shown—(x, y) and (ϕ, ψ)—are simply examples. The panel (a) highlights two regions "i" and "j," while the panel (b) shows schematic trajectories that could occur in either real or configuration space.

we know that dynamics are occurring ceaselessly, some molecules must be leaving and entering the sub-volume all the time. The flux-balance idea simply means that, on average, the number of particles leaving and entering the sub-volume must be equal for equilibrium.* More specifically, the flux will be balanced (equal and opposite) between any pair of regions. This idea is illustrated in Figure 4.5.

We can also apply the flux-balance idea to more abstract spaces, like the space of all possible protein configurations. (This is a high-dimensional space where, for N atoms in a protein, $3N$ numbers are needed to specify the full configuration—x, y, and z for every atom; thus it is a $3N$-dimensional space. See again Figure 4.5). Focus on any subset of configuration space—for instance, all configurations where the angle between two helices falls within a certain range of values. Again, we know that protein configurations will be constantly fluctuating due to finite temperature and collisions with solvent. However, in equilibrium, if we consider a large number of proteins, the average number leaving the specified angle range (per unit time) must equal the number entering it.

Note that the balance of flux must hold regardless of the relative sizes of the regions under consideration. For instance, we can consider an arbitrarily small sub-volume of the test tube and a much larger neighboring volume.

By probing a bit more deeply, we can infer a slightly more descriptive picture, which explains why the flux-balance principle applies to spaces of any size. Denote the two regions of our space—whether concrete or abstract—as A and B, with populations N_A and N_B, that are constant in time by the definition of equilibrium. (Note that other populated regions C, D, . . . may be present but do not alter our discussion.) To make progress, we must define the microscopic rate k_{ij}, the conditional probability per unit time for a given molecule to change to state j starting from state i, in equilibrium. Then the macroscopic rate—the flux or the total number of molecules moving

---

* Our use of the word flux is slightly different from the traditional physics use, where the net number of particles is considered. In traditional physics terminology, then, our notion of "flux balance" corresponds to zero (net) flux.

from i to j per unit time—is the number of molecules times the microscopic rate:

$$\text{Flux}(i \rightarrow j) = N_i\, k_{ij}. \tag{4.4}$$

We switched to using i and j instead of A and B only to emphasize the generality of our discussion.

Finally, we can state the flux balance principle in a simple equation,

$$\text{Flux}(i \rightarrow j) = \text{Flux}(j \rightarrow i), \tag{4.5}$$

which holds for every ij pair in equilibrium.

It is critical to note that we have not assumed equality of the microscopic rates, k. Rather, it is the overall flux that maintains the constant populations N_i and N_j, which occur in equilibrium.

The powerful and general idea of flux balance greatly facilitates the understanding of binding (and unbinding!), which we will study in Chapter 9.

4.4.1 "Detailed Balance" and a More Precise Definition of Equilibrium

We said above that equilibrium is defined by a situation in which, on average, there is no flow of probability. Equivalently, but perhaps more carefully, we can say that equilibrium is defined by a balance of probability flow (flux) between every pair of microstates or configurations. This condition is called "detailed balance." Not only is it a fundamental property of any physical system in equilibrium but it is the starting point for deriving a host of interesting Monte Carlo simulation algorithms (see Chapter 12).

PROBLEM 4.1

Detailed balance implies "coarse balance." Show that if there is balanced flux between every pair of configurations, that there is a also a balance of flux between any pair of arbitrarily defined states. Proceed by defining states as arbitrary sums of microstates/configurations, noting that the overall flux will be dictated by the relative probabilities of the configurations.

4.4.2 Dynamics Causes Equilibrium Populations

Equilibrium populations result from ratios of rates, and this is one of the basic principles of biophysics—or indeed of statistical physics in general. In fact, the flux-balance Equations 4.4 and 4.5 can be rewritten to express this principle via the fundamental relation

$$\frac{N_i}{N_j} = \frac{k_{ji}}{k_{ij}}. \tag{4.6}$$

The ratio of equilibrium populations on the left is fully determined by the rates. A simple intuitive understanding of Equation 4.6 comes from looking again at Figure 4.4: because the barrier from A to B is higher than the reverse, we can expect the microscopic rate k_{AB} to be lower than k_{BA}, implying the larger population of state A. It's easier to enter A than to leave. Note that we reach this conclusion without direct reference to the energies of states A and B (though of course they are implicit in the rates).

If you are feeling contrary and picky, you may object that the rates determine only the ratios of populations, as in Equation 4.6, rather than their absolute values. However, the absolute values are usually unimportant. And when they are, they are determined from the ratios with a single additional piece of information: the overall population. That is, the absolute populations can be determined from the rates along with the total population $N = \sum_i N_i$.

4.4.3 THE FUNDAMENTAL DIFFERENTIAL EQUATION

It is very useful to recast the statistical discussion of dynamics in terms of differential equations that will describe the way the populations of two states, A and B, will change, assuming that they have well-defined rate constants. The question then becomes: given the current (possibly nonequilibrium) populations of the states, how will they change/evolve over time? For two states, we can answer fully via a set of two equations that describe the incremental change occurring in each state's population (dN) in a time dt. Quite simply, we follow the definition of rate constants, which tells us how many transitions to expect in our time interval—namely, $N_i k_{ij} dt$—for the two possible transitions.

Thus, state A can gain population via transitions from state B to A and it can lose by transitions from A to B, and similar logic applies to state B. Therefore,

$$\frac{dN_A}{dt} = N_B k_{BA} - N_A k_{AB},$$

$$\frac{dN_B}{dt} = N_A k_{AB} - N_B k_{BA}. \qquad (4.7)$$

We can convert these descriptions to concentrations [A] and [B] simply by dividing the volume of our system. Note that $N_A = N_A(t)$ and $N_B = N_B(t)$.

Where is the equilibrium in the differential equations? Actually, it's easy to find. One basic rule of equilibrium is that populations are constant in time. Therefore, one can set the time derivatives on the left-hand sides of the Equations 4.7 to zero. It is easy to show that either equation leads immediately to the correct ratio of populations as given by Equation 4.6.

PROBLEM 4.2

For the case of two states, A and B, derive Equation 4.6 from the differential equations (Equations 4.7), using the time-invariance of equilibrium.

4.4.4 ARE RATES CONSTANT IN TIME? (ADVANCED)

Perhaps it is time for an important caveat. Our whole discussion has assumed that rates are constant in time—and this is true in equilibrium, by definition. Certainly

FIGURE 4.6 Two examples of flux balance. States are represented by circles, with the size denoting the equilibrium population. The arrow thickness represents the microscopic rate. Thus, two equi-sized (equally probable) states must have equal rates in the two directions. For a large state to be in equilibrium with a small one, the small-to-large rate must greatly exceed its counterpart.

such time-invariance makes all the mathematics and the overall "picture" simple to understand. But this assumption, built into Equations 4.7, corresponds to an underlying physical picture. Specifically, in the energy landscape, each state should not have significant internal barriers. If any barriers internal to a state are similar in height to the barrier separating the two states, then the time for equilibration within a state could become comparable to that for barrier crossing. For instance, if there is an internal barrier inside state A, then in the "substate" of A closest to B, there could be zero population at some point in time—leading to an observed flux of zero—even though there might be population in a more distant part of state A. Thus, the basic rule is that equilibration within a state must be much more rapid than transitions between states. You can think of this as a sensible rule for defining states. Again, these considerations are critical only for considering nonequilibrium phenomena, when populations initially exhibit nonequilibrium values (Figure 4.7).

4.4.5 Equilibrium Is "Self-Healing"

The "self-healing" or "self-correcting" property of equilibrium is both pretty and powerful. Think of it as one of the luxuries of the equilibrium scenario.

To understand self-correcting behavior, we must imagine a perturbation to our equilibrium situation in which one or more of the state populations is changed. (This can occur in real life if a sample is "pulsed" with light, temperature, or a change in chemical composition.) Let's assume the perturbation increases the ratio of populations N_i/N_j over its equilibrium value, as in Figure 4.7. In other words, state i is more populated than it should be, compared to j. But if the rates k are fixed (see below), then the fluxes will not be balanced, and will tend to restore the equilibrium

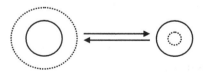

FIGURE 4.7 A schematic example of the self-healing tendency of a system perturbed away from equilibrium. Two states with equal equilibrium populations (solid-line circles) have been perturbed to other nonequilibrium populations (dashed lines), while the microscopic rates k (arrows) remain equal. Based on Equation 4.4 more flux will flow from the overpopulated to the underpopulated state, and this flow imbalance will continue until equilibrium is restored—at which time there is no net flow of population/probability.

populations. In our example, the flux from i to j—see Equation 4.4—will be higher than before and hence will tend to deplete the state i population relative to that of j.

Remember, the self-healing property of equilibrium does not mean that fluctuations about equilibrium values—in any non-infinite system—will not occur. Rather, it means that fluctuations should tend not to grow, but to decrease.

Echoing Section 4.4.4, microscopic rates k could change along with a perturbation to the populations. As an example, a given state might have an internal barrier and the populations of the two basins might be perturbed differently. Nevertheless, after a period of time, the rates will gradually revert to their equilibrium values. This will occur as the internal populations of individual states readjust to their equilibrium values.

4.5 SIMPLE DIFFUSION, AGAIN

Diffusion is remarkably simple, remarkably important, and just-plain remarkable. It is blind and random but "senses" gradients and smoothes them out. It uses no energy but provides critical transport in living cells. Understanding this fundamental mechanism will provide a very useful window into the means and ends of stochastic processes.

Diffusion is the basis for what might be called "passive" transport, where no special forces are applied and no energy is expended. Such passive transport is critical in biology, where signaling molecules, drugs, and ions must move around "on their own." ("Active" transport, where energy is used, is equally important in biology, but is outside the scope of this chapter.) Of course, motion of any kind, including passive diffusion, requires some type of energetic processes. Yet diffusion, amazingly enough, relies only on the energy already present in our finite-temperature environment.

Because of its "passivity," all diffusion can do is to disperse a set of molecules. For instance, a drop of ink or dye placed in the middle of a cup of water will spread out. At the same time, diffusion is indeed a mechanism for producing equilibrium. In the case of ink in a cup, the equilibrium condition is evenly dispersed color, which is what diffusion will yield after enough time. Although any individual dye molecule diffuses "blindly" without any preference of direction, the net effect of diffusion leads to transport in opposition to gradients (high-concentration regions tend to get depleted).

A semiquantitative microscopic picture of diffusion is very useful. You should imagine a single dye molecule far from all other dyes, which is colliding on all sides with water molecules in a completely random way. These collisions are driven by thermal energy (i.e., all molecules have average translational kinetic energy of $\sim k_{\mathrm{B}} T$) and lead to random displacements. In a short time interval Δt, let us call the displacement that occurs $\overrightarrow{\Delta x}$. Because we are considering "isotropic" diffusion, the vector $\overrightarrow{\Delta x}$ is equally likely to point anywhere in space, and we can expect that longer displacements are less likely than shorter ones. For simplicity, we will focus solely on displacements Δx in the x direction of motion (as opposed to y and z directions), which will give us as much of a picture as we need.

Consider again the drop of dye or ink in a glass of water. Dye molecules are equally likely to move to the left as to the right—that is, there is a 50% probability of each. If we imagine crudely that every molecule moves either left or right a fixed amount in each time interval Δt, then consider what happens as time proceeds. After one Δt, half the population has moved left and half right. Focusing on the half that has moved right, after two Δt, half of these have moved left and back to their starting location; the other half is now two steps to the right. The behavior of the half that originally moved left is a mirror image. Hence, the original location now has only half the population, with a quarter displaced two units to each side. Two key aspects of diffusion have been illustrated: (1) as time passes, some dye molecules tend to move further and further away; and (2) regions with initially high concentrations tend to have their concentrations reduced.

PROBLEM 4.3

As in the text above, consider a simplified discrete model of one-dimensional diffusion where the population of molecules is restricted to $x = 0, \pm1, \pm2, \ldots$. Assume that in every time interval Δt, molecules move one unit left or right, with equal probabilities. Starting with 100% of the population at $x = 0$ at time $t = 0$, plot the populations as a function of x for $t = 0, \Delta t, 2\Delta t, 3\Delta t, 4\Delta t$.

PROBLEM 4.4

In the continuous case, imagine a distribution of dye molecules that starts out very narrow at $t = 0$. Sketch how you think this evolves with time.

We can now understand how diffusion—which is "blind" and nondirectional—"knows" how to even out the distribution of solute molecules to make it uniform. If we focus on a small element of volume (v_a) with a higher momentary concentration of ink than its neighbors, we know that a fraction of the molecules will be moving left. In the next volume over to the right (v_b), we expect an identical *fraction* will be moving left and into the first volume. Because there are fewer solute molecules to begin with in v_b, however, more molecules will move out of v_a than into it. Note that the arguments here are completely general for any isotropic diffusion process, and in fact, are identical to those presented in the "self-healing" discussion.

PROBLEM 4.5

(a) By considering displacement probabilities for individual molecules, explain why the *fractions* of populations moving left and right in the volumes v_a and v_b, discussed above, must be identical. (b) In turn, explain why this means that more molecules will leave the volume with higher concentration.

4.5.1 THE DIFFUSION CONSTANT AND THE SQUARE-ROOT LAW OF DIFFUSION

Diffusion was just described (correctly) as consecutive random displacements. If we assume these displacements are uncorrelated, then the concatenation of these random increments is exactly what is embodied in the convolutions described in Chapter 2.

To be more precise, let us assume that in a given time increment, Δt_i, each diffusing molecule suffers a random displacement Δx_i (due to collisions), which is distributed according to $\rho(\Delta x_i)$. (We limit ourselves to one dimension for simplicity and note that ρ is the same for all i.) We now want to know about the convolution of N such increments, that is, what are the characteristics of the distribution $\Delta x_{\text{tot}} = \sum_i \Delta x_i$?

What, precisely, can we learn? Unfortunately, to calculate the exact distribution of Δx_{tot}, we would need to know the exact functional form of ρ (which we will not assume is known). We only assume that all molecules behave the same at all times and hence there is a single ρ. However, we did learn in Chapter 2 that the variance of a sum (i.e., of Δx_{tot} in this case) is exactly equal to the sum of the variances of the increments Δx_i. That is,

$$\langle \Delta x_{\text{tot}}^2 \rangle = \sum_{i=1}^{N} \langle \Delta x_i^2 \rangle = N \sigma_\rho^2, \tag{4.8}$$

where σ_ρ^2 is the variance of any of the individual Δx_i distributed according to ρ.

Two important facts are embodied in Equation 4.8. First of all, as in Section 2.3.4, the classic "square-root-of-time" behavior becomes evident by taking the square root of both sides and noting that N is proportional to the number of time increments—that is, $t = N\Delta t$. This is more clear if we substitute $t = N\Delta t$ and rewrite the result:

$$\langle \Delta x_{\text{tot}}^2 \rangle^{1/2} = \sqrt{N} \sigma_\rho = \frac{\sqrt{2N\Delta t} \sigma_\rho}{\sqrt{2\Delta t}} \equiv \sqrt{2Dt}, \tag{4.9}$$

where we have defined the diffusion constant,

$$D \equiv \frac{\sigma_\rho^2}{(2\Delta t)}. \tag{4.10}$$

Unlike conventional "ballistic" or inertial motion, where the distance traveled is proportional to time, the expected magnitude of the displacement of a diffusing particle grows with $\sqrt{t}$. This square-root-of-time behavior is generic to any diffusive system.

Secondly, the diffusion constant D of Equation 4.10 is also fundamental, but in quite a different way, since it describes the specific properties of the diffusing particle in the particular medium or solvent. Note that because the variance σ_ρ^2 is expected to grow linearly with Δt, the ratio defining D should be independent of Δt. We expect that D (and σ_ρ^2) will be smaller for more viscous solvents, for instance, since it is harder to move in such solvents. More precisely, D is found to be proportional to the temperature and inversely proportional to the particle diameter and solvent viscosity.

PROBLEM 4.6

We know that diffusive behavior results from averaging over random collisions with solvent molecules. Given this picture, try to describe in words a time increment that would be too small for a diffusive description.

It is useful to know typical values of diffusion constants. For small molecules in water, it turns out $D \sim 1 \; \mu\text{m}^2/\text{ms}$, whereas for proteins, the value is about a tenth of that (see the book by Nelson).

4.5.2 DIFFUSION AND BINDING

Binding in biomolecular systems is very dependent on diffusion, in two ways. The first is almost trivial: in order to bind, molecules must first be able to "find" one another, and they do so by random diffusion. That is, to visualize molecules present in a solution at a certain concentration, you should not imagine a regular lattice of locations, but rather numerous random three-dimensional trajectories, constantly evolving. Sooner or later receptors and ligands will meet. Thus diffusion, which embodies random thermal motion, guarantees that receptors and ligands can find one another sooner or later.

The second aspect of diffusion in binding is best understood by distinguishing two parts of the binding process: (1) collision—that is, the molecules "finding" one another, and (2) complexation—the molecules actually joining together in a meta-stable complex. Above, we merely stated that diffusion enables collision. But now, we can go further and note that diffusion governs the specific timescales for binding: Given the diffusion constant (the measure of typical displacements per second), one can estimate the time between collisions, given the molecular concentrations. Perhaps equally importantly, the diffusion constant can help provide an estimate of the "opportunity time" a receptor and ligand possess. That is, once two molecules have found each other, how much time do they have to form a stable complex, before they will "lose" each other by diffusing away? Camacho and coworkers have estimated that this window of opportunity typically lasts only a few nanoseconds (which in turn has implications for the types of structural behavior that can occur during complexation). For more information, see the article by Rajamani et al. and also the book by Phillips et al.

4.6 MORE ON STOCHASTIC DYNAMICS: THE LANGEVIN EQUATION

There are several important reasons to become familiar with the Langevin equation, which describes the dynamics of stochastic systems. First, as we argued above, it is impossible, in principle, to describe a system of interest in a fully deterministic way—because interactions with the environment will affect any system. Second, if we want to consider toy models in one or two dimensions, we really are forced to use stochastic dynamics, since Newtonian dynamics yield pathological results (see below). Third, even for molecules simulated with molecular dynamics, there is often a stochastic component hidden in the "thermostat," which adjusts atomic velocities to maintain consistency with temperature. One can therefore argue that there is little harm in studying dynamics that are explicitly stochastic.

It is easy to see why Newtonian dynamics are not suitable in a thermal context, particularly for low-dimensional models. Imagine a system undergoing energy-conserving Newtonian dynamics in a simple one-dimensional harmonic (parabolic) potential. As we all learned in high school, the system exchanges potential and kinetic energy to maintain constant overall energy. But, perversely, this means that the system moves slowest (and spends the most time!) at highest potential energy—apparently in direct opposition to the Boltzmann factor!

PROBLEM 4.7

One-dimensional Newtonian dynamics. (a) If a particle moves at velocity $v = dx/dt$, approximately how much time does it spend in a small increment Δx. (b) Now consider a particle undergoing Newtonian motion with constant total energy E (potential plus kinetic) moving in a harmonic potential with spring constant κ. Derive an equation that gives the amount of time spent at every x value, $\hat{t}(x)$. Explain where the minima and maxima of the function $\hat{t}(x)$ are. (c) Can you think of a way to generate a true thermal ensemble—consistent with the Boltzmann factor for the total energy—using Newtonian dynamics? Hint: Consider using more than one initial condition.

How can we add thermal behavior and stochasticity directly into Newton's equation of motion? In our heuristic discussion, simplifying that from Reif's book, we will imagine a particle that "feels" the one-dimensional potential energy $U(x)$ but also feels collisions with solvent molecules in motion at a temperature T. The collisions will have two effects: (1) they will lead to random forces on our particle (and hence random displacements), which will increase with higher temperature, and (2) they will lead to "friction" and always tend to slow down a moving particle. These two phenomena cannot be separated, and are intrinsic to the behavior of any molecule in solution. However, both result from collisions with the solvent.

More quantitatively, we would like to add two corresponding terms to Newton's equation. The first will be a "drag" or frictional force that always opposes the particles motion: $f_{drag} = -\gamma m v$, where γ is a friction constant, m is the particle mass, and $v = dx/dt$ is the velocity. (The friction constant γ has units of s^{-1} and can be understood as the approximate frequency of collisions with the solvent.) Second, this slowing down will be counterbalanced by a random thermal force, f_{rand}, which operates even at zero velocity and increases with temperature. Using the derivative definitions for acceleration, force, and velocity, we therefore have

$$\text{Newton:} \quad m\frac{d^2x}{dt^2} = -\frac{dU}{dx}, \tag{4.11}$$

$$\text{With solvent:} \quad m\frac{d^2x}{dt^2} = -\frac{dU}{dx} + f_{drag} + f_{rand},$$

$$\text{Langevin:} \quad m\frac{d^2x}{dt^2} = -\frac{dU}{dx} - \gamma m \frac{dx}{dt} + f_{rand}. \tag{4.12}$$

The distribution of the random forces, not surprisingly, is determined by the temperature, as described in detail in the book by Reif.

4.6.1 OVERDAMPED, OR "BROWNIAN," MOTION AND ITS SIMULATION

We also want to consider an important simplification of the Langevin equation, which will occur in the case where there is high friction—perhaps from a dense solvent. In a dense solvent, where there are frequent collisions, the instantaneous velocity will change frequently and drastically, resulting in large fluctuations in the acceleration. Nevertheless, if we imagine viewing our particle at lower time resolution, we know that a high-friction solvent leads to slow *average* motion. In other words, despite frequent collisions, the changes in the time-averaged velocity—that is, changes in the velocity averaged over a window of time greater than the inverse collision frequency γ^{-1}—will be negligible in the high-friction limit. (The classic example is a falling particle reaching terminal velocity.) This is called overdamped or Brownian motion, and is mathematically approximated by setting the acceleration to zero in Equation 4.12, yielding

$$\text{Overdamped/Brownian:} \quad \frac{dx}{dt} = -\frac{1}{m\gamma}\frac{dU}{dx} + \frac{1}{m\gamma}f_{\text{rand}}. \tag{4.13}$$

The temperature-dependent balance (or competition) among the drag and random forces remains.

The overdamped dynamics described by Equation 4.13 is not only pertinent to molecular behavior in aqueous solutions, but is also ideal for simulating low-dimensional systems. In a simulation, we need to know the rule for generating configuration x_{j+1} from configuration x_j when these are separated by a time Δt. The simulation rule can be derived by changing the derivative to an incremental form $(dx/dt \rightarrow \Delta x/\Delta t)$ in Equation 4.13, with $x_{j+1} - x_j = \Delta x$. In one dimension, the equation for simulating overdamped motion is therefore

$$x_{j+1} = x_j - \frac{\Delta t}{m\gamma}\frac{dU}{dx}\bigg|_{x_j} + \Delta x^{\text{rand}}, \tag{4.14}$$

where $\Delta x^{\text{rand}} = (m\gamma)^{-1}\Delta t f_{\text{rand}}$ is the random displacement or "noise" resulting from the random force. As we mentioned earlier, the random collisions must balance the frictional drag force (embodied in γ) in a temperature-dependent way. In overdamped dynamics, this leads to the requirement that the random displacements Δx^{rand} be chosen from a distribution with zero mean and variance $2k_{\text{B}}T\Delta t/(m\gamma)$. Because the random displacements are the net result of many collisions, they are typically modeled as having a Gaussian distribution. While it is possible to model successive Δx^{rand} values as correlated in time, they are usually considered independent, which is termed "white noise." (Time-correlated random forces or displacements are called "colored noise." These artistic names originate from the Fourier-transform analysis of the time correlations.)

PROBLEM 4.8

In general, physical functions vary continuously in space in time. (a) What physical quantity is assumed to be constant in Equation 4.14 over the interval

Δt? (b) Imagine you have performed a simulation based on Equation 4.14. How could you check whether the assumption from (a) is valid?

PROBLEM 4.9

Write down the explicit form of the conditional probability $\rho(x_{j+1}, t + \Delta t | x_j, t)$ assuming simulation by Equation 4.14 with Δx^{rand} Gaussian distributed.

4.7   KEY TOOLS: THE CORRELATION TIME AND FUNCTION

In Chapter 2, we studied correlations among different coordinates—for example, whether the value of x affects the values of y likely to be seen in a certain distribution. We touched briefly on correlations in time, to which we now return. Correlations in time are, in a sense, different from other cases, since they encompass the possibility for a coordinate (or any observable, really, such as the energy) to be correlated with itself.

Beyond its "theoretical value," an understanding of time correlations can be directly applied, in many cases, to analyzing both experimental and simulation data. Quite simply, correlations in time indicate that behavior at any point in time depends on previous behavior. This should immediately remind you (by contrast!) of the notion of independent variables discussed in the context of probability theory. The butane trajectory shown in Figure 4.8 is a prototypical illustration of time-correlated behavior. The dihedral angle tends to remain the same for lengths of time (called the correlation time, t_{corr}), rather than immediately varying among the possible values ranging over $\sim 300°$.

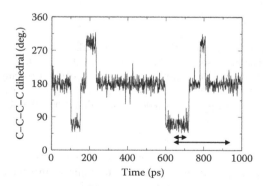

FIGURE 4.8   Correlations in the dynamics of butane's main dihedral. For short times (short arrow), values of the dihedral tend to be similar to previous values. For longer times (long arrow), the molecule "forgets" its earlier conformation, as it typically is found in a completely different state. Based on the figure, the "memory" of previous configurations typically appears to endure for the correlation time of ~ 100 ps in the case of butane.

4.7.1 QUANTIFYING TIME CORRELATIONS: THE AUTOCORRELATION FUNCTION

The basic idea is to quantify the difference from statistically independent behavior, which only occurs for time points beyond the correlation time. Recall from Chapter 2 that if two variables x and y are independent, then we can factorize averages, so that, for instance, $\langle xy \rangle = \langle x \rangle \langle y \rangle$, which we can trivially rewrite as $\langle xy \rangle - \langle x \rangle \langle y \rangle = 0$. Now, however, we want to consider the same variable (or observable) at two different times, say, $x(t_1)$ and $x(t_2)$. If this variable is correlated at the two times, then we expect statistical independence will not hold and result in a nonzero value: $\langle x(t_1)x(t_2) \rangle - \langle x(t_1) \rangle \langle x(t_2) \rangle \neq 0$. The degree to which this measure of correlation differs from zero will be the basis for defining the correlation function.

To define the correlation function itself, note first that we are really interested in how the difference between the two times $\tau = t_2 - t_1$ affects the degree of correlation, so we should consider all pairs of times, $x(t)$ and $x(t + \tau)$. That is, for a given time-difference τ, we will want to average over all possible t values. Putting these ideas together, it makes sense to consider the correlation function

$$
\begin{aligned}
C_x(\tau) &\equiv \frac{\langle x(t)\, x(t+\tau) \rangle - \langle x(t) \rangle \langle x(t+\tau) \rangle}{\langle x(t)x(t) \rangle - \langle x(t) \rangle \langle x(t) \rangle} \\
&= \frac{\langle x(t)\, x(t+\tau) \rangle - \langle x \rangle^2}{\langle x^2 \rangle - \langle x \rangle^2},
\end{aligned} \tag{4.15}
$$

where the averages are to be performed over all t values and the normalizing denominator ensures that $C_x(0) = 1$. This is called an "autocorrelation" function since "auto" means self and it measures the self-correlation of the x variable/observable, instead of between, say, two different variables. Equation 4.15 can be compared to the Pearson coefficient (Equation 2.29), noting that $\sigma_x^2 = \langle x^2 \rangle - \langle x \rangle^2$.

PROBLEM 4.10

The definition of C_x given in Equation 4.15 is difficult to understand the first time around. Therefore you should (a) explain why the variable t appears on the right-hand side, but not the left; (b) explain why $C_x(0) = 1$ for any variable or observable x; (c) explain why $\langle x(t) \rangle = \langle x(t + \tau) \rangle = \langle x \rangle$; (d) explain, similarly, why $\langle x(t)x(t) \rangle = \langle x^2 \rangle$; and (e) based on the preceding, show that the second line of Equation 4.15 is correct.

How will an autocorrelation function behave? It is defined to start at one, for $\tau = 0$, but the more important aspect of C is to quantify the decay of correlations. Indeed, the further separated in time the x values are (i.e., the larger the τ value), the less correlated we expect them to be. We can therefore anticipate two limits of the correlation function, namely,

$$
C_x(0) = 1 \quad \text{and} \quad C_x(\tau \to \infty) = 0. \tag{4.16}
$$

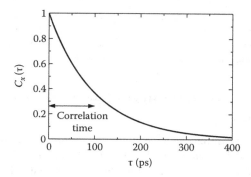

FIGURE 4.9 An idealized correlation function, with an exponential form and correlation time $t_{corr} = 100$ ps. With real data, the correlation function becomes increasingly noisy at larger τ values. A real correlation function can also become negative.

In many circumstances, the correlation function will behave exponentially—in the simplest case as $\exp(-\tau/t_{corr})$. Therefore, a rough estimate of the correlation time t_{corr} can be obtained by the τ value for which C has decayed to a value of e^{-1}.

A schematic simple correlation function is shown in Figure 4.9.

PROBLEM 4.11

Extend the definition given in Equation 4.15 to the case of two different variables. That is, construct a suitable definition for $C_{xy}(\tau)$ that describes the degree of correlation between x at a certain time and y at a time τ later. Explain whether or not you expect $C_{xy} = C_{yx}$? [Hint: try to think of a counterexample, such as one involving two cars on a road.]

4.7.2 DATA ANALYSIS GUIDED BY TIME CORRELATION FUNCTIONS

Imagine studying either a computer model or experimental assay that yields a time trace showing when a particular ion channel is open (conducting current) or closed; your goal is to determine the fraction of time the channel is open given a specific set of conditions. As another example, imagine you are performing a molecular dynamics simulation of a protein and you want to measure the average angle between two helices. In both of these cases, the correlation time, as perhaps measured by analyzing the a correlation function, can be useful.

Since the correlation time should tell you how long it takes for the system to lose its memory of its previous state—at least for the variable under consideration—you want to take experimental or computational "measurements" for a period greatly exceeding t_{corr}, ideally by a factor of 10 or 100. Recall that your precision will improve as the square root of the number of independent measurements.

PROBLEM 4.12

In the example given above of studying an inter-helical angle, it is fairly clear that one should use the autocorrelation of the angle itself. However, to determine the correlation time associated with the opening of an ion channel, one should use an "indicator function." Explain what this is (by looking it up if necessary) and why it is the appropriate variable for assessing correlation in the case of measuring the fractional population of a state of the ion channel.

4.7.3 THE CORRELATION TIME HELPS TO CONNECT DYNAMICS AND EQUILIBRIUM

The previous section referred to estimating averages based on time traces (trajectories)—but is this really valid? The correlation time is the key to letting us know when such averaging of dynamic data is legitimate. The correlation time indicates how long it takes a system to forget its previous state—that is, to generate an independent, new state. On the other hand, statistical averages are based on sufficient quantities of independent samples. Thus, averages based on timescales much longer than the correlation time(s) are truly equilibrium results, while behavior occurring on a timescale equal to or less than the correlation time must be considered dynamical (i.e., nonequilibrium).

It is important to recognize that complex systems like proteins possess many correlation times. Larger-scale motions (e.g., the motion of an entire helix) are likely to be slower than those at small scales (e.g., the motion of solvent-exposed side chain). Which timescales matter will depend on what is being calculated, but it is always safer to worry about longer correlation times.

4.8 TYING IT ALL TOGETHER

Understanding dynamics and the consequences is not easy. In this chapter, we have focused on descriptions of nature that may seem at odds with one another: equilibrium vs. dynamics, and deterministic vs. stochastic dynamics. However, at bottom, there are no contradictions, just different perspectives on the same phenomena.

Most basically, dynamics causes equilibrium. After all, nature does not perform partition-function integrals. Nature causes dynamics, which can lead to (apparent) equilibrium over time. No finite-temperature system is truly static, but rather, constant balanced flows are the heart of equilibrium. These same flows cause equilibrium to be self-healing so that temporary "nonequilibrium" perturbations to populations will relax back to equilibrium values.

The true dynamics of molecules are always governed by the basic laws of physics (quantum mechanics, which we can think of in our simplified picture as Newton's laws of classical mechanics). Nevertheless, in this chapter we primarily discussed stochastic processes, like diffusion. What's the connection to deterministic Newtonian dynamics? In fact, diffusion and stochastic dynamics do not represent different dynamics or phenomena from those predicted by Newton's laws. Rather, stochastic dynamics should be seen as a somewhat simplified description (projected, in the probability sense) of the more fundamental and detailed Newtonian description. Recall, indeed, that the diffusive description of a single particle's motion is actually a simplified version of the huge number of collisions with solvent molecules that the particle actually experiences.

On the other hand, since many biological outcomes and experiments are genuinely stochastic (reflecting the average behavior of a large number of events at finite temperature), a stochastic description is genuinely useful. The stochastic descriptions we have considered include both diffusion—which provides limitations on protein–ligand binding based on concentrations of the species—as well as the general flux picture, where rates represent average stochastic descriptions. When so many molecules (repeated events) are involved, simple stochastic descriptions, like flux-balance, are very powerful.

How do time correlations fit into the big picture? All physical dynamics, including realistic stochastic descriptions, will exhibit time correlations: after all, the present evolves from the past. For example, if a large number of molecules in a system were somehow initialized in the same state, we will not immediately have equilibrium since it takes time to forget the initial state. The correlation time t_{corr} provides a measure of how long it takes to relax back to equilibrium from a "perturbed" state. At times much longer than t_{corr}, we can expect an equilibrium description to be accurate—even though we know that a (flux) balance among never-ceasing dynamical processes actually underlies "equilibrium."

4.9 SO MANY WAYS TO ERR: DYNAMICS IN MOLECULAR SIMULATION

We have often referred to molecular dynamics computer simulations, sometimes as an idealization of dynamics. Of course, simulations are far from perfect, and it is a useful theoretical exercise to probe the intrinsic limitations of simulations. There are at least three ways in which every (classical) molecular dynamics simulation is approximate. See if you can name them before reading on.

1. *The potential is wrong.* Every molecular mechanics forcefield is a classical approximation to quantum mechanics. In reality, the electron density in a molecule will shift in response to changes in configuration or the environment—in contrast to the fixed parameters employed in molecular mechanics.
2. *The Lyapunov instability.* Computers are digital and can only approximate continuum mathematics. Because computers have finite precision, even if the forcefield and dynamics are treated exactly, simulated trajectories must deviate from the true solution of the equations. In fact, this deviation grows exponentially with the duration of the trajectory, and is termed the Lyapunov instability. (This instability also implies that two trajectories started from initial conditions differing very slightly will soon become greatly different, but this is not an error in itself.)
3. *The environment cannot be modeled exactly.* No system can truly be isolated. And, as we discussed earlier, the "environment" of a system of interest, such as the solvent, can never be modeled exactly. The local environment is always in contact with an outer environment that ultimately includes the whole universe! Stochastic descriptions approximate these effects, but

it is not obvious which atoms of a molecular systems should be modeled stochastically—for example, perhaps just an outer layer?

The list above does not include the challenge of obtaining enough sampling, which is a far-from-solved problem in large biomolecular systems. However, at least the sampling problem can be solved in principle one day! More simulation ideas are discussed in Chapter 12.

Should you despair of ever running a simulation? Well, you should be careful, but also be aware that there is a phenomenon called "shadowing," which indicates that even approximate treatments of dynamics can create trajectories that may be quite similar to real trajectories. In other words, if you have the right basic ingredients in your simulation, you can hope to get physically reasonable behavior. (Shadowing is quite a technical subject, however, and the details are well beyond the scope of our discussion.)

Also, the example below will show that you can understand important physical principles even from simple simulations.

4.10 MINI-PROJECT: DOUBLE-WELL DYNAMICS

Consider the toy potential function $U(x) = E_b[(x/d)^2 - 1]^2$, where x denotes the value of a ("reaction") coordinate in a molecule in units of angstroms. Let $d = 1$ Å and $E_b = 5k_BT$. The coordinate will be assumed to evolve according to overdamped stochastic dynamics as given in Equation 4.14 with $\gamma = 10^9 \text{s}^{-1}$. Let the particle's mass be given in strange-looking units as $m = 10^{-9}k_BTs^2/\text{Å}^2$, which emphasizes the key role of the thermal energy scale in stochastic dynamics. Assume the random fluctuation Δx^{rand} at any time step is either of the two values $\pm\sqrt{2k_BT\Delta t/m\gamma}$. In other words, Δx^{rand} is not chosen from a Gaussian distribution, but for simplicity takes only one of these two values chosen with equal random probabilities. Answer or complete all of the following items.

a. Sketch the potential and force by hand on the same plot. Check that the force curve makes physical sense.

b. Why is the choice $\Delta t = 0.001$ s justified? (This is the value to be used below.) See Problem 4.8.

c. Write a computer program to simulate overdamped stochastic dynamics of the coordinate x.

d. Starting from $x = -1$, run your program for one million time steps. Save the output to a file, but you probably do not want to print out every time step to save disk space.

e. Plot the trajectory—that is, plot x as a function of time. Print out the plot, or part(s) of it, on the best scale or scales to illustrate any interesting features. Comment on the connection to the potential function you sketched. [*Hint:* if your trajectory does not switch between the two minima, look for a bug in your computer program.]

f. Now increase E_b by $1k_B T$ and repeat the simulation. Based on the Arrhenius factor, what do you expect the ratio of transition rates to be in the two cases? Does this seem consistent, approximately, with your trajectories?

g. Make a histogram of the distribution of x values. With what function can you compare this to see if your simulation makes sense?

h. Estimate, roughly, the length of simulation that would be required to obtain a good estimate of the equilibrium distribution for the potential under consideration. Justify your estimate, perhaps by suggesting the rough error/uncertainty in it.

FURTHER READING

Berg, H.C., *Random Walks in Biology*, Princeton University Press, Princeton, NJ, 1993.

Dill, K.A. and Bromberg, S., *Molecular Driving Forces*, Garland Science, New York, 2003.

Leach, A., *Molecular Modelling*, Prentice Hall, Harlow, U.K., 2001.

Ma, S.-K., *Statistical Mechanics*, World Scientific, Hackensack, NJ, 1985.

Nelson, P., *Biological Physics*, W.H. Freeman, New York, 2008.

Phillips, R., Kondev, J., and Theriot, J., *Physical Biology of the Cell*, Garland Science, New York, 2009.

Rajamani, D., Thiel, S., Vajda, S., and Camacho, C.J., *Proceedings of the National Academy of Sciences*, 101:11287–11292, 2004.

Reif, F., *Fundamentals of Statistical and Thermal Physics*, McGraw-Hill, New York, 1965.

5 Molecules Are Correlated! Multidimensional Statistical Mechanics

5.1 INTRODUCTION

We can now extend our understanding of statistical mechanics to much more realistic systems. Most systems of interest, of course, are "macroscopic" and contain not just complicated molecules, but many, many copies of them. You can guess that mathematical complexity is about to set in. But don't panic—all of the intuition we developed in one-dimensional systems remains highly pertinent. Free energy still "means" probability, and entropy still describes the size of a state. It's true even in high-dimensional systems that we should expect an average energy of $k_B T/2$ per degree of freedom. Part of this chapter will be devoted to showing you that one-dimensional reasoning still works.

This chapter and the next will also explain how to cope with intrinsically multidimensional phenomena, like correlations and projections down to lower dimensional descriptions. For these, we will draw on our experience with probability theory.

Large molecules are naturally described by "internal coordinates" and we will describe these and explain how they are used in statistical mechanics calculations. Nevertheless, most of your understanding can come from Cartesian coordinates, which are much simpler to use in partition functions.

As usual, we will proceed by starting with the simplest examples. This will further develop our intuition and hopefully make the equations less intimidating.

5.1.1 MANY ATOMS IN ONE MOLECULE AND/OR MANY MOLECULES

For better or worse, we are faced with the twin tasks of studying a single molecule with many atoms (hence many dimensions), as well as understanding systems with many molecules, perhaps of different types. For instance, think of a ligand-binding system: it will contain proteins ("receptors"), ligands, and solvent. In fact it's not hard to write down partition functions for such systems. Rather, the hard thing is knowing what to do with them.

Despite the fact that we will never be able to "solve" a partition function for a real system, it is nevertheless critical to understand multidimensional statistical mechanics. In this chapter, we will develop the theory that will be the beginning of our understanding of such fundamental phenomena as binding and hydrophobicity.

Further, the theory is the basis for performing and understanding computer simulations. We will make as much contact with biophysics as possible throughout the chapter.

5.1.2 WORKING TOWARD THERMODYNAMICS

In case you haven't noticed yet, this book has strong bias for statistical descriptions, rather than using thermodynamics. Nevertheless, thermodynamics does matter, particularly since most experimental measurements are thermodynamical. Eventually, we must take the plunge into thermodynamics. To do so as painlessly as possible (i.e., statistically!), we must have a very firm grasp of multidimensional statistical mechanics.

5.1.3 TOWARD UNDERSTANDING SIMULATIONS

Although the author is guilty of writing papers on one-dimensional systems, it's fair to say that practically every computer simulation of biophysical interest investigates a high-dimensional system. Often, a protein is studied, perhaps along with a ligand molecule—and worse still, perhaps along with so much water that it constitutes the majority of atoms in the system. How can we get a grip on such systems with tens of thousands, or even up to millions of degrees of freedom? How do the scientists performing the work even look at the data? The PMF (potential of mean force) is a key part of the answer, which you will readily understand based on our study of projection (marginalization) in probability theory. This chapter presents the preliminaries necessary for understanding the PMF, which will be discussed in the next chapter.

5.2 A MORE-THAN-TWO-DIMENSIONAL PRELUDE

A two-dimensional (2D) "volume" presents great opportunities for carefully considering high-dimensional spaces.

In a series of baby steps, we will study a system of two atoms (A and B) constrained to two dimensions—imagine two balls on a table—because key mathematical and physical aspects of multidimensional problems can be seen. Thus, we have an eight-dimensional phase space since there are eight degrees of freedom (coordinates), namely, x^A, y^A, v_x^A, and v_y^A, for atom A and similarly for B. The configuration space, however, is four-dimensional since there are four positional coordinates. We'll first consider the two balls as noninteracting, but later we will connect the balls by a spring, as shown in Figure 5.1.

For the most part, we'll focus on constructing the correct partition functions. But we'll also consider some physical properties to get our intuition primed for the more complex systems to come.

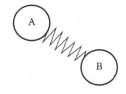

FIGURE 5.1 A simple description of a diatomic molecule, namely, two atoms connected by a spring. For even more simplicity, we assume the two atoms can only move in two dimensions—that is, they are constrained to a surface.

5.2.1 One "Atom" in Two Dimensions

As always, the partition function normalizes the probability density, and hence is an integral over all possible configurations of the system. From Chapter 3, we know that every ensemble average of interest can be written as a ratio of integrals: in the numerator, there is one integral quite like the partition function, normalized by the partition function in the denominator. And the free energy, the most important physical quantity of all, is just the log of the partition function. So the partition function is the place to start.

Let us assume we have a single atom of mass m that moves freely ($U = 0$) in a 2D square box of side L. The partition function considers all possible positions and velocities (really, momenta $p = mv$ to get the detailed physics right):

$$Z = \frac{1}{h^2} \int\limits_{-\infty}^{+\infty} dp_x \int\limits_{-\infty}^{+\infty} dp_y \int\limits_{0}^{L} dx \int\limits_{0}^{L} dy \, \exp\{-\left[p_x^2/2m + p_y^2/2m + U(x,y)\right]/k_B T\}. \quad (5.1)$$

The factors of Planck's constant, h^{-2}, are part of the classical approximation to quantum mechanics described in Chapter 3, and conveniently serve to make Z dimensionless. As usual, U is the potential energy.

Note that in Equation 5.1 the velocity/momentum integrals are "decoupled" from the rest of the integral. That is, without approximation, we can separate each of these and evaluate it as in Chapter 3—that is, as a Gaussian integral. We therefore write

$$Z = \left(\frac{1}{h} \int\limits_{-\infty}^{+\infty} dp_x e^{-\left[p_x^2/2m\right]/k_B T}\right) \left(\frac{1}{h} \int\limits_{-\infty}^{+\infty} dp_y e^{-\left[p_y^2/2m\right]/k_B T}\right)$$

$$\times \int\limits_{0}^{L} dx \int\limits_{0}^{L} dy \, e^{-U(x,y)/k_B T}$$

$$= \frac{1}{\lambda^2} \int\limits_{0}^{L} dx \int\limits_{0}^{L} dy \, e^{-U(x,y)/k_B T}$$

$$= \left(\frac{L}{\lambda}\right)^2. \quad (5.2)$$

Because both temperature and mass affect the width of the Gaussian velocity integrals, λ depends on both (see Chapter 3). In deriving the last equality, we have set $U = 0$ for all x and y, which is equivalent to saying there are no external "fields" exerting forces on our atom. For instance, in principle, gravity or electrostatics could affect the energy of the atom, but we have excluded such effects.

Is there physical meaning in the trivial-looking equation $Z = (L/\lambda)^2$? In fact, there is, and it's our old friend the entropy. To see this, first note that since the potential energy is always zero, we can immediately say $\langle U \rangle = 0$. You can easily show by Gaussian integration, as in Chapter 3, that $\langle KE \rangle = k_B T$, which is just $k_B T/2$ for each momentum/velocity component. This means that $S = (\langle E \rangle - F)/T = k_B - F/T = k_B(1 + \ln Z)$ or

$$S = k_B \left[\ln \left(\frac{L}{\lambda} \right)^2 + 1 \right] \simeq k_B \ln \left(\frac{L}{\lambda} \right)^2, \tag{5.3}$$

where we have used the fact that $L \gg \lambda$ since λ is a tiny quantum-mechanical lengthscale.

Equation 5.3 shows that the entropy is just the log of the available configurational volume, which happens to be an area in our 2D case. The same basic rule as in one-dimension carries over to 2D, as promised!

5.2.2 TWO IDEAL (NONINTERACTING) "ATOMS" IN 2D

The next baby step is to consider two atoms that do not interact, as shown in panel (a) of Figure 5.2. In other words, $U = 0$ for all configurations, which of course will keep our calculation nice and simple. Physically, $U = 0$ means that neither particle can "feel" the other (no forces). In fact, they wouldn't even know if they were on top of each other, which is part of the standard "ideal" assumption. But such overlaps represent a tiny part of the configuration space since we will assume the system size L is much greater than the particle size.

Based on our experience above, we can really just write down the exact value for the partition function. All four momentum integrals will factorize and simply lead to powers of the "thermal wavelength" λ^{-1}—except that now we may have two different wavelengths because the masses of A and B may be different. The configurational coordinates are also easy: each will lead to a power of L. In other words, we can evaluate the partition function for two ideal particles "by inspection,"

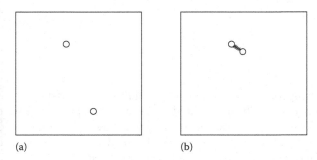

FIGURE 5.2 Two-particle systems for which we calculate partition functions: (a) two noninteracting "atoms" in a 2D box and (b) a diatomic "molecule" confined to a 2D box.

as the mathematicians would say:

$$Z = \left(\frac{1}{h} \int\limits_{-\infty}^{+\infty} dp_x^A \, e^{-\left[(p_x^A)^2 / 2m_A \right] / k_B T} \right) \left(\frac{1}{h} \int\limits_{-\infty}^{+\infty} dp_y^A \, e^{-\left[(p_y^A)^2 / 2m_A \right] / k_B T} \right)$$

$$\times \left(\frac{1}{h} \int\limits_{-\infty}^{+\infty} dp_x^B \, e^{-\left[(p_x^B)^2 / 2m_B \right] / k_B T} \right) \left(\frac{1}{h} \int\limits_{-\infty}^{+\infty} dp_y^B \, e^{-\left[(p_y^B)^2 / 2m_B \right] / k_B T} \right)$$

$$\times \int\limits_0^L dx^A \int\limits_0^L dy^A \int\limits_0^L dx^B \int\limits_0^L dy^B \, \exp\left[- U(x^A, y^A, x^B, y^B) / k_B T \right] \qquad (5.4)$$

$$= \frac{L^4}{\lambda_A^2 \lambda_B^2}. \qquad (5.5)$$

High-dimensional integrals don't get much easier than this, and we will indeed exploit such factorizability extensively in ideal systems with more than two particles. Remember that when "A" and "B" are used as superscripts here, they are just labels, and not exponents! Also note that Equation 5.4 still includes U and is valid for nonideal systems— unlike Equation 5.5 where we have set the ideal value $U = 0$.

Note that Equations 5.4 and 5.5 apply for the case when the A and B atoms are different. The case when the atoms are identical requires a correction for the fact that interchanging the positions of identical atoms leads to a configuration indistinguishable from the original. This is discussed in Section 5.5.

Once again, we can calculate the entropy of our system, which contains the most important physical information from our two-atom exercise. Using the same procedure that produced the single-atom entropy, Equation 5.3, we find

$$S \simeq k_B \ln \frac{L^4}{\lambda_A^2 \lambda_B^2} = k_B \left[\ln \left(\frac{L}{\lambda_A} \right)^2 + \ln \left(\frac{L}{\lambda_B} \right)^2 \right]. \qquad (5.6)$$

For two atoms in two dimensions, L^4 represents the full configuration-space volume, which is available to two ideal atoms. Each atom has L^2, as in Equation 5.2, and these factors get multiplied together in the two-atom partition function. Because the entropy is based on the log of the available configuration volume, the single-atom entropy has essentially doubled by the addition of a second ideal atom (of roughly equal mass). You can compare Equations 5.3 and 5.6.

5.2.3 A DIATOMIC "MOLECULE" IN 2D

Finally, time for a real molecule! Well, maybe that's an exaggeration, but we will now study a 2D balls-on-a-spring model, which includes many important features—and complications—that one encounters in nonideal molecular models.

We will study the diatomic "molecule" shown in Figure 5.1 confined to a box as in Figure 5.2b. The atoms, of masses m_A and m_B, are connected by an ideal spring of spring constant k, meaning the energy of every configuration is given simply by $U = k(r - r_0)^2 / 2$, where r is the separation between the atoms and r_0 is the equilibrium bond length. We assume the molecule is much smaller than our box, by having $r_0 \ll L$ and also having k large enough so that the spring can't stretch much.

PROBLEM 5.1

Using your knowledge of one-dimensional springs, define a criterion for a "stiff" spring without doing any integration. Your criterion should be an inequality relating k, r_0, and T.

It's now time for the partition function, which is exactly of the form of Equation 5.4. First note that even though the atoms are connected by a spring, their momentum integrations are still independent in the classical partition function. Physically, this means that the atoms are indeed free to move at high speeds, even though they can never get far from one another. Mathematically, it means that our momentum integrations remain trivial and yield powers of $1/\lambda_i$, with $i = $ A or B. The potential U depends only on the interatom separation r, but of course, r itself is a function of the x and y values of both A and B. Given these points, the partition function takes the form

$$Z = \frac{1}{\lambda_A^2 \lambda_B^2} \int_0^L dx^A \int_0^L dy^A \int_0^L dx^B \int_0^L dy^B \, e^{-U(r)/k_B T}. \qquad (5.7)$$

The difficulty with this expression is that it's not obvious how to do the x and y integrations.

5.2.3.1   Internal Coordinates to the Rescue!

Fortunately, there is an old solution to the problem, and it is one that clarifies both the mathematics and the science. "Internal coordinates" (sometimes called "generalized coordinates") are essentially relative coordinates, as opposed to absolute ones. They are perfect for descriptions of molecules. Indeed, the interatom distance r is a prototypical internal coordinate, which does not depend on the absolute location of the molecule or its orientation, but only on the "internal" configuration. The other internal coordinates typically used are bond and dihedral angles, which we'll discuss below.

Using internal coordinates, we can change our integration variables to match those in the potential energy. We do so by observing that every "Cartesian" configuration (x^A, y^A, x^B, y^B) can be described equivalently by the set of coordinates (x^A, y^A, r, ϕ), where ϕ is an angle giving the orientation of the molecule (based on any reference value you like). Note that we chose A as the reference atom, but this was arbitrary—it could just as easily have been B.

But—be warned!—it is not completely trivial to change to internal coordinates in a partition function. We must preserve "meaning" of Z, which will automatically preserve its mathematical value. If you simply integrate over the variables (x^A, y^A, r, ϕ), you will make an error. The meaning of a partition function like (5.7) is that *all possible Cartesian configurations*—all (x^A, y^A, x^B, y^B) values—*are examined on an equal basis, and then weighted by the Boltzmann factor.* If you imagine evaluating the integral by sampling the whole space, you would pick an equal number of points in every little "volume" in the four-dimensional configuration space and evaluate the Boltzmann factor at each point. Equivalently, you could check all locations for A—one at a time—while for each A location, examining every possible configuration for B. (This is essentially the internal coordinate prescription.)

The subtlety comes in examining all possible B locations given some fixed A location. The Cartesian x^B and y^B values must be checked on a uniform basis (prior to Boltzmann factor weighting). But this is absolutely not equivalent to uniformly sampling r and ϕ values. To see why, note that if you sampled the Cartesian coordinates of B for a fixed A location, you would necessarily sample more values further away— that is, at larger r. On the other hand, if you sample uniformly in r, then by definition, you always get the same number of configurations at every r. This is the origin of the infamous Jacobian factors that you may have studied in changing variables in calculus.

Once the Jacobian is included in our 2D problem, the bottom line is that

$$\int \int \int \int dx^A \, dy^A \, dx^B \, dy^B = \int \int \int \int dx^A \, dy^A \, r \, dr \, d\phi. \tag{5.8}$$

If the integrand (not shown) does not depend on ϕ, as in our case (since U depends only on r), then one can further simplify this to $2\pi \int \int \int dx^A \, dy^A \, r \, dr$ by performing the trivial ϕ integration.

PROBLEM 5.2

By drawing a picture, give a simple geometric argument for why the factor of r is necessary in Equation 5.8. You may treat the case when the integrand is independent of ϕ for simplicity.

You should be able to "read" Equation 5.8. It says that we will integrate over all absolute locations of atom A while, for each A position, we will integrate over all possible relative positions of atom B—that is, of B relative to A—described by a distance and an orientation angle. We will properly account for the increasing area of rings of thickness dr as r increases.

PROBLEM 5.3

(a) Write down the equation analogous to Equation 5.8 for a system of three particles in two dimensions. (b) Draw a picture to illustrate your coordinates. (c) Explain in words why you expect your equation in (a) to be correct.

Finally, we are ready to evaluate the partition function, but not without one last "trick" (which, fortunately, we already used in Chapter 3). We have

$$Z \overset{(a)}{=} \frac{1}{\lambda_A^2 \, \lambda_B^2} \int_0^L dx^A \int_0^L dy^A \int_0^{r_{max}} r \, dr \int_0^{2\pi} d\phi \, e^{-U(r)/k_B T}$$

$$\overset{(b)}{\simeq} \frac{2\pi}{\lambda_A^2 \, \lambda_B^2} \int_0^L dx^A \int_0^L dy^A \int_0^{r_{max}} dr \, r \, \exp\left[- k(r - r_0)^2 / 2 k_B T\right]$$

$$\overset{(c)}{\simeq} \frac{2\pi L^2}{\lambda_A^2 \lambda_B^2} \int_0^\infty dr\, r\, \exp\left[-k(r-r_0)^2/2k_B T\right]$$

$$\overset{(d)}{\simeq} \frac{2\pi L^2}{\lambda_A^2 \lambda_B^2} r_0 \int_{-\infty}^\infty dr\, \exp\left[-k(r-r_0)^2/2k_B T\right]$$

$$\overset{(e)}{\sim} \frac{2\pi L^2}{\lambda_A^2 \lambda_B^2} r_0 \sqrt{2\pi} \sqrt{\frac{k_B T}{k}} \propto T^{5/2}, \tag{5.9}$$

where in the second line (b) we have performed the ϕ integration and also substituted the explicit form for U. In (b), we have assumed that all ϕ values are equivalent, but this is not quite true when the molecule is near the edge of the box. However, the large box ($L \gg r$) means that edge configurations constitute a negligible fraction of the total. In the third line (c), we performed the x^A and y^A integrations and assumed the value of the integral would be essentially unchanged if we set $r_{max} = \infty$.

How can we justify an infinite r_{max} in the partition function (5.9)? At first glance, it seems r must be limited by $L/2$ or certainly by L. But then the situation seems even worse: What if atom A is quite near the wall of the box? Indeed, in principle, the maximum r depends on the orientation (i.e., on ϕ) and could be quite small for some A locations. This is where our two earlier conditions come in—the stiffness of the spring (large k) and the small equilibrium size of the molecule ($r_0 \ll L$). The stiffness means that for any position of A, only a small range of r near r_0 is important (i.e., contributes to the integral). The smallness means that the fraction of A positions that are close to the wall (on the order of r_0 distant) is negligible. Together, these assumptions tell us we can ignore the ϕ dependence in (b), set r_{max} to infinity and the lower limit to $-\infty$ in line (c), and set the linear r factor in the integrand to r_0 in (d), making only very small errors.

Given all the explanation necessary to perform a reasonable calculation on even this simplest of molecular systems, you should recognize the intrinsic difficulty of performing statistical mechanics calculations for molecules. Indeed, it is fair to argue that, even for our model, we have somewhat oversimplified. The symbol "$\sim$" was used in line (e) of Equation 5.9 to emphasize that the stiff spring assumption is somewhat inconsistent with our earlier assumption that the velocities were fully independent. After all, if the spring gets infinitely stiff, at least one of the four components of the momenta (represented in suitable relative coordinates) is not fully independent. While this book relies solely on classical mechanics to underpin statistical mechanics, diatomic molecules are usually treated quantum mechanically—see the discussions in the book by McQuarrie and Simon—where the coupling between configurational and momentum degrees of freedom is handled more naturally.

5.2.3.2 Loss of Translational Entropy

The most important physical implication of the diatomic partition function (5.9) is a loss of entropy, as compared to noninteracting atoms. Further, we will see that translational entropy is lost, which is a generic feature of "tying" together two or more particles, as we did with our spring.

The calculation of the entropy proceeds similarly to the ideal case, except that now we have $\langle U \rangle = k_B T$ because there are effectively two interacting configurational coordinates (the relative x and y positions embedded in r and ϕ). Thus, $\langle E \rangle = \langle U \rangle + \langle KE \rangle = 2k_BT$. On the other hand, because $L \gg \lambda$, we expect $F = -k_BT \ln Z \gg 2k_BT$. Therefore, it is again a reasonable approximation to set

$$S \simeq k_B \ln Z = k_B \left[\ln \left(\frac{L}{\lambda_A} \right)^2 + \ln \left(\frac{r_0 \sqrt{k_BT/k}}{\lambda_B^2} \right) \right] \simeq k_B \ln \left(\frac{L}{\lambda_A} \right)^2, \qquad (5.10)$$

where we used our stiff-spring and small r_0 assumptions. (Note that in the full entropy, both λ_A and λ_B correctly appear in a symmetric way.)

The diatomic entropy (5.10) is only marginally larger than the *single*-atom entropy (Equation 5.3) and much smaller than that for two noninteracting atoms (Equation 5.6). Physically, this makes sense. Because the two atoms are tied together, they really can only explore the configuration space available to one of the atoms. They cannot perform "translational" motion independently, and hence the entropy loss is specified in this way.

5.2.4 LESSONS LEARNED IN TWO DIMENSIONS

Later, you'll see that the 2D calculations we just performed are excellent models for more advanced calculations. Indeed, we have seen both types of multidimensional systems: a multi"molecular" system (two atoms) and a nontrivial diatomic "molecular" system. Four things are clear already: (1) noninteracting systems are easy, (2) internal coordinates make a lot of sense for describing the configuration of a molecule, and (3) molecular calculations are tricky in general, and with internal coordinates, one particular subtlety is the Jacobian factors needed in configurational integrals. Furthermore, (4) the loss of translational entropy occurring in our simple diatom qualitatively indicates what happens in protein–ligand binding.

5.3 COORDINATES AND FORCEFIELDS

5.3.1 CARTESIAN COORDINATES

"Cartesian" coordinates are just a fancy way of calling the x, y, z coordinates you are already familiar with. The only generalization from what you may be used to is that we can have many more than three coordinates defining a single point configuration space. After all, to specify the configuration of a molecule with N atoms requires $3N$ numbers—x, y, and z for each atom. We'll often denote the full set of coordinates by $\mathbf{r}^N$, which is meant to remind you that there is a long list of numbers, something like in exponentiation (though N here is not an exponent). This notation builds on the common use of $\mathbf{r}$ for the vector describing the location of a point in 3D.

To put it concretely, we have

$$\begin{aligned} \mathbf{r}^N &= (\mathbf{r}_1, \mathbf{r}_2, \ldots, \mathbf{r}_N) \\ &= (x_1, y_1, z_1, \ x_2, y_2, z_2, \ \ldots, \ x_N, y_N, z_N), \end{aligned} \qquad (5.11)$$

where $\mathbf{r}_i$ denotes the location of the atom i and similarly for x_i, y_i, z_i.

Momenta in Cartesian space will be denoted in direct analogy to the configurational coordinates, via $\mathbf{p}^N$ and $\mathbf{p}_i$.

You should be aware that there is no standard notation for multidimensional systems. Each book and paper will adopt its own convention, and some may be clearer than others about the dimensionality hidden in a given symbol.

5.3.2 INTERNAL COORDINATES

Internal coordinates, as we have already seen in Section 5.2.3, provide a much more intuitive description of a molecule than do their Cartesian counterparts. Bond lengths, bond angles, and dihedral/torsion angles (defined below) essentially tell you how to construct a molecule starting from one reference atom. Even if the molecule is branched, such as a protein, internal coordinates can be (and are) used with suitable conventions.

There are always fewer internal coordinates than Cartesian, since the internal description leaves out the absolute location and orientation of a molecule. As we saw in the case of the diatomic molecule in 2D, above, there is really only one internal coordinate, the bond length r, though there are four Cartesian coordinates. For a molecule in three dimensions, there are always six fewer internal coordinates ("degrees of freedom") because three are required to specify the location and three more are required for the orientation. All these are omitted from the internal description. To see that three numbers are required to specify the orientation of an object in three-dimensional space, you can think of possible rotations about the x, y, and z axes.

Defining internal coordinates can be tricky, particularly if the molecule is not linear. However, in a linear molecule, like an idealized polymer, the conventions are fairly obvious, as illustrated in Figure 5.3. The bond lengths, r, and bond angles, θ, are almost self-explanatory. The only point of caution is to note that the index usually refers back to the first possible coordinate—for example, r_1 connects atoms 1 and 2, while θ_1 is the angle formed by atoms 1, 2, and 3. A dihedral angle, ϕ, among

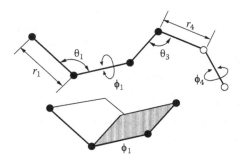

FIGURE 5.3 Selected internal coordinates for a hypothetical linear molecule based on a reasonable indexing convention. The two planes that define the first dihedral angle, ϕ_1, are shown below the molecule. Atom numbers are not shown to avoid clutter, but start with 1 at left.

four consecutive atoms is defined to be the angle between the two planes formed by the first three and last three atoms, as sketched in the figure. Note that sometimes a "torsion" angle is defined to be $360° - \phi$, while at other times the words dihedral and torsion are used synonymously. So be careful!

While we will not describe naming conventions for branched molecules (a category that includes peptides and proteins), it is interesting to consider a concrete example showing that a configuration is still described by $3N - 6$ internal coordinates. The next problem does this in the simple methane molecule.

PROBLEM 5.4

(a) Sketch methane, CH_4. Note that $3N - 6 = 9$ in this case, and also note that there are no dihedral angles in the molecule. (b) How many bond lengths and angles are there, considering all possibilities? (c) One bond angle is redundant—that is, is not necessary to specify the configuration. Explain in words and/or sketches why only five bond angles are necessary to specify the configuration.

5.3.3 A Forcefield Is Just a Potential Energy Function

In this section, we will briefly describe typical potential energy functions used in molecular mechanics computations. Such functions are termed "forcefields," which should be considered as a synonym for $U(\mathbf{r}^N)$. The statistical mechanics of molecules we will discuss will not rely on the particular functional form of the forcefield below, but it is useful to have in mind a concrete potential energy function.

Perhaps the most important thing for you to recognize—and hopefully think about—is that the forcefields used in protein and nucleic acid computations are not necessarily the most accurate. They may be used because they are simpler than better approximations, or simply because they represent a historical standard. You should be aware that the "standard" forcefield equation given below does not represent the best classical approximation to interatomic interactions by any means. More advanced forcefields have been developed, but are beyond the scope of this book. The bottom line is that every classical (non-quantum-mechanical) forcefield is approximate.

The typical forcefield used in biomolecular computation contains both "bonded" terms, embodying interactions between covalently bonded atoms and "nonbonded" van der Waals and electrostatics terms. The van der Waals and electrostatic interactions between bonded atoms are considered to be built in to the bonded terms, and therefore are not represented separately.

$$U(\mathbf{r}^N) = \sum_{i=1}^{N-1}\left(\frac{k_{b,i}}{2}\right)(r_i - r_{i0})^2 + \sum_{i=1}^{N-2}\left(\frac{k_{\theta,i}}{2}\right)(\theta_i - \theta_{i0})^2 + \sum_{i=1}^{N-3} u_i(\phi_i)$$

$$+ \sum_{i<j}\left\{\frac{q_i q_j}{4\pi\epsilon_0 r_{ij}} + 4\epsilon_{ij}\left[\left(\frac{\sigma_{ij}}{r_{ij}}\right)^{12} - \left(\frac{\sigma_{ij}}{r_{ij}}\right)^{6}\right]\right\} + \cdots. \quad (5.12)$$

The first line includes the bonded terms, while the second line includes the nonbonded terms. The typical u_i function for dihedrals is a sum of a few cosine functions, and

r_{ij} is the distance between the nonbonded atoms i and j. It is the set of parameters $(k, r_{i0}, \ldots,$ plus those hidden in $u_i)$—as opposed to the variables r_i, θ_i, ϕ_i—which distinguishes the different forcefields (e.g., CHARMM, AMBER, OPLS, GROMOS) from one another, since each will use different parameter values. Typically, there will be hundreds of parameters.

Common forcefields may well contain additional explicit or implicit terms; hence the "$\cdots$" in Equation 5.12. For instance, "improper dihedral" terms among four nonconsecutive atoms are often used to maintain the planarity of ring groups. Further, van der Waals interactions—and sometimes electrostatics—are typically not assumed to extend to infinity, but are tapered off to zero at some cutoff distance. Hydrogen bonds are sometimes treated separately from the standard interactions of Equation 5.12—and sometimes not. Details like this may differ among forcefields and even individual users of computer software when an option is given.

At first glance, it seems that forcefields include only interactions among atom pairs. However, Equation 5.12 indeed contains "three-body" and "four-body" terms. After all, a bond angle depends on the location of three atoms, and a dihedral on the location of four.

5.3.3.1 Modeling Water

In fact, water is often modeled using the same basic forcefield form (Equation 5.12). However, there are different approaches to setting up the atomic interaction centers within the molecule. Common models used in biomolecular simulation involve fixed point charges, typically three (one for each hydrogen and one for oxygen) embedded in a single van der Waals sphere. In such cases, there are no nonbonded terms, since the "atoms" are not free to move relative to one another. Other water models include four interaction centers (representing two hydrogens and two "lone pairs" of electrons, see Chapter 8), and still others are flexible and/or polarizable.

You should also be aware that forcefield parameters for proteins and nucleic acids are typically developed to be consistent with a particular water model.

5.3.3.2 Where Is the Quantum Mechanics?

The forcefields embodied in Equation 5.12 are purely classical. There is no allowance for covalent bond formation or breakage, for instance. Such exclusion of chemical reactions may be reasonable in many situations, but other fundamentally quantum behaviors can be important in general. In particular, the quantum "wave function," which specifies the electron density (the spatial distribution of negative charge), will vary as the locations of atoms (nuclei) change during dynamics. In other words, the electrostatic properties of an atom will depend on the configuration of the whole molecule! For Equation 5.12, this means the charges q_i generally should change with configuration, although the sum of the charges should not. These "polarization" effects are an important topic of current research, but as of 2009, the charges typically are held constant.

To keep the discussion of this book simple, we will assume that atomic parameters do not change with molecular configuration. Thus, we ignore degrees of freedom associated only with electrons. The statistical principles we describe, nevertheless, are applicable to more complex descriptions of molecules.

5.3.4   JACOBIAN FACTORS FOR INTERNAL COORDINATES (ADVANCED)

To write down a correct partition function for a molecular system, we need to know the correct Jacobian factors for multiple atoms in a three-dimensional space. In other words, we want to develop the analog of Equation 5.8 for multiatomic molecules embedded in a three-dimensional space. Certainly, the same logic holds: The partition function written in internal coordinates must be the same as that in Cartesian coordinates. Translated into the language of sampling, this means that if we generate points in internal coordinates with a distribution proportional to the Jacobian factors, this should yield a uniform distribution in Cartesian space.

In terms of equations, we want to determine the correct function J so that the relation

$$
\int dx_1\, dy_1, dz_1 \ldots dx_N\, dy_N\, dz_N f(\mathbf{r}^N)
$$

$$
= \int dx_1\, dy_1\, dz_1\, d\theta_0\, d\phi_{-1}\, d\phi_0\, J_0(\theta_0)
$$

$$
\times \int dr_1 \ldots dr_{N-1}\, d\theta_1 \ldots d\theta_{N-2}\, d\phi_1 \ldots d\phi_{N-3} f(\mathbf{r}^N)
$$

$$
\times J(r_1 \ldots, r_{N-1},\ \theta_1, \ldots, \theta_{N-2},\ \phi_1, \ldots, \phi_{N-3}) \tag{5.13}
$$

holds true for any function f. This is complicated. Since the left-hand side of the equation includes rotation and translation, so too must the right-hand side (RHS). The first six integrations on the RHS are exactly the six omitted degrees of freedom. To understand these six integrations, we should consider them for a single rigid molecular configuration—that is, a single set of internal coordinates. It is useful to consider Figure 5.3, which depicts the indexing scheme followed in Equation 5.13.

The first three integrations on the RHS describe the translational degrees of freedom (which could have been imputed to the center of mass, but here we have just used the first atom). But even if we know the location of the first atom, we still have not specified the orientation of our molecule. To do so, we must consider it to be embedded in some Cartesian (x, y, z) reference frame, with the first atom placed at the origin. Then, extrapolating backward from the indices shown in Figure 5.3, the variable θ_0 describes the angle the first bond forms with the z axis. (The choice of z is conventional. You may want to look up "spherical polar coordinates" if you are not familiar with them.) The angle ϕ_{-1} (really an "azimuthal" and not a dihedral angle) describes the rotation of the second atom about the z axis, relative to the x–y plane—or equivalently, the angle between the projection/shadow of the first bond in the x–y plane and the x axis. Finally, ϕ_0 is very much like a dihedral: given the orientation of the bond between atoms 1 and 2, the location of the third atom can be specified by the amount of rotation necessary about the first bond relative to some arbitrary reference.

Let's summarize the logic in an "operational" way: Imagine you are integrating over all possible internal coordinates—that is, over all molecular configurations.

For each configuration, you will also need to consider all possible translations and rotations. Thus, for every internal configuration (considered fixed), you integrate over all possible locations of the first atom. The next three integrations consider all orientations, by establishing the first atom as the origin of a set of local Cartesian coordinates. Integration over θ_0 and ϕ_{-1} considers all possible locations of atom 2 (since the first bond length is considered fixed). For every location of atom 2, the "pseudo-dihedral" ϕ_0 describes the orientation of the whole molecule via the degree of rotation about the first bond (between atoms 1 and 2). You'll need to scribble on Figure 5.3 to understand this.

To be fully general, our function J must apply for any integrand f in Equation 5.13, whether the Boltzmann factor or any other function. This means that we are not only getting the volume/sampling equivalence correct globally, but also locally—in any and every microscopic volume. Note that the angle θ_0 (which is not an internal coordinate) requires a Jacobian to ensure equivalence of the partition functions, although the potential energy may not depend on it in many cases.

A detailed justification for the exact form of the Jacobian can be found in any discussion of spherical-polar coordinates. See for instance the book by McQuarrie and Simon. The ideas are an extension of what we discussed above for 2D systems. Basically, the volume of a tiny trapezoidal section of a spherical shell (radius r, thickness dr, and subtending the angles $d\theta$ and $d\phi$) centered at (r, θ, ϕ) is the product of the thickness and the lengths of the two sides, namely, $dr \cdot r \sin \theta \, d\phi \cdot r \, d\theta$. This implies the Jacobians are given by

$$J = \prod_{i=1}^{N-1} r_i^2 \prod_{j=1}^{N-2} \sin \theta_j \quad \text{and} \quad J_0 = \sin \theta_0. \tag{5.14}$$

Finally, note that for internal coordinates, we imagine integrating over all possible spherical-polar coordinates for a given atom assuming that all "previous" atoms have been considered. This is equivalent to integrating over all Cartesian coordinates.

PROBLEM 5.5

By considering the "local Cartesian frame" described above, explain why the Jacobian J_0 is necessary. Draw a sketch to justify your answer. Note that the same logic applies to the true internal coordinates $\theta_1, \ldots, \theta_{N-2}$ since internal coordinates implicitly rely on local frames referenced to the previous two bond angles.

5.4 THE SINGLE-MOLECULE PARTITION FUNCTION

Well, it's finally time for real statistical mechanics, the kind that researchers face every day. This is the painful kind, in a way, because nothing can be calculated exactly. Even the approximations tend to be poor, especially for the kinds of systems (e.g., biomolecules) in which we are primarily interested.

So why are we bothering? Obviously, the goal here is not for you to perform calculations that have frustrated physicists for more than a hundred years. Rather, the goals are as follows: (1) to understand concretely why molecular calculations are hard, (2) to understand the explicit connections with the probability theory we described earlier, and (3) to look at some special examples that can be solved exactly in order to gain a bit of intuition.

In fact, it is straightforward to spell out why systems with more than two interacting atoms are difficult to handle mathematically. With two atoms interacting by typical forces, the problem essentially reduces to only one nontrivial coordinate, namely, the distance between the atoms—as we saw earlier in the chapter. But now add a third atom, and suddenly, you need at least three interatom distances to describe the state of the system, and this is assuming we're ignoring gravity and all other "external fields." Worse still, you can't change one distance without changing at least one another. With three atoms, then, we face three nontrivially correlated coordinates. In general, you will not be able to calculate a partition function exactly that depends on more than two interacting particles. If you don't believe this now, just wait and see for yourself in the next section.

Mathematically, the issue is related to statistical independence. Because the coordinates are not independent, the fundamental multidimensional integral (the partition function) cannot be factorized into easy one-dimensional integrals. Rather, many coordinates need to be simultaneously integrated, and this tends to be impossible.

The ultimate solution, although an imperfect one, is the use of computations to evaluate integrals representing averages of interest. Indeed, one way to do so is to perform the molecular dynamics simulations referred to in Chapter 1. But such numerical calculations typically are inadequate, because they are too short to generate a representative sample (ensemble) that truly reflects the underlying Boltzmann factor distribution. Chapter 12 describes essential elements of computer simulations so you can get a feel for how it all works.

5.4.1 THREE ATOMS IS TOO MANY FOR AN EXACT CALCULATION

Let's get our hands dirty by considering three argon-like "atoms" interacting only by van der Waals forces, which we will model by a Lennard-Jones potential. The Lennard-Jones potential is the same as that used earlier in the forcefield

$$u^{LJ}(r; \epsilon, \sigma) = 4\epsilon \left[\left(\frac{\sigma}{r} \right)^{12} - \left(\frac{\sigma}{r} \right)^{6} \right], \tag{5.15}$$

where r is the interparticle distance and the real argument of the function u^{LJ}, and ϵ and σ are parameters of the potential. You can write the function as $u^{LJ}(r)$ so long as you remember about the two parameters lurking in there, which may be specific to the types of atoms that are interacting. Recall that we studied a dimensionless version of this potential in Section 3.2.4.

We'll look at the partition function with the momentum integrals already performed to yield factors of the thermal wavelengths. In our case, we'll assume that all three atoms are different, so that there will be three wavelengths, $\lambda_1, \lambda_2, \lambda_3$. Further, the particles will occupy three-dimensional space, so we should expect to see λ_i^3.

We have

$$Z = \frac{1}{\lambda_1^3 \lambda_2^3 \lambda_3^3} \int_V d\mathbf{r}_1 \int_V d\mathbf{r}_2 \int_V d\mathbf{r}_3 \, \exp\left\{ \frac{-\left[u^{LJ}(r_{12}) + u^{LJ}(r_{23}) + u^{LJ}(r_{13}) \right]}{k_B T} \right\}. \quad (5.16)$$

In this expression, the notation r_{ij} refers to the distance between atoms i and j, which of course depends on both $\mathbf{r}_i$ and $\mathbf{r}_j$.

The integral in (5.16) simply cannot be performed exactly. We could try the usual trick of fixing one atom at the origin—that is, $\mathbf{r}_1 = 0$—and then performing the simple volume integral for $\mathbf{r}_1$ afterward. However, even with $\mathbf{r}_1$ fixed, we still need to integrate over all possible values of both $\mathbf{r}_2$ and $\mathbf{r}_3$. Even if there were some way to fix the positions of two of the atoms, integrating over all positions of the third would involve changing two r_{ij} distances (to the two fixed atoms). While this could be possible for very simple integrands, our Boltzmann factor is far from simple—even though we have greatly "stripped down" the forcefield of Equation 5.12.

The lesson here is that no chemically reasonable model of almost any molecule of interest is amenable to an exact statistical mechanical treatment. As for proteins or DNA molecules with thousands of atoms—just forget about it!

In fact, there is a highly simplified model of a triatomic model that can be solved exactly, but it no longer works once even more atoms are added. You can explore this model in the following problem.

PROBLEM 5.6

A linear triatomic molecule can be described as three impenetrable hard spheres connected by freely rotating fixed-length bonds. Thus, $u^{HS}(r_{ij}) = 0$ except when two atoms overlap, in which case the potential is infinite. (a) Excluding translational and rotational degrees of freedom, sketch the molecule and illustrate the basic degree of freedom. (b) Explain in words the geometric quantity calculated in the partition function, again without including translation and rotation. Do not calculate the integral. (c) Explain why calculation is not so simple if a fourth atom is added to the chain.

5.4.2   THE GENERAL UNIMOLECULAR PARTITION FUNCTION

We now want to consider the partition function of a general molecule, containing N atoms. (By the way, "unimolecular" just means pertaining to one molecule.) A complicating factor, however, is that some of the atoms may be of the same type (element) while others may be different. For better or worse, the partition functions will be different depending on whether atoms of the same type (e.g., two hydrogen atoms) can be distinguished from one another. This leads to an overall prefactor in the partition function, $\mathcal{N}_D^{-1}$, where $\mathcal{N}_D$ is a duplication factor or "degeneracy number" explained below. By now, you should know that an overall constant multiplying the partition function—which leads to an additive constant in the free energy—does not affect relative probabilities. So feel free to ignore $\mathcal{N}_D$ for the moment.

The basic form of the partition function should be obvious by now. If we once again use $\mathbf{r}^N$ to denote the set of x, y, z coordinates for all N atoms, then the partition function is

$$
Z = \frac{1}{\mathcal{N}_D}\left[\prod_{i=1}^{N}\lambda_i^{-3}\right]\int dx_1\, dy_1, dz_1 \ldots dx_N\, dy_N\, dz_N\, e^{-U(\mathbf{r}^N)/k_B T}
$$

$$
= \frac{1}{\mathcal{N}_D}\left[\prod_{i=1}^{N}\lambda_i^{-3}\right]\int d\mathbf{r}^N\, e^{-U(\mathbf{r}^N)/k_B T}, \tag{5.17}
$$

where the second line simply uses our short-hand notation. Note that Equation 5.17 is the correct partition function in general. It is correct even if the potential energy U depends on dihedral angles or distances of any kind. After all, every angle and distance can be computed from the set of x, y, z values in a Cartesian description of a configuration—and thus the potential energy can be calculated. (Internal coordinates lack absolute position and orientation information, as you should remember, and therefore contain less information than Cartesian coordinates.)

5.4.3 BACK TO PROBABILITY THEORY AND CORRELATIONS

It's worth emphasizing that the probabilistic interpretation of the partition function (Equation 5.17) is exactly analogous to that for the one-dimensional systems we studied in Chapter 3. Once again, the probability density is proportional to the Boltzmann factor—that is,

$$
\rho(\mathbf{r}^N) \propto e^{-U(\mathbf{r}^N)/k_B T}. \tag{5.18}
$$

There's no more to it than that. Even the degeneracy number is an overall constant and so does not affect relative probabilities of states or configurations.

Of course, there's a physical interpretation to the mathematics. The physics and chemistry is fully embodied in the energy function, and most importantly in the potential energy U. To emphasize this, consider the following "translation" problems—from math to physics and back.

PROBLEM 5.7

Explain the physical meaning of the potential $U(\mathbf{r}^N) = \sum_{i=2}^{N} u^{SW}(r_{i-1,i})$, where $r_{i,j}$ is the distance between atoms i and j. Here "SW" stands for square well: the function $u^{SW}(r)$ is defined to be infinite for all r except between r_0 and $1.1 \cdot r_0$. Also, sketch the molecule so described.

PROBLEM 5.8

Write a potential function that could describe the interactions among the atoms of methane, CH_4.

5.4.3.1 Correlations Come from Energy Cross-Terms

Way back in Chapter 2, we learned that if two variables are correlated, then the value of one affects the (distribution of) values of the other. And what could be more correlated than a molecule? Covalent bonds are the most obvious source of correlations: if two atoms are bonded then clearly the position of one affects the position of the other. But the situation is actually much worse—much more complicated—since *any* interaction among two atoms causes correlations. Mathematically, any energy term that depends on the simultaneous positions of two or more atoms leads to correlations, because the multiatomic integral cannot be factorized. To be concrete, consider only two atoms, i and j. If the potential energy depends on the interatomic distance, then $\rho(\mathbf{r}_i, \mathbf{r}_j) \neq \rho(\mathbf{r}_i) \cdot \rho(\mathbf{r}_j)$.

Indeed, take a second look at the integrals (5.7) and (5.16). So long as U depends on interatomic distances, the integrals cannot be factorized into independent integrals depending only on the coordinates of individual atoms. This is correlation. Furthermore, bond angles depend on the simultaneous positions of three atoms, and dihedrals on four. Such dependence can be described as a direct correlation among three (or four) bodies. Of course, correlations are "transitive" in that if atom pairs 1-2 and 2-3 are each correlated, then the 1-3 pair should also be correlated.

5.4.4 TECHNICAL ASIDE: DEGENERACY NUMBER

While this section is not advanced in the sense of being very difficult, it is perhaps not a truly essential part of most biomolecular calculations. For readers with a chemistry background, however, the distinction between our "degeneracy number" and the conventional symmetry number may be of interest. The conventional symmetry number can be overly restrictive in describing molecules that lack rotational symmetry.

Look again at the partition function (Equation 5.17). Each atom is numbered, and clearly Z entails the integration over all $3N$ configurational coordinates. That is, every possible value of each x_i, y_i, and z_i must be considered.

Nevertheless, from a physical point of view, there is redundancy in the integral (Equation 5.17) for some molecules. Consider one of the simplest and most important, H_2O. To evaluate the integral for Z (for a single water molecule), we would integrate over all possible locations of $\mathbf{r}_1$, say the position of oxygen, $\mathbf{r}_2$, for the first hydrogen, and $\mathbf{r}_3$ for the second hydrogen. But consider any three specific positions for the atoms, say $\mathbf{r}_1 = \mathbf{r}_a$, $\mathbf{r}_2 = \mathbf{r}_b$, and $\mathbf{r}_3 = \mathbf{r}_c$. Physically, this configuration is no different from the configuration resulting from swapping the labels "2" and "3," namely, $\mathbf{r}_1 = \mathbf{r}_a$, $\mathbf{r}_2 = \mathbf{r}_c$, and $\mathbf{r}_3 = \mathbf{r}_b$, since each hydrogen is bonded only to the oxygen atom. In other words, for every configuration integrated over, there is a physically equivalent, label-swapped ($2 \leftrightarrow 3$) configuration that is also included in the integral: the integral exactly double-counts physically distinct conformations. Hence, water should be assigned the degeneracy number, $\mathcal{N}_D = 2$.

There is a subtlety here. While the test for degeneracy/redundancy can be stated as invariance of the physical configuration upon swapping atom labels—as in the case of water, above—it is easy to make an error. The invariance that matters is not the way the configuration appears to your eye, but how the configuration would be evaluated in the potential energy function. For degeneracy to hold, the label-swapped configuration must also have the same potential energy. (It should also have the same "stereochemistry" although this may not be explicitly in the force-field. Stereochemistry is the relative ordering of atoms about a central atom to which they are bonded. Refer to your favorite chemistry book for details.) Consider ethane, which is C_2H_6 or CH_3–CH_3. On a drawing of ethane, we could swap labels of two hydrogens from the different—that is, opposite—carbons and only see a labeling (rather than a physical) change. However, the potential energy function "knows" which hydrogens belong to which carbon because of the bond length terms that depend on labeled distances r_{ij}. Such a swap, therefore, yields a physically different configuration and fails the degeneracy test. On the other hand, rotating the labels of three hydrogens in a single methyl group (CH_3) keeps the potential energy happy, even when it specifically encodes stereochemistry.

Degeneracy issues certainly are fundamental in the statistical mechanics of molecular systems—and they help in setting the stage for the "distinguishability" issue in multimolecular systems—but the value of $\mathcal{N}_D$ will not affect any of the probabilities or ensemble averages we are concerned with in this book. Also, as perhaps you can imagine, degeneracy issues can get quite complicated for some molecules.

5.4.4.1 Degeneracy Number vs. Symmetry Number

Chemistry books often discuss the "symmetry number," which is intended to play the same role as our degeneracy number in the partition function—that is, to compensate for overcounting. However, the symmetry number is defined solely with regard to whole-molecule rotations. In typical complex molecules, and especially in biomolecules, there will rarely be rotational symmetry. Nevertheless, there is still degeneracy in the partition function based on the label-swapping picture described above. If one only accounts for rotational symmetries, the partition function can fail to have the correct overall prefactor. Said another way, the conventional factorization of chemical partition functions based on a rotational part may only be useful in considering the simplest molecules.

5.4.5 SOME LATTICE MODELS CAN BE SOLVED EXACTLY

Some highly simplified models of polymers can be treated exactly. Perhaps we should say these models are ridiculously simplified—at least they look that way at first glance—but they turn out to be very useful in gaining conceptual understanding (see Figure 5.4). For instance, Chan and Dill introduced the "HP" (hydrophobic-polar) model to study sequence-dependent aspects of protein folding. Many lessons have been learned from this model, and it is still studied today. See the book by Dill and Bromberg for further information.

Lattice models can be easy to solve for relatively small systems, because the integral in a usual partition function gets converted to a sum. Atoms can only occupy discrete sites. There is also no kinetic energy. Specifically, then, the partition function of a lattice model can be written as a sum over configurations.

$$Z = \frac{1}{\mathcal{N}_D} \sum_{\text{configs}} e^{-U(\text{config})/k_BT}. \tag{5.19}$$

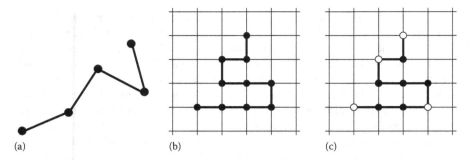

(a)   (b)   (c)

FIGURE 5.4   Three polymer models. Model (a) is a freely jointed chain, with unrestricted bond angles, which is not exactly solvable for more than three self-avoiding (hard-core) "atoms" or beads. Model (b) is a simple lattice homopolymer, in which all beads are identical. Model (c) shows a lattice heteropolymer with two bead types. The right configuration is low in energy for the HP model, in which the black hydrophobic beads are self-attractive.

In two dimensions, it is easy to enumerate all possible configurations by hand for very short polymers, as illustrated in the following problems, which are based on the book by Dill and Bromberg.

PROBLEM 5.9

Consider a polymer on a 2D square lattice, such as in Figure 5.4b. Further, exclude translations and rotations by fixing the first two beads of the polymer horizontally and growing to the right. (a) Assuming all beads are distinguishable, draw all possible configurations for polymers containing $N = 3$, 4, and 5 beads. (b) Repeat part (a) but now assuming a homopolymer of indistinguishable beads, so that mirror images are excluded.

PROBLEM 5.10

Consider now a self-attractive homopolymer on a 2D square lattice. Whenever two nonbonded beads are separated by a single vertical or horizontal lattice-spacing, an attractive energy $-\epsilon < 0$ results. (a) For the four-bead homopolymer, sketch the configurations and energies. (b) Write down the partition function. (c) Describe in words the states that are exhibited at very high and low temperatures.

5.5   MULTIMOLECULAR SYSTEMS

In a biomolecular context, it's rare that we are interested in a single isolated molecule. Rather, biomolecules tend to be solvated in aqueous solution, whether in experiments or in the cell. Thus, the isolated molecules (in "gas phase," as the chemists would say) considered in the previous section are not the most complicated and pertinent case. Having said that, there are no qualitatively new mathematical complications that will arise when we consider multimolecular systems—the basic difficulty remains correlations, or energy cross-terms.

Much remains the same in multimolecular systems, compared to the unimolecular case. For instance, the same basic forcefield, Equation 5.12, can be used. To see this, consider the case of a "solute" molecule in water. In addition to the interactions internal to the solute, we must also consider solute–water and water–water interactions. However, all of these are typically treated by the same Lennard-Jones and electrostatic interactions already included in Equation 5.12.

Indeed, the equation (5.17) for the partition function is fully correct for a multimolecular system so long as the degeneracy number is treated properly. Below, we'll see how to treat the degeneracy number.

Finally, although it would seem that multimolecular systems should be intrinsically more difficult than unimolecular ones, this is not always the case. Indeed, dilute systems of very many atoms or simple molecules can be treated very accurately when approximated as noninteracting or "ideal." Ideal systems are extremely useful learning tools for thermodynamics, as we will see in Chapter 7. By contrast, it is very difficult to make a good approximation for any single large molecule that is exactly solvable.

5.5.1 PARTITION FUNCTIONS FOR SYSTEMS OF IDENTICAL MOLECULES

We already said that the partition function remains the same as in the unimolecular case. The thing we need to do is specify the degeneracy number correctly—which tends to be easier with a large number of identical small molecules than for a single large molecule.

The essential point is that with N identical molecules, the labels of the molecules can be switched around (permuted) in exactly $N!$ ways without changing the physical configuration. If the molecules are just simple atoms, or beads in some toy model, that is the end of the story. However, if the molecule has its own nontrivial degeneracy number (the number of ways to relabel the atoms without changing the configuration), $\mathcal{N}_D^{mol} > 1$, then for every labeling of the molecules, the atomic labels can also be changed. This leads to an overall degeneracy number of $\mathcal{N}_D = N!(\mathcal{N}_D^{mol})^N$. If there are two types of molecules, A and B, then the degeneracy number will have a factor $N_A!N_B!$ and separate $\mathcal{N}_D^{mol}$ factors for A and B.

PROBLEM 5.11

Consider three identical molecules with internal degeneracy number $\mathcal{N}_D^{mol} = 2$, which could be a diatomic molecule. For a single configuration, sketch all the possible permutations to check that $\mathcal{N}_D = N!(\mathcal{N}_D^{mol})^N$.

To be concrete, let's write out a few partition functions.

$$N \text{ identical atoms: } \quad Z = \frac{\lambda^{-3N}}{N!} \int d\mathbf{r}^N \, e^{-U(\mathbf{r}^N)/k_B T}, \tag{5.20}$$

$$N_A \text{ A atoms, } N_B \text{ B atoms: } \quad Z = \frac{\lambda_A^{-3N_A}\lambda_B^{-3N_B}}{N_A!N_B!} \int d\mathbf{r}^{N_A} d\mathbf{r}^{N_B} \, \exp\left[-U(\mathbf{r}^{N_A}, \mathbf{r}^{N_B})/k_B T\right], \tag{5.21}$$

where $d\mathbf{r}^{N_X}$ indicates integration over all Cartesian coordinates of all atoms of type X and $(\mathbf{r}^{N_A}, \mathbf{r}^{N_B})$ represents the coordinates of all atoms of both types.

5.5.2 IDEAL SYSTEMS—UNCORRELATED BY DEFINITION

While (nearly) ideal systems, where the constituent molecules are uncorrelated, are common enough in everyday life (e.g., the air), such ideality is essentially nonexistent in biology. Everything is correlated with everything else, to a good approximation. Nevertheless, ideal gases form the key example in connecting statistical mechanics to thermodynamics, as we will see. It is therefore worth the effort to understand ideal systems—and fortunately, these are the easiest systems around.

If we consider first a monatomic gas (one of the noble elements, say), and we say the atoms are uncorrelated, then no configuration is favored (energetically) over any other. That is, the potential energy is a constant, which we can safely consider to be zero: $U(\mathbf{r}^N) = 0$. We can then evaluate the partition function exactly, since the Boltzmann factor is always one. Thus, Equation 5.20 becomes

$$N \text{ identical ideal atoms:} \quad Z = \frac{\lambda^{-3N}}{N!} V^N, \tag{5.22}$$

where V is the volume of the system. If the system consists of N ideal molecules, each containing n atoms, the partition function becomes

$$N \text{ identical molecules:} \quad Z = \frac{\lambda^{-3N}}{N!} \left(q^{\text{mol}} V\right)^N, \tag{5.23}$$

where q^{mol} is the partition function of one molecule and λ is the whole-molecule thermal wavelength.

> **PROBLEM 5.12**
>
> Starting with the partition function for $N \cdot n$ atoms of a system N ideal molecules, derive the partition function (5.23).

These ideal systems will be essential for understanding thermodynamics in Chapter 7, and therefore for our general understanding of biophysics.

5.5.3 NONIDEAL SYSTEMS

In the real world, there are no ideal systems, since atoms and molecules always exert forces on one another. In biological cells, which are crowded with solvent and macromolecules, this is certainly true. Here, we will very briefly describe some key qualitative aspects of typical nonideal systems of interest.

5.5.3.1 Solute in Solvent

A common type of system with two molecule types is when solute molecules are dissolved in a solvent. If you think of the forcefield equation (5.12), in this case, there are three basic types of interactions: between pairs of solvent molecules, between pairs of solute molecules, and between solvent–solute pairs. Although the average energy can rigorously be decomposed into these three parts, the entropy generally cannot, since this would require that the partition function be factorizable into different components. Some simplifications may be possible, however. For instance, if the solution is dilute—contains few solute molecules—then solute–solute interactions are unimportant and the other two interaction types predominate.

5.5.3.2   A Look Ahead at Binding: Receptor and Ligand in Solvent

Consideration of quasi-biological binding typically requires a protein receptor, a ligand, plus the solvent (which often contains further components like salts). Thus, there are a minimum of six types of interactions: those of every component with itself, plus the three different pairings of the three species. In the dilute limit, it may be possible to ignore receptor–receptor and ligand–ligand interactions. This idea will be pursued in Chapter 9.

> **PROBLEM 5.13**
>
> (a) Write down a partition function for a volume V containing one receptor, one ligand, and N solvent molecules. (b) Discuss a possible definition for the bound state and explain how to determine the fraction of time the receptor–ligand pair are bound.

5.6   THE FREE ENERGY STILL GIVES THE PROBABILITY

We've said several times that the intuition we gained in one-dimensional systems would carry over to more complex systems. Now is the time to make good on that claim.

Here we want to show two things: (1) the free energy gives the probability, and (2) the entropy gives the configuration-space volume of the system. In one dimension, the entropy gave the log of the effective width, and in higher dimensions, the analog of width is the volume.

First, the free energy. As in one dimension, our real interest is always in observable phenomena—that is, in populations, or equivalently, probabilities. And just as in one dimension, a partition function is the sum of all probabilities in the state for which the integration is performed. Thus, the ratio of probabilities for two states is the ratio of the partition functions. We can simply rewrite our previous equation (3.12) to reflect the higher dimensionality. Specifically,

$$\frac{p(\text{state A})}{p(\text{state B})} = \frac{\int_{V_A} d\mathbf{r}^N \, e^{-U(\mathbf{r}^N)/k_B T}}{\int_{V_B} d\mathbf{r}^N \, e^{-U(\mathbf{r}^N)/k_B T}} = \frac{Z_A}{Z_B} \equiv e^{-(F_A - F_B)/k_B T}. \qquad (5.24)$$

The difference from one dimension is that the integrals are over $3N$-dimensional configuration volumes. Since the same system is considered in the numerator and denominator of Equation 5.24, both degeneracy numbers and kinetic energy factors of λ cancel out.

The configuration-space volumes V_A and V_B define the states for which the probability ratio is given. States can be defined in a number of ways. For binding, state A might be the bound state and state B unbound—with the latter possibly defined as when the ligand is further than a certain distance from the binding site. It is also possible to define states based purely on the configuration of one molecule. For instance, butane could be in the *trans* or *gauche* states depending on the value of the central dihedral angle. A protein could be defined as in a folded or unfolded state based on a measure of the "distance" from a specific known structure.

5.6.1 THE ENTROPY STILL EMBODIES WIDTH (VOLUME)

In our study of 2D atomic systems (Section 5.2), we saw that entropy was the log of volume. That is, the translation of "width" for multidimensional systems is "volume" in configuration space. For molecules, this volume will not be three dimensional. For instance, configuration space for methane (5 atoms) is 15 dimensional and for butane (14 atoms), it's 42 dimensional. Typical proteins have thousands of atoms and accordingly huge configuration spaces.

We still use exactly the same definition for entropy as we did for one-dimensional systems: entropy will be the quantity that makes up the difference between the average energy and the log of the probability (a.k.a. free energy). More precisely, we still have $S \equiv (\langle E \rangle - F)/T$ and therefore

$$e^{-F/k_B T} = e^{+S/k_B} \, e^{-\langle E \rangle / k_B T}, \qquad (5.25)$$

where all quantities must refer to the same state of interest. Thus, the entropy indeed accounts for the size of the state considered in the partition function, as in one dimension.

5.6.2 DEFINING STATES

It is rare that we are interested in the absolute free energy of a state. Typically, we are interested in knowing the free energy difference between two states so that we can estimate probability (population) differences via Equation 5.24. In one dimension, there is basically one way to define a state: via start and end points, or possibly a set of these.

In high-dimensional molecular systems like macromolecules, many definitions are possible. In fact, from a theoretical point of view, a state can be defined in any way you like. Thus, a state could be defined as when exactly half of the molecules in a box occupy the box's left side. One can also use more microscopic definitions such as when a certain dihedral angle is within a certain range, or when two atoms of a protein are closer than a certain cutoff.

However, when the goal is to develop a theory—or interpret a simulation—so that an experiment can be explained, more care must be taken. While it may never be possible to correlate a state definition exactly with an experimental measurement, one can at least try to define theoretical states in a way that the predicted results are fairly insensitive to small changes in the definition of the state. In rough terms, one would want to draw boundaries between states along points of high energy (low probability) so that a slight shift of the boundary would tend to include or exclude only a tiny amount of probability.

5.6.3 Discretization Again Implies $S \sim -\sum p \ln p$

In Chapter 3, we showed that the validity of the relation $S = -k_B \sum_j p_j \ln p_j + \text{const.}$, where the sum was over states of equal sizes in configuration space. In higher dimensions, the derivation of Chapter 3 remains correct, so long as the lengths l_j are now traded for configuration-space volumes V_j. Thus, p_j is the probability of the state occupying the (fixed) volume V_j, where again V_j is small enough so that the energy is nearly constant over the whole volume.

5.7 SUMMARY

In this chapter, we discussed two types of multidimensional systems: systems of many "particles" and also molecular systems with many atoms. The single most important conclusion we reached is that the basic ideas about the meaning of free energy (in terms of probability) and entropy (in terms of state volume) remain the same as in one dimension. A complication intrinsic to multiatomic systems is that correlations among coordinates must occur: there are no truly ideal systems and interactions cause correlations. At the simplest level, the position of one atom dictates that other atoms cannot occupy the same location. But any interatomic forces mean that the "behavior"—that is, distribution—of one atom is affected by others. Aside from correlations, other technical complications arise in multidimensional systems like the need to consider molecular degeneracy numbers, distinguishability, and Jacobians for internal coordinates.

FURTHER READING

Dill, K.A. and Bromberg, S., *Molecular Driving Forces*, Garland Science, New York, 2003.
McQuarrie, D.A. and Simon, J.D., *Physical Chemistry*, University Science Books, Sausalito, CA, 1997.

6 From Complexity to Simplicity: The Potential of Mean Force

6.1 INTRODUCTION: PMFS ARE EVERYWHERE

The potential of mean force (PMF), sometimes called the free energy profile (or landscape) without doubt is one of the key statistical mechanical objects for molecular biophysics. It is right up there with entropy and correlation. The PMF provides a rigorous way to visualize high-dimensional data, and consequently is seen in countless papers. Even a simple molecule like butane, as discussed previously, has 36 internal degrees of freedom. If we want to know the "energy profile" appropriate to a single coordinate, we must use the PMF, as shown in Figure 6.1.

A thorough understanding of the PMF will also prevent us from the very real danger of overinterpretation: just because the PMF is rigorously derived from the Boltzmann-factor distribution does not mean that it will reveal the physics—or the biology—we want to know!

The PMF will also help us to understand a key aspect of thermodynamics, in the next chapter.

6.2 THE POTENTIAL OF MEAN FORCE IS LIKE A FREE ENERGY

High-dimensional systems are difficult not only because the calculations are hard, but also because of our limited ability to visualize and understand dimensionality higher than three. Even if we could calculate partition functions exactly, we would not fully understand the statistical implications. Consider again the three-atom system of the previous chapter (Section 5.4.1). Although you can easily imagine any single configuration of such a system, how about the distribution? Indeed, if you look at the integrands in the partition functions in Chapter 5, we already "know" the exact form of the distribution. But—and this is even more valid for more complex systems—it is difficult to understand the statistical behavior when we cannot imagine the way the distribution looks.

Having studied projection (marginalization) in Chapter 2, you should already know the practical, if imperfect, solution to this conundrum. One needs to examine distributions projected onto subsets of coordinates. If the system is huge (e.g., a protein, with thousands of coordinates), then great care must be taken in choosing the coordinates to be examined explicitly. More on this later.

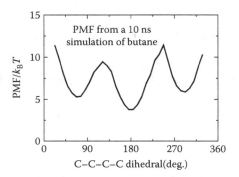

FIGURE 6.1   The PMF based on the 10 ns butane trajectory from Figure 1.3. Simply put, the PMF is the energy whose Boltzmann factor yields the probability distribution—in this case, as shown in Figure 2.3.

6.2.1 THE PMF IS EXACTLY RELATED TO A PROJECTION

Let's quickly remind ourselves of projection basics. In essence, one obtains the correct distribution for a subset of coordinates by performing a suitable weighted average (integration) over all other coordinates. A projection can be performed from an original configuration space of any size down to any smaller size space. Simple examples include

$$\rho(x) = \int dy\, \rho(x, y), \tag{6.1}$$

$$\rho(x, z) = \int dy\, \rho(x, y, z), \tag{6.2}$$

$$\rho(x) = \int dy\, dz\, \rho(x, y, z). \tag{6.3}$$

In statistical mechanics, the configuration-space distribution is defined by the Boltzmann factor of the potential energy, $\rho(\mathbf{r}^N) \propto \exp[-U(\mathbf{r}^N)/k_\mathrm{B}T]$. Let's say we are interested solely in the distribution of distances between atoms 2 and 6 in some protein. There's no such thing as $U(r_{26})$ to use in a Boltzmann factor. Even though there may be a forcefield term depending on r_{26}—such as a Lennard-Jones potential—such a term by itself does not determine the distribution of r_{26} values. In a folded protein, for instance, two atoms may rarely come into contact even though they attract one another.

Basically, the statistical mechanical analog to the projected distribution is a free-energy-like quantity called the potential of mean force. It is similar to a free energy because the Boltzmann factor of the PMF is defined to give the probability distribution on the coordinates of interest. Put another way, the PMF is the energy landscape corresponding to the distribution in a subset of coordinates. Thus,

$$e^{-\mathrm{PMF}(x, y, \ldots)/k_\mathrm{B}T} \propto \rho(x, y, \ldots). \tag{6.4}$$

Sometimes, a PMF is casually referred to as a free energy, or a bit more appropriately, as a "free energy profile"—profile in the sense of the plot of a distribution.

More concretely, consider a single particle moving in an external potential $U(x, y, z)$, so that $\rho(x, y, z) \propto \exp\left[-U(x, y, z)/k_\mathrm{B}T\right]$ is the fundamental distribution. We can then write analogs of Equations 6.1 through 6.3 using the PMF. It will actually be clearest to use the reverse order, and we have

$$e^{-\mathrm{PMF}(x)/k_\mathrm{B}T} \propto \int dy\, dz\, e^{-U(x,y,z)/k_\mathrm{B}T}, \tag{6.5}$$

$$e^{-\mathrm{PMF}(x,z)/k_\mathrm{B}T} \propto \int dy\, e^{-U(x,y,z)/k_\mathrm{B}T}, \tag{6.6}$$

$$e^{-\mathrm{PMF}(x)/k_\mathrm{B}T} \propto \int dy\, e^{-\mathrm{PMF}(x,y)/k_\mathrm{B}T}, \tag{6.7}$$

The analogies to statistical projection should be clear. Of the three equations above, Equation 6.7 may be a bit more difficult. You should therefore complete the following problem.

PROBLEM 6.1

Use the definition of $\mathrm{PMF}(x, y)$ to show the validity of Equation 6.7.

Because of the proportionality symbols ($\propto$) in Equations 6.5 through 6.7, we know there are overall factors missing. Nevertheless, the overall prefactors are physically irrelevant because they do not affect the relative probabilities for the coordinates of interest on the left-hand sides. In almost every case, then, it is safe to substitute "=" for "$\propto$" in Equations 6.5 through 6.7 (if you ignore issues of dimensionality).

6.2.1.1 Unless Coordinates Are Independent, the PMF Is Needed

In general, when atoms interact, none of their coordinates will be statistically independent—and we need to use the PMF. This is evident in the example above where we considered the r_{26} coordinate in a multiatomic system.

6.2.1.2 The PMF for a General Coordinate

What if you're interested in a special coordinate that is not one of the usual variables of integration in the partition function? As with r_{26}, you might want to know the distribution of the distance between two atoms of interest. Another example is the RMSD "value" of a protein configuration, which depends on all coordinates. (Really RMSD is a distance between two conformations but typically one is a fixed reference such as the x-ray structure—see Equation 12.11).

A PMF can give a distribution for any coordinate, but the forms exemplified in Equations 6.5 through 6.7 really apply only to Cartesian coordinates. The most general PMF requires the use of a delta function and a messier equation. Say we're interested in some coordinate R, like a distance, which can be determined from any single configuration; so we have $R = \hat{R}(\mathbf{r}^N)$, where $\hat{R}$ is the mathematical function and R is the value it yields. We would like to know the distribution of R values (up to a proportionality constant), perhaps in order to see the most likely R and the degree of variability. We can use a delta function (Section 2.2.8) to selectively include probability (i.e., "original" Boltzmann factors for the full set of coordinates $\mathbf{r}^N$) that are consistent with a given R. That is, we want to sum up Boltzmann factors for all configurations such that $\hat{R}(\mathbf{r}^N) = R$ and do this for all R to get the full distribution. Mathematically, this is equivalent to

$$e^{-\mathrm{PMF}(R)/k_B T} \propto \rho(R)$$

$$\propto \int d\mathbf{r}^N \, \delta\left(R - \hat{R}(\mathbf{r}^N)\right) e^{-U(\mathbf{r}^N)/k_B T}. \qquad (6.8)$$

You can readily extend this type of formulation to PMFs of multidimensional subsets of general coordinates.

The Equation 6.8 looks different from the preceding Equations 6.5 through 6.7 because all variables are integrated over in the present instance. However, the two formulations are equivalent if you consider the special properties of a delta function (Section 2.2.8).

PROBLEM 6.2

Show the equivalence of Equations 6.8 and 6.5—when projecting from three dimensions to one—by setting $\hat{R}(x, y, z) = x$ and integrating Equation 6.8 over x.

PROBLEM 6.3

Consider the two-dimensional potential $U(x,y) = U_1(x) + U_2(x, y)$, where $U_1(x) = E_b(x^2 - 1)^2$ and $U_2(x, y) = [\kappa(x)/2]y^2$ with $\kappa(x) = \kappa_0 + \kappa_1 x^2$. This is a double-well potential in x with a harmonic potential in y whose width depends on x. Assume that at the temperature, T, of interest, the constants are equal to $E_b = 5k_B T$, $\kappa_0 = k_B T$, and $\kappa_1 = 3k_B T$. For simplicity, x and y are dimensionless in this problem.

    a. Roughly sketch a contour plot of the potential.
    b. Given that the potential is harmonic in y, what is the (x-dependent) standard deviation of the corresponding Gaussian Boltzmann factor.
    c. Use the result from (b) to calculate the necessary Gaussian integral and determine $\mathrm{PMF}(x)$ up to a constant.
    d. Plot U_1 and the PMF on the same graph and compare their barrier heights.
    e. Explain what differs between the two curves and why.

6.2.2 PROPORTIONALITY FUNCTIONS FOR PMFS

What we have described already about PMFs embodies the essential physical understanding you should have. As Equation 6.4 indicates, simply enough, the Boltzmann factor of the PMF gives the probability distribution for the coordinates it includes. The distribution will be correct if the PMF is properly computed using standard statistical projection, as illustrated in Equations 6.5 through 6.7. Again, if you ignore constants of proportionality, you will still get PMFs that yield the right relative probabilities and are sufficient for most purposes.

For completeness, nevertheless, we should specify the proportionality constants that are required in the preceding equations. Your first guess might be (should be!) to use the inverse configuration integral $\hat{Z}^{-1} = 1/\lambda Z$. Inserting $\hat{Z}^{-1}$ on the right-hand sides of Equations 6.5 through 6.7 indeed yields properly normalized probability densities. However, a density, by definition, has dimensions—some power of an inverse length in configuration space—while the left-hand sides of the PMF equation are dimensionless Boltzmann factors. We therefore require something more.

Following the historically original description of a PMF (for the pair correlation function described below)—and indeed following the spirit of physics description—we want our dimensionless quantity to reflect a sensible physical comparison. In this case, the only choice would seem to be to compare to a uniform density in configuration space. Since we assume we are dealing with some physical system confined to a volume $V \equiv L^3$ (i.e., L is defined to be the cube root of V), a uniform density in configuration space is $1/L^n$, where n is the number of spatial dimensions considered in the PMF.

Putting all this together, the general definition of a PMF for the subset of coordinates $x_1, x_2, \ldots, x_n$—from a total of N—is given by

$$
e^{-\mathrm{PMF}(x_1,\ldots,x_n)/k_B T} \equiv \frac{\rho(x_1,\ldots,x_n)}{\rho^{\mathrm{uni}}(x_1,\ldots,x_n)}
$$
$$
= \frac{\hat{Z}^{-1} \int dx_{n+1} \cdots dx_N \, e^{-U(x_1,\ldots,x_N)/k_B T}}{\rho^{\mathrm{uni}}(x_1,\ldots,x_n)}. \tag{6.9}
$$

Note that the second line simply writes out the projection needed for $\rho(x_1,\ldots,x_n)$. The function ρ^{uni} represents a uniform distribution over its arguments, and we have not written it out explicitly because its mathematical form depends on the variable. For a single Cartesian coordinate—for example, x_5—we have $\rho^{\mathrm{uni}}(x_5) = 1/L$, whereas for a dihedral angle, it is $\rho^{\mathrm{uni}}(\phi) = 1/2\pi$. For more general coordinates—such as those requiring delta functions as in Equation 6.8—you will have to employ the correct ρ^{uni}.

6.2.3 PMFS ARE EASY TO COMPUTE FROM A GOOD SIMULATION

In many cases, the PMF is incredibly easy to compute, even for complicated variables like distances or RMSDs. (And in cases where it's hard to get a PMF, it's usually hard to calculate anything of value.)

Why is the PMF so easy to calculate from a (good) simulation? You have to recall the basic idea of sampling: a good simulation will sample the desired distribution, which in our case is proportional to a Boltzmann factor. Once a set of configurations has been sampled, all we have to do is count—make a histogram. For every configuration $\mathbf{r}^N$, we simply calculate R or whatever variable interests us, and we bin the values to make a histogram. Since the configurations are distributed properly, so too will be the R values. The histogram will therefore be proportional to $e^{-\mathrm{PMF}(R)/k_B T}$, and the PMF can be obtained from the log of the histogram.

PMFs from our two favorite simple molecules are shown in Figures 6.1 and 6.2.

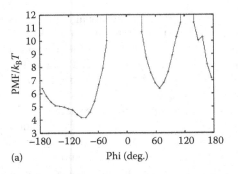
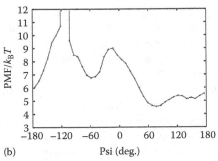

(a)   Phi (deg.)   (b)   Psi (deg.)

FIGURE 6.2 Two different PMFs for alanine dipeptide (Ace-Ala-Nme) estimated from a single Langevin dynamics simulation. Alanine dipeptide is a small molecule that models a single amino acid in a peptide (Section 1.8.8). Panel (a) shows the PMF for the dihedral angle ϕ, while panel (b) shows the PMF for ψ. Note that it's not completely clear how many free energy minima have been found in total. (Just because there are two minima for each coordinate does not mean there are four minima in total—there could be only two or three.) In this case, we could examine the combined ϕ-ψ PMF, but for general complex systems, one does not know which coordinates to examine for the "real answer."

6.3 THE PMF MAY NOT YIELD THE REACTION RATE OR TRANSITION STATE

At the beginning of the chapter, we pointed out that the PMF provides one of the few ways to rigorously analyze a high-dimensional system in a way that our low-dimensional minds can understand it. There was also a cryptic statement about overinterpretation of the PMF. It is now time to clarify the "dangers."

Two key "answers" often sought by both computational and experimental biophysicists are transition states and barrier heights controlling kinetics via the Arrhenius relation that we considered in Chapter 4. Here, we are imagining a system that makes a transition between two states. In one-dimensional systems, the transition state is simply the configuration of highest energy between the states of interest, and the barrier height controlling kinetics is the energy difference between the initial and transition states. Because a PMF can project a high-dimensional landscape onto a single coordinate—and a two-state landscape may indeed appear in such a projection—it is tempting to reason by analogy in obtaining the transition state and barrier height.

Nevertheless, it is wrong, in principle, to obtain a transition state or barrier height pertinent to kinetics via a PMF. This point was emphasized in the paper by Dellago et al., and is illustrated in Figure 6.3. The figure contains the essential points of this section: the quality of physical conclusions based on (a fully correct) PMF is highly sensitive to the specific topography of the underlying landscape. Figure 6.3 shows a particularly pathological case, but the fact is that it's not really known what "typical" molecular landscapes look like because they are difficult to study and visualize. The figure is a very important one in statistical biophysics and you should etch it in your memory.

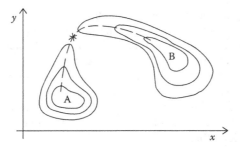

FIGURE 6.3   The PMF may yield the wrong transition state and an irrelevant barrier height if the energy landscape is complicated. In the landscape depicted here, the asterisk (*) denotes the saddle point of the landscape—also called the transition state. However, a PMF projected onto either x or y clearly would give a very poor indication of the location of the transition state. (Adapted from Dellago, C. et al., *J. Chem. Phys.*, 108, 1964, 1998.)

6.3.1   IS THERE SUCH A THING AS A REACTION COORDINATE?

"Reaction coordinates" are discussed all the time in chemical contexts. The basic meaning is that such a coordinate (which could be multidimensional) describes all the essential aspects of a reaction or structural transformation. That is, the free energy along this coordinate—that is, the PMF—yields a good picture of the transition state and presumably embodies an estimate of the reaction rate via the barrier height. Of course, these claims are exactly what we are warning against when using a PMF.

Does this make reaction coordinates a fantasy? Not necessarily, at least in principle. After all, in Figure 6.3, if one could somehow project a PMF along the dashed line passing through the true transition state, then the resulting free energy profile would indeed exhibit the desired characteristics. In other words, it is not crazy to expect that a good reaction coordinate exists—even if it is hard to find.

However, it is important to recognize that, in principle, there could be more than one "channel" connecting two states of interest, as shown schematically in Figure 6.4. In such a case, it could be essentially impossible to define an effective reaction

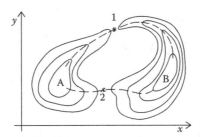

FIGURE 6.4   Two stable states connected by two paths. High-dimensional systems like macromolecules, in general, are expected to have a multiplicity of pathways connecting (meta)stable states. Whether a single path predominates in biological systems is an open question. Asterisks (*) 1 and 2 indicate saddle points of the two paths.

coordinate, even in principle. For instance, in the transition from the unfolded to the folded state of a protein, many biophysicists believe that many channels/pathways are possible in typical cases. Even though there may be coordinates that can effectively describe the progress from the unfolded to the folded state—such as the overall size of the protein, perhaps measured by the radius of gyration—this does not mean that such "progress coordinates" fulfill the requirements of a true reaction coordinate.

Certainly, reaction coordinates are conceptually useful and likely to exist in many cases of importance, yet it is important to recognize that such coordinates may be difficult or impossible to find.

6.4 THE RADIAL DISTRIBUTION FUNCTION

The PMF often focuses on specific coordinates, but it can also be used to study the correlation occurring among pairs of atom types. For instance, a polar amino acid may sit on the surface of a protein, and it could be of interest to see how often a water molecule is in close proximity (see Figure 6.5). A similar query could be made concerning groups in a protein's binding site and atoms of a ligand. In the case of water, one could imagine constructing a PMF for the distance between a polar hydrogen atom on the side chain and the oxygen of a certain water molecule. But in an experiment or a simulation, undoubtedly the water molecule that is nearby at one time will drift off and be replaced by another water molecule. This would make the PMF analysis, as described above, somewhat confusing.

The natural tool for studying such correlations between atom types—rather than specific atoms—is the radial distribution function (which is essentially equivalent to something known as the pair correlation function). Interestingly, the notion of a

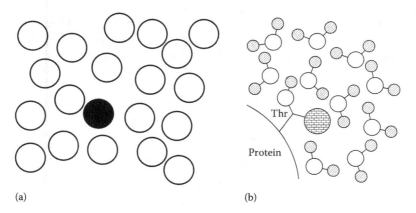

(a)   (b)

FIGURE 6.5 Cartoons of two systems for which a radial distribution function $g(r)$ might be calculated. Panel (a) shows a monatomic fluid of identical atoms, where the reference atom has been filled. Panel (b) depicts a protein side chain, threonine (Thr), surrounded by water molecules. Oxygen atoms are shown as open circles, hydrogens are diagonally hatched, and the bricked circle is a methyl group.

PMF was first developed in the context of distribution functions. These functions are usually used for liquids but could be applied to almost any system.

In essence, the radial distribution function $g_{ab}(r)$ gives the probability to find an atom of type "b" at a distance r from an atom of type "a." More precisely, it gives the relative probability compared to a uniform distribution of "b" particles. To write this mathematically requires ordinary "number densities" n (as opposed to pdfs), which represent the number of particles per unit volume or per unit area or length—depending on what is of interest. In our case, we are interested in the distance r from our reference atom "a," and the defining equation is

$$g_{ab}(r) = \frac{n_{ab}(r)}{n_b^{uni}(r)} = \frac{n_{ab}(r)}{N_b/V} \equiv e^{-PMF(r)/k_B T}. \qquad (6.10)$$

Before we specify the symbols, you should see that g_{ab} has the usual structure of a dimensionless definition for the Boltzmann factor of a PMF—namely, the ratio of two distributions.

The main ingredient in the definition of g_{ab} is $n_{ab}(r)$, the conditional number density—given the condition that the center of an "a" particle is the reference point—to find a "b" particle at a distance r. A "number density" is just what it sounds like: $n(r)$ gives the average number of particles per unit volume between r and $r + dr$ for small dr. In n_{ab}, the "a" particle could refer to a specific atom (e.g., the amide hydrogen on a particular asparagine), while "b" could mean oxygen atoms in water. In a homogeneous system, such as a liquid, it is expected that there is no difference between the different molecules. Thus, if we were interested in n_{OH} for water, the distribution of hydrogen atoms around one oxygen atom should be the same as for any other oxygen.

Note that $n(r)$ is defined to require a Jacobian, so that the total number of particles between r and $r + dr$ is $4\pi r^2 n(r)\, dr$. By this convention, the total number of "b" particles is obtained via

$$N_b = \int_V dr\, 4\pi r^2\, n_{ab}(r). \qquad (6.11)$$

The denominator of Equation 6.10 is just the uniform number density, N_b/V.

6.4.1 What to Expect for $g(r)$

The radial distribution function g_{aa} for a monatomic fluid has a typical shape, as depicted in Figure 6.6. You should expect that as r gets very large, correlations with the reference particle will decay. Given the mathematical definition of g, this means that g will approach unity at large distances. Closer in, the exact form of g will reflect the full behavior of the system—that is, correlations among all particles. The basic point is that close layers of neighboring particles will be clearly distinguished. Indeed, radial distribution functions can reveal much about the structure and thermodynamics of a system—but this is beyond the scope of this book. See the book by McQuarrie for further details.

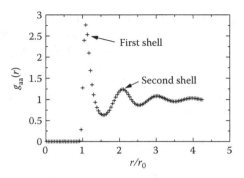

FIGURE 6.6 The radial distribution function $g_{aa}(r)$ for a typical monatomic fluid. With the center of an atom as the reference, you can expect higher than normal density in the first "shell" of neighboring atoms, followed by a depletion zone, followed by the next shell, and so on (see Figure 6.5). At increasing distances, the correlation with the reference atom becomes weaker and hence g decays to one. The data are from a simulation of a dense fluid of Lennard-Jones particles with radius r_0.

PROBLEM 6.4

Even though a crystal is ordered, $g(r)$ can still be calculated. (a) Sketch g for cubic lattice of a single type of atom, without calculating it exactly. (b) Describe the key qualitative differences between the crystalline g and that for a fluid.

6.4.2 $g(r)$ Is Easy to Get from a Simulation

One of the nice things about simulation is that the result—the trajectory of configurations produced—embodies the Boltzmann-factor distribution, assuming the sampling was good. Echoing Section 6.2.3, to calculate probabilities of interest from a simulation requires merely counting.

6.4.2.1 The Simplest Case: A Fluid

An instructive first example is the simplest fluid you can imagine: a system of N monatomic (argon-like) atoms, with every atom the same, contained in a volume V. We'll assume we have obtained M configurations of the system—each one containing the x, y, z coordinates of every atom—in our trajectory. These configurations are the "snapshots" of a molecular dynamics "movie."

We are interested in $g(r)$, the relative probability of finding an atom a distance r from a reference atom, as compared to the probability in a uniform fluid of the same overall density, N/V. To proceed, we should recall the "experimental" definition of probability: the expected outcome based on a large number of independent realization of an event. Since our trajectory has M configurations, we can assume these configurations are our repeated realizations. (Whether they are really independent—uncorrelated—will be discussed in Chapter 12. However, if the trajectory is long enough, it will still yield correct probabilities—for example, for computing $g(r)$.)

The one trick to the counting is that binning is necessary—or call it "discretization" to make it sound fancier. Even though you may have many thousands of configurations in the simulation trajectory (say, $M \sim 10^4$), this is still only a finite statistical sample. Therefore, if you choose a particular distance such as $r = 4.19$ Å without first looking at the trajectory, you cannot expect to have even one distance that precisely matches this value. Instead, construct a bin, perhaps $4.0 < r < 4.2$, in which you have many distance values. Having chosen a set of bins spanning the range of interesting r values, g is calculated simply by counting the number of configurations in the bin and suitably normalizing based on the r value and the bulk density.

In other words, you discretize Equation 6.10 by using the ratio of counts in a bin of width Δr—the ratio of the observed counts to the expected reference case. We also employ the standard unsubscripted notation for a monatomic fluid (all of type "a"), $g(r) \equiv g_{aa}(r)$, to obtain

$$g(r) = \frac{n_{aa}(r)\Delta r}{(N_a/V)\Delta r} = \frac{\langle \text{counts}(r; \Delta r)\rangle / (4\pi r^2)}{(N/V)\Delta r}, \qquad (6.12)$$

where the numerator is the average number of counts (per configuration) in the bin which extends from $r - \Delta r/2$ to $r + \Delta r/2$. The number of counts must be normalized by the Jacobian-like factor accounting for the growing volume of the bins as r increases.

Estimation of g via Equation 6.12 is simple to implement in a computer program. For instance, after choosing an arbitrary "colored" atom as the reference atom (see Figure 6.5), one calculates r values to all other atoms and bins them. This is repeated for every configuration for better averaging. The final result (Equation 6.12) is based on the average number of counts per bin seen in a configuration, with the density of the monatomic fluid being simply N/V.

PROBLEM 6.5

Explain why you can compute $g(r)$ from a single snapshot of good simulation for a homogeneous fluid—that is, a fluid consisting of just one type of molecule.

PROBLEM 6.6

Using the meaning of g, derive an equation that determines the total number of particles in a certain range—from r_{min} to r_{max}.

6.4.2.2   Slightly Harder: Water around a Protein

Imagine now that we have simulated a protein along with the surrounding water, and we are interested in the distribution of water molecules around a particular side chain of the protein, a threonine (see Figure 6.5). We can quantify this by considering the (more precisely defined) distribution of oxygen atoms from water around the partially positively charged hydrogen of threonine's hydroxyl group. The corresponding radial distribution function, $g_{HO}(r)$, measures the relative probability of finding an oxygen atom around threonine's hydroxyl hydrogen—compared to that for the oxygen in bulk water. The only tricky part here, compared with Equation 6.12, is that the bulk density of the "b" atoms (oxygen from water) is not simply N_b/V because not all of the volume is occupied by the water. Therefore, we must estimate bulk density of oxygen atoms some other way—for instance by using n_{ab} for large r, yielding

$$g_{ab}(r) = \frac{n_{ab}(r)}{n_b^{uni}(r)} = \frac{\langle \text{counts}(r; \Delta r) \rangle}{\langle \text{counts}(r \to \infty; \Delta r) \rangle}, \tag{6.13}$$

where for our example, "a" is H and "b" is O.

6.4.3 THE PMF DIFFERS FROM THE "BARE" PAIR POTENTIAL

Just one look at $g_{aa}(r)$ in Figure 6.6 should tell you that the PMF derived from the relation $\text{PMF}(r) = -k_B T \ln g(r)$ is qualitatively different from the (nonbonded) pairwise interaction potential u^{pair}—that is, from the potential that generates forces in a system where the total potential is

$$\text{Only pairwise interactions:} \quad U(\mathbf{r}^N) = \sum_{i<j} u^{pair}(r_{ij}). \tag{6.14}$$

(This potential energy implies there are no three-body or higher terms.)

The PMF corresponding to a multi-peaked g will have multiple minima. However, a typical interatomic potential (whether in reality or computer simulations) possesses only a single energy minimum, as in the Lennard-Jones potential. Nevertheless, we know that atoms interacting under a simple potential can form dense fluids, such as that depicted in Figure 6.5.

In other words, a single-well nonbonded interaction potential should be expected to lead to a PMF with multiple minima. One atom attracts its neighbors, which attract their own neighbors, and so on.

A key consequence of this is that *it is wrong, in principle, to equate u^{pair} and the* PMF. Certainly, interactions outside the first shell of neighbors will be badly mistaken this way. (The PMF for the first shell will also include subtle correlations among members of the first shell and also with their neighbors.) On the other hand, sometimes the PMF is the only tool available for estimating an interaction potential, and the PMF does form the basis of several approximate "knowledge-based" potentials for proteins, described below.

Just to build your intuition, consider the case of densely packed but purely repulsive particles. Qualitatively, the radial distribution function for this system will look the same as Figure 6.6: peaks of decreasing height for the various shells of neighbors. These peaks correspond to minima in the PMF. Of course, the true u^{pair} function will have no attractive part at all.

PROBLEM 6.7

Consider a system of hard particles in one dimension—that is, where u^{pair} is either zero or infinity depending on whether the particles are closer than some hard-core distance "*d*." Think of beads on a string. (a) If one hard particle is in a fixed position, and it has one neighbor that can move freely in a space of length $2d$, sketch the system and the function $g(r)$. Note that g should correspond to the probability of the distance between the centers of these one-dimensional, hard particles. (b) Sketch the corresponding PMF on the same graph as u^{pair}, and comment on the differences.

6.4.4   FROM $g(r)$ TO THERMODYNAMICS IN PAIRWISE SYSTEMS

One reason why the radial distribution function receives so much attention is that it forms a direct connection between the "microscopic" world of statistical mechanics and the "macroscopic" world of thermodynamics. Concretely, for a monatomic system, both the average total potential energy and the pressure can be calculated from g when the interactions occur only between pairs of atoms. (For molecules, one expects to experience interactions among three atoms—bond angles— and even four—dihedrals—as in typical forcefields.)

Here, we will only discuss the relation of g to the average potential energy. We are assuming the total potential energy for N atoms is just a sum of identical pairwise terms u^{pair}, as in Equation 6.14. If this is the case, then we can determine the potential energy associated with any single central reference atom using g, and then multiply the result by N. To accomplish this, we can use the definition of g to note first that around a reference atom, there are on average $4\pi r^2\, dr\, g(r)\, N/V$ other atoms in a range of size dr at a distance r. Each of the atomic pairs separated by r contributes $u^{\text{pair}}(r)$ to the total potential energy—of which half should be associated with each atom. Putting all this together, we find

$$\text{Pairwise:} \quad \langle U \rangle = N\frac{N}{2V} \int dr\, 4\pi r^2\, g(r)\, u^{\text{pair}}(r), \tag{6.15}$$

which applies to systems with only pairwise interactions.

As promised, a macroscopic thermodynamic average can indeed be derived from the average microscopic behavior embodied in g. This is not really surprising, since we already know the relationship between free energy and the partition function, but it is a nice statistical thermodynamical validation.

6.4.5 $g(r)$ IS EXPERIMENTALLY MEASURABLE

You may have learned already that in determination of protein structures from x-ray experiments, the measured scattering data is the Fourier transform of the electron density. Similarly, for fluids, scattering data can reveal (the Fourier transform of) $g(r)$. In a crystal, there are repeated structural features that give rise to special scattering patterns, while in a fluid there is a well-defined average structure embodied in g. For further details, you are referred to McQuarrie's book.

The key point is that both structural and thermodynamic information can be measured experimentally, since both are built in to g.

6.5   PMFS ARE THE TYPICAL BASIS FOR "KNOWLEDGE-BASED" ("STATISTICAL") POTENTIALS

We have focused on the PMF as an analysis tool—a tool for understanding data or a model that has already been created. However, the PMF can also be used to construct models.

The so-called knowledge-based or statistical potential energy functions (force-fields) are derived from PMF-style analyses of structural databases, such as the protein data bank. The idea is to assume the set of known experimental structures represents a statistical sample, and then derive effective interactions from there. Such an analysis is very flexible because effective interactions can be derived for "particles" of arbitrary definition—for example, atoms, amino-acid side chains, or entire amino acids.

The assumption that an experimental database represents a good statistical sample is highly doubtful, but knowledge-based potentials should still be expected to capture the most important interactions—for example, hydrophobic tendencies. Furthermore, traditional forcefields have many known imperfections despite the care taken in parameterizing them.

Examples of knowledge-based potentials include the following references:

- S. Tanaka and H. Scheraga, *Macromolecules* 9:945 (1976).
- S. Miyazawa and R. Jernigan, *J. Mol. Bio.* 256:623 (1996).
- I. Bahar and R. Jernigan, *J. Mol. Bio.* 266:195 (1997).
- D. Tobi and R. Elber, *Proteins* 41:40 (2000).

PROBLEM 6.8

Given the differences between the PMF and the underlying potential discussed in Section 6.4.3, do you expect knowledge-based potentials to be more accurate at short or large distances?

6.6   SUMMARY: THE MEANING, USES, AND LIMITATIONS OF THE PMF

The PMF is exactly analogous to a free energy: it is the energy that gives the probability in a Boltzmann factor. The PMF, however, is explicitly defined to give not just the probability but a probability *distribution* as some coordinate of interest varies. It is often called a "free energy profile."

There are many types of PMF, which is good and bad. The PMF can represent probability distributions for different—indeed, arbitrary—coordinates. This is its great strength and the basis of its importance. However, caution is required because of the freedom in defining a PMF. For instance, the PMF for a radial distribution function $g(r)$ is defined with an implicit factor of r^2 to scale out effects of radial distance. To be as careful as possible: given that the Boltzmann factor of the PMF should correspond to a dimensionless ratio of distributions, be sure to know what the normalizing (denominator) distribution is.

The key limitation of a PMF is the information it hides—which generates the temptation to overinterpret the PMF as an energy landscape. In (high-dimensional) molecular systems, it is often desirable to estimate kinetics—and one naturally turns to Arrhenius-like Boltzmann factors of barrier heights. However, different PMFs— that is, projections onto different coordinates—can yield different apparent barrier heights. The location of the transition state, moreover, can easily be mistaken based on a PMF. Other intrinsic limitations characterize "knowledge-based" potentials for proteins, which are derived from PMF analyses of structures. Be careful!

FURTHER READING

Dellago, C. et al., *Journal of Chemical Physics*, 108: 1964–1977, 1998.
McQuarrie, D.A., *Statistical Mechanics*, University Science Books, Sausalito, CA, 2000.

7 What's Free about "Free" Energy? Essential Thermodynamics

7.1 INTRODUCTION

If this were a "normal" book, we would already have discussed classical thermodynamics. From my point of view, however, thermodynamics is actually a harder subject than statistical mechanics. Maybe much harder. As we have seen, "stat mech" can be understood very well from a probability theory perspective.

Now we will see that the aspects of thermodynamics that are most important for biophysics can be understood in terms of the statistical mechanics we already know. This subset of thermodynamics can be called "statistical thermodynamics," and we will discuss that first. In essence, we will derive many key thermodynamic results by taking certain derivatives of partition functions or free energies.

"Classical thermodynamics"—that is, the laws and their consequences—is better appreciated from a fully macroscopic perspective. In other words, we will discuss the classical topics in a quite similar way to an undergraduate physics text. Nevertheless, you will gain valuable intuition when we use a statistical treatment of the ideal gas to illustrate the classical laws.

By the way, the "free" in free energy means available. The free energy is the energy available to do work. As in statistical mechanics, the free energy is not the average energy—and we will see why explicitly.

7.1.1 AN APOLOGY: THERMODYNAMICS DOES MATTER!

While this book has taken the perspective that molecular behavior is best understood from a microscopic/statistical point of view, this doesn't mean thermodynamics is irrelevant. Far from it. First, note that thermodynamics provides the basic language of biophysics—terms like entropy, enthalpy, free energy, and specific heat. To communicate with other scientists, you must know this language well. (And you may not realize it yet, but communication is a huge part of science.) Secondly, thermodynamics enables quite simple descriptions of a system's behavior, at least in a number of important cases. Finally, being a macroscopic description, thermodynamics is quite useful to know in ordinary life: I promise you'll never look at your refrigerator in the same way!

Historically, thermodynamics was developed before statistical mechanics, which is why thermodynamics terminology dominates both fields.

What does it mean to say thermodynamics is "macroscopic"? For the most part, up until this point we have only been considering in detail systems of one or a few degrees of freedom. Even though a single molecule may possess tens to thousands of degrees of freedom, this is a tiny number compared to what is studied experimentally. Even a small volume like $(1 \text{ mm})^3 = 10^{-9} \text{ L}$ contains about $10^{-9} \text{ mol} \sim 10^{15}$ molecules! We could never perform statistical mechanics calculations on such experimental-sized systems, except under extreme idealizations. Therefore, thermodynamics is key for understanding experiments.

We should also mention that there are limitations to the probabilistic interpretation we have focused on. Specifically, the Boltzmann factor distribution $\rho(\mathbf{r}^N) \propto \exp[-U(\mathbf{r}^N)/k_B T]$ provides a complete probabilistic description at any constant T, but it does not tell you everything you need to know in the context of common temperature-dependent experiments. For instance, free energy changes as a function of temperature are important in the protein-folding field. Of course, free energy differences (i.e., relative populations) of states can be compared at two different temperatures—and this seems the most intuitive thing to do—but whole-system free energy differences at two temperatures cannot be interpreted in such a way.

7.2 STATISTICAL THERMODYNAMICS: CAN YOU TAKE A DERIVATIVE?

Our initial discussion of thermodynamics will continue our focus on systems maintained at constant volume and temperature. Under these conditions, the symbol $F(T, V) = -k_B T \ln Z(T, V)$ is called the Helmholtz free energy. Later, we will also study the Gibbs free energy $G(T, P)$ appropriate to constant temperature and pressure P.

7.2.1 QUICK REFERENCE ON DERIVATIVES

You already know all this calculus, of course. It's just here for your reference.

Differentiating under integral:
$$\frac{d}{d\alpha} \int dx f(x; \alpha) = \int dx \left[\frac{d}{d\alpha} f(x; \alpha) \right], \quad (7.1)$$

where the key point is that the derivative is applied to some variable (a "parameter") that is not being integrated over. The relation is not true when the derivative is to be taken with respect to the variable of integration (x in this case).

Logarithm:
$$\frac{d \ln [f(x)]}{dx} = \frac{f'(x)}{f(x)}, \quad (7.2)$$

Logarithm of composition:
$$\frac{d \ln [f(g(x))]}{dx} = \frac{f'(g(x)) g'(x)}{f(g(x))}, \quad (7.3)$$

where $f'(g(x))$ indicates that one should differentiate the function f with respect to its argument and substitute in $g(x)$ after—that is, $f'(g(x)) = f'(u)|_{u=g(x)}$.

Product rule: $\dfrac{d}{dx}[f(x)g(x)] = f'(x)g(x) + f(x)g'(x),$ (7.4)

Total derivative: $\dfrac{d}{dx}f(x,y) = \dfrac{\partial f}{\partial x} + \dfrac{\partial f}{\partial y}\dfrac{dy}{dx},$ (7.5)

for the case where $y = y(x)$ is not independent of x.

7.2.2 Averages and Entropy, via First Derivatives

If you have studied thermodynamics before, you have probably seen derivatives of free energies and perhaps wondered what they could possibly mean. Here, we will see that because of the mathematical structure of the partition function, such derivatives lead directly to the statistical mechanical averages we studied in Chapter 3.

Watch what happens with the temperature derivative of the partition function Z. We will continue to denote the total energy by E, so that, in one dimension, we have $E(x, p) = U(x) + \text{KE}(p) = U(x) + p^2/2m$, with momentum $p = mv$. Making use of the differentiation rules of Section 7.2.1, we find

$$\frac{\partial Z}{\partial T} = \frac{\partial}{\partial T}\, h^{-1}\int dx\, dp\, e^{-E(x,p)/k_B T}$$

$$= h^{-1}\int dx\, dp\left(\frac{+1}{k_B T^2}\right) E(x,p)\, e^{-E(x,p)/k_B T}.$$ (7.6)

Then using the log rule,

$$\frac{\partial \ln Z}{\partial T} = \frac{1}{k_B T^2}\frac{\int dx\, dp\, E(x,p)\, e^{-E(x,p)/k_B T}}{\int dx\, dp\, e^{-E(x,p)/k_B T}}.$$ (7.7)

In other words,

$$\langle E \rangle = k_B T^2 \frac{\partial \ln Z}{\partial T}.$$ (7.8)

Recall that $\ln Z = -F/k_B T$. If you trace back through the math, you will see that this simple relation results from the fact that the Boltzmann factor (the statistical weight) is an exponential, so the T-derivative brings down the energy, and ultimately yields the weighted average.

Is it thermodynamics or probability theory? You can decide for yourself.

Note that the partial derivative $\partial/\partial T$ (as opposed to the total derivative d/dT) is a reference to the fact that the partition function also depends on the available configuration "volume" V—that is, $Z = Z(T, V)$. In one dimension, V is just the length of the system, but that's not so important at the moment.

Note further that the differentiation "trick" used above has nothing to do with the dimensionality of the integral. Equation 7.8 is a general result as you will see in the next problem.

PROBLEM 7.1

Show that Equation 7.8 holds for systems of arbitrary dimensionality.

In terms of notation, we will find ourselves writing Boltzmann factors so often that it is convenient to define the inverse thermal energy $\beta = 1/k_B T$ (sometimes referred to as the inverse temperature by physicists who are cavalier about units). Then the Boltzmann factor is $e^{-\beta E}$, and in fact, Equation 7.8 simplifies to

$$\langle E \rangle = -\frac{\partial}{\partial \beta} \ln Z = +\frac{\partial}{\partial \beta} \beta F \qquad \left(\beta = \frac{1}{k_B T} \right). \qquad (7.9)$$

PROBLEM 7.2

Derive Equation 7.9.

It is often important to consider only the potential energy, and it should be fairly obvious from the derivations above that the average U can be written in terms of a derivative of the configuration integral:

$$\langle U \rangle = -\frac{\partial}{\partial \beta} \ln \left(\frac{\hat{Z}}{l_0} \right), \qquad (7.10)$$

where the configurational partition function is $\hat{Z} = \int dx \, e^{-U(x)/k_B T}$ in one dimension, and l_0 is an arbitrary constant length to keep our math straight. The relation (7.10) holds for systems of arbitrary dimensionality with $l_0 \to l_0^{3N}$ for N atoms.

Later, when we study other thermodynamic conditions (e.g., constant pressure instead of constant volume as we have been doing implicitly), we'll see that other first derivatives of the appropriate free energy lead to other averages.

7.2.2.1 The Entropy

Now that we've seen the average energy extracted from the free energy by differentiation, and since we know that F is obtained from the log of Z, you may not be surprised to learn that $S = (\langle E \rangle - F)/T$ can also be calculated from differentiation. In fact, simply using the product rule, one obtains

$$\frac{\partial F}{\partial T} = \frac{\partial}{\partial T} (-k_B T \ln Z) = -k_B \ln Z - k_B T \frac{\partial}{\partial T} \ln Z. \qquad (7.11)$$

The first term is F/T by definition, and the second is $-\langle E \rangle/T$ from Equation 7.8, thus implying

$$S = -\frac{\partial F}{\partial T}. \qquad (7.12)$$

7.2.3 FLUCTUATIONS FROM SECOND DERIVATIVES

The relationship between fluctuations (i.e., variances) and second derivatives turns out to be one of the most important and practical in thermodynamics. For instance, in the study of protein folding and unfolding, the specific heat (which is a second temperature derivative, as we'll see) is often measured experimentally. The meaning of such measurements in terms of the "microscopic" molecular behavior (as opposed to the "macroscopic" heat input/output) can only be grasped by knowing the connection between fluctuations and second derivatives. Other second derivatives are equally fundamental in other physical measurements.

For now, we will continue our focus on temperature derivatives. To simplify the math, we will again employ the variable $\beta = 1/k_B T$. We then have

$$
\frac{\partial^2 \ln Z}{\partial \beta^2} = \frac{\partial}{\partial \beta} \left[\frac{1}{Z} \frac{\partial Z}{\partial \beta} \right]
$$

$$
= \frac{\partial}{\partial \beta} \frac{\int dx\, dp\, [-E(x,p)]\, e^{-\beta E(x,p)}}{\int dx\, dp\, e^{-\beta E(x,p)}}
$$

$$
= \frac{\int dx\, dp\, [E(x,p)]^2\, e^{-\beta E(x,p)}}{\int dx\, dp\, e^{-\beta E(x,p)}} - \left[\frac{\int dx\, dp\, [-E(x,p)]\, e^{-\beta E(x,p)}}{\int dx\, dp\, e^{-\beta E(x,p)}} \right]^2
$$

$$
= \langle E^2 \rangle - \langle E \rangle^2 = \sigma_E^2. \tag{7.13}
$$

This is exactly the variance of the energy!

7.2.4 THE SPECIFIC HEAT, ENERGY FLUCTUATIONS, AND THE (UN)FOLDING TRANSITION

The specific heat or heat capacity, c, is commonly measured in biophysical experiments. As you learned in high school, c quantifies the amount of heat energy needed to raise the temperature of a sample. But from a molecular/statistical point of view, we will see that the specific heat measures the variance in the energy. It may not be obvious right now how important this is, but the energy fluctuations can reveal when a transition (e.g., unfolding) is taking place: compared to low or high temperatures where a single conformational state is important, at a transition temperature, both states are occupied leading to a large variance in the energy distribution. In essence, the distribution is bimodal (and hence very broad) only near the transition. At temperatures above and below an unfolding transition, as we saw in Section 3.7.1, only a single state is occupied and the energy variance is much smaller.

PROBLEM 7.3

Using the square-well model of protein folding/unfolding studied in Section 3.7.1, (a) sketch the distribution of energy values at temperatures above, below, and equal to the transition temperature. (b) Explain why the variance is highest at the transition.

True "phase transitions," such as the melting of ice, occur only in the "thermo-dynamic limit" of very large systems. For such a macroscopic system undergoing a phase transition, some free energy derivative will become infinite. This is the case for the specific heat of ice. By contrast, a protein is very finite, even when placed in a large number of solvent molecules and other proteins. The specific heat of a very large solution of proteins never becomes infinite at the folding "transition."

7.2.4.1 The Formal Connection between Specific Heat and Energy Fluctuations

How exactly is the specific heat related to energy fluctuations? By definition, the specific heat is obtained from the entropy via

$$c_v \equiv T\frac{\partial S}{\partial T} = -T\frac{\partial^2 F}{\partial T^2}. \tag{7.14}$$

Later, we'll see explicitly that this definition is indeed related to heat.

To determine the correct constant of proportionality between the specific heat and σ_E^2, you could take two temperature derivatives of F. Alternatively, you can use our result (7.13) for the second β derivative. Either way, one finds

$$c_v = \frac{\sigma_E^2}{(k_B T^2)}. \tag{7.15}$$

We will examine the specific heat further below because of its importance in molecular biophysics.

PROBLEM 7.4

Derive the relation between specific heat and energy variance (7.15). Use the standard constant-volume ensemble where $Z = h^{-1} \int_V dx \int dp \, \exp[-\beta E(x,p)]$ in 1D. Your calculation should not depend on the system being one dimensional, however, and hence it should apply to multi-atom systems confined to an ordinary three-dimensional volume.

By the way, the subscript v denotes constant volume, and c_v is correspondingly derived from the constant-volume (Helmholtz) free energy F. However, the ther-modynamic relation $c \equiv T\partial S/\partial T$ applies in other conditions as well: for example, at constant pressure $c_p = T\,\partial S/\partial T$ also, although S can no longer be derived by differentiating F. More on this below.

7.3 YOU LOVE THE IDEAL GAS

The ideal gas is your friend. Its partition function can be computed exactly (ana-lytically) at any temperature and it lends itself to illustrating thermodynamic results starting from first-principles statistical mechanics. A dream come true!

An ideal gas is defined to consist of particles that do not interact—as we saw in Chapter 5. The easiest way to achieve this is by setting the potential energy function to be exactly zero, $U(\mathbf{r}^N) = 0$ for every configuration $\mathbf{r}^N$. To keep the present discussion as simple as possible, we will also assume that the particles have no internal degrees of freedom (such as the bond length or orientation angles of a diatomic molecule), which means we are essentially considering a monatomic gas.

We already calculated the ideal gas partition function in Equation 5.22. The result was

$$Z^{\text{idl}} = \frac{1}{N!} V^N \left[\frac{1}{\lambda(T)} \right]^{3N}, \qquad (7.16)$$

where $\lambda(T) = h/\sqrt{2\pi m k_B T} \equiv V_0^{1/3}(T_0/T)^{1/2}$—that is, $\lambda(T) \propto 1/\sqrt{T}$. The symbols V_0 and T_0 are constants that are relevant only for their dimensions, as we'll see later. Showing the thermodynamic variables explicitly, we have

$$Z^{\text{idl}}(T, V) = \frac{1}{N!} \left(\frac{V}{V_0} \right)^N \left(\frac{T}{T_0} \right)^{3N/2}. \qquad (7.17)$$

This form will be useful for taking a logarithm (to get the free energy), since if we're careful, we should only use logs of dimensionless expressions.

The average energy of the ideal gas can be derived by one of the differentiation rules given above. Alternatively, recall from Chapter 3 that there is $k_B T/2$ of average energy per degree of freedom for each of the harmonic (in velocity) degrees of freedom, and there is no potential energy. Either way, one obtains

$$\langle E \rangle^{\text{idl}} = \langle \text{KE} \rangle = \left(\frac{3}{2} \right) N k_B T \qquad (7.18)$$

for an ideal gas of N particles in a three-dimensional space.

7.3.1 FREE ENERGY AND ENTROPY OF THE IDEAL GAS

Using the definition of F and Stirling's approximation ($\ln N! \simeq N \ln N - N$), the free energy is given by

$$F^{\text{idl}} = -k_B T \ln Z = -k_B T \left[-N \ln N + N + N \ln \frac{V}{V_0} + \frac{3N}{2} \ln \frac{T}{T_0} \right]. \qquad (7.19)$$

Note that in the "thermodynamic limit" ($N \to \infty$), Stirling's approximation is essentially exact, and so too is our expression for F.

The entropy is simply derived by differentiation

$$S^{\text{idl}} = -\frac{\partial F^{\text{idl}}}{\partial T} = N k_B \left[\ln \frac{V}{N V_0} + \frac{5}{2} + \frac{3}{2} \ln \frac{T}{T_0} \right]. \qquad (7.20)$$

7.3.2 The Equation of State for the Ideal Gas

The "equation of state" is another of those phrases that sounds like it was intended to intimidate beginners. In fact, it's just the relationship between the thermodynamic variables and, for the ideal gas, you learned the equation of state in high school: $PV = nRT$. Here P is the pressure, n is the number of moles, and R is the gas constant. By simply noting that $nR = Nk_B$, we can cast the relation more statistically to obtain

$$PV = Nk_B T \tag{7.21}$$

We haven't really discussed pressure yet. Of course, as you were taught in freshman physics, the pressure is just the force per unit area. In the case of the ideal gas, the force is based on the momentum transferred per unit time to the walls of the container. If you think about it, we have all the necessary information to compute the momentum transfer: the distribution of velocities (Maxwell–Boltzmann, as in Chapter 3), the size of the container (V), and the average kinetic energy ($k_B T/2$ in each direction). From this information, the ideal gas equation of state (7.21) can be derived. We won't perform the derivation carefully here, but you can find it in a number of textbooks (e.g., by Phillips et al.).

We will examine a quick-and-dirty (and inexact) derivation of Equation 7.21. In the x direction, for example, we know $\langle mv^2/2 \rangle = k_B T/2$. We consider a single particle moving at the average speed $\langle v \rangle = \sqrt{k_B T/m}$ in the x direction, and contained in a volume $V \equiv L^3$. (The average velocity is zero.) There is a collision with each wall every time the particle goes back and forth—that is, in a time $t = 2L/\langle v \rangle$. Each collision changes the momentum by $2m \langle v \rangle$, which is transferred to the wall as an impulse, so the rate of momentum transfer to a given wall is $2m \langle v \rangle/t = m \langle v \rangle^2/L = k_B T/L$. This is the definition of the average ("impulsive") force. Since each wall has area L^2, the average force per area (the pressure!) due to a single particle is $k_B T/L^3 = k_B T/V$. The total pressure due to N particles is $P = Nk_B T/V$. Voila! In a derivation like this, you may have to worry that you have missed constant factors of 2 or π from exact integrals (although we were lucky to get the right result), but you can often expect to obtain the right dependence on the variables.

The key point is that Equation 7.21 is a statistical result. It can't be derived from thermodynamics unless the form of the free energy (from statistical mechanics) has been used. The reason we want the (statistically derived) equation of state is that it will help us illustrate the laws of thermodynamics from a purely statistical point of view.

7.4 BORING BUT TRUE: THE FIRST LAW DESCRIBES ENERGY CONSERVATION

The first law of thermodynamics is real simple: energy is conserved. That's it, you're done.

Alas, from a pedagogical point of view, the author would be remiss in his obligations to you if a beautiful example were not provided. Fortunately, we are masters of

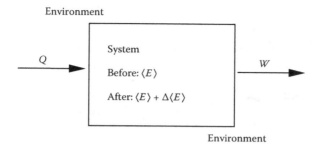

FIGURE 7.1 An illustration of the definitions used in a process governed by the first law of thermodynamics. The environment is everything external to the figure.

the ideal gas. The plan is to consider a simple situation and show that the work done goes into heat. And not abstract, entropy-something-or-other, but heat that increases the temperature of our favorite gas.

We first need to agree on some names (and symbols), and then we can apply the simple energy conservation idea. We divide the universe into two parts—our system of interest and everything else ("the environment"). The symbol Q is defined to be the heat (energy) put in to a system—from the environment, of course, as there's no other choice. The internal energy of our system will be the total average energy $\langle E \rangle = \langle U \rangle + \langle KE \rangle$. Finally, the work done by our system on the environment will be called W. Here, Q and W will refer to any process we want to consider, and $\langle E \rangle$ describes the instantaneous energy, suggesting we may want to consider the possibly differing values of $\langle E \rangle$ before and after the process under consideration.

We can now apply conservation of energy, as illustrated in Figure 7.1. In words, what goes in must either stay in or go out, which implies

$$\text{First law:} \quad Q = \Delta \langle E \rangle + W, \tag{7.22}$$

where $\Delta \langle E \rangle$ is the final energy less the initial. By the way, the work is performed in standard ways—for example, mechanical (piston moving against pressure) or electrostatic (charge moving across potential difference). If work is done on the system, then $W < 0$.

7.4.1 Applying the First Law to the Ideal Gas: Heating at Constant Volume

The reason to apply the first law to a process in the ideal gas is to convince ourselves that the broader thermodynamics "formalism" (all the derivatives) really mean something. Specifically we'll focus on the (constant-volume) specific heat c_v by imagining that we heat our gas so that the temperature changes by ΔT (Figure 7.2). Until now, we've seen that the specific heat has a statistical meaning in terms of the variance of the energy distribution. But what about the "heat" part of c_v?

Our strategy will be to calculate $Q/\Delta T$ from purely statistical results, and show that it is exactly the same as the specific heat we have already defined

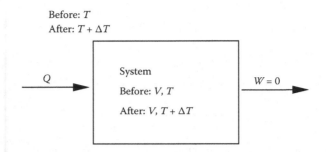

FIGURE 7.2 Heating a gas at constant volume, a process in which no mechanical work is done. Heat is added as the system temperature increases by ΔT.

thermodynamically. Thus, we will show that $c_v = Q/\Delta T$ really does describe both heat exchange and energy fluctuations in the ideal gas.

So we heat our ideal gas from temperature T to $T + \Delta T$, keeping it in a fixed-size box (constant volume). The heat added to raise the temperature is Q. Is work done? Since our gas particles don't interact with anything (except occasionally the walls of the enclosing box), the only way for them to do work is mechanically by changing the volume of the box. However, the volume is constant in this case, so no work is done ($W = 0$). Thus, from the first law, $Q = \Delta \langle E \rangle$, or from Equation 7.18, $Q = (3/2)Nk_B \Delta T$.

In other words, based on purely statistical results, we find that $Q/\Delta T = (3/2)Nk_B$ for any ΔT, in the ideal gas. (You should not make the mistake of assuming the specific heat is constant—independent of T—for any nonideal system.) But what if we use the thermodynamical definition of specific heat, namely, $c_v \equiv T\partial S/\partial T$? Well, you can differentiate our result (7.20) above for the entropy yourself and find exact agreement with the statistical result.

We have therefore shown that, at least in the ideal gas, the traditional meaning of the specific heat—the heat needed to raise the temperature—just happens to be proportional to the variance of the energy. This is an important result that you should think about.

PROBLEM 7.5

Show that you can "weigh" an ideal gas (determine the total mass) using a calorimeter, which is a machine that can simultaneously measure Q and ΔT. Assume you know the mass of each particle (atom) ahead of time, but not the number of particles.

7.4.2 Why Is It Called "Free" Energy, Anyway? The Ideal Gas Tells All

We can start with the punch line: "Free" means available. We can even give away the secret behind the punch line: a constant-temperature system exchanges energy

Slowly decreasing external pressure

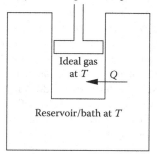

FIGURE 7.3   A process that reveals the meaning of "free" in free energy: slow expansion of an ideal gas. The gas is permitted to expand only very slowly, by a gradual decrease of the external pressure, so that the temperature remains constant (i.e., the gas is in equilibrium with the bath). The volume increases as the gas expands and does mechanical work by pushing the piston up against the external pressure.

(i.e., the heat Q) with its bath/environment, and this energy can be available to do work (Figure 7.3).

A constant-temperature expansion in the ideal gas tells the story. The first thing you should be thinking when you hear "constant temperature" for an ideal gas is that the energy is also constant—in other words, $\Delta \langle E \rangle = 0$ in any equilibrium process at constant T because $\langle E \rangle = (3/2)Nk_BT$. On the other hand, in an expansion, by definition, the volume is changing and so some work is being done. So your first thought might be that there's a paradox here: the energy is constant but work is being done.

But the first law rescues us, since there's still a Q term. Interestingly, this heat exchange remains fundamental even when the temperature is constant. We can calculate the heat from the first law, namely, $Q = \Delta \langle E \rangle + W$ or $Q = W$ when T is constant in the ideal gas. The total work is simply the integral of force over distance, which translates to the integral of pressure over volume:

$$ W = \int_{V_i}^{V_f} P\,dV = Nk_BT \ln \frac{V_f}{V_i}, \tag{7.23} $$

where V_i and V_f are the initial and final volumes. The work performed was obtained by using $P = Nk_BT/V$.

So what does this have to do with free energy? Well, let's check. Compare F^{idl} before and after the expansion, using Equation 7.19. Remarkably, $F^{idl}(V_f, T) - F^{idl}(V_i, T) = -Nk_BT \ln V_f/V_i$, which means that the free energy has decreased by exactly the amount of work done!

That's worth saying again: the gas did work, while the energy $\langle E \rangle$ did not change. However, the "free" energy correctly accounted for the work done in the expansion, since less free energy is now available. If we ask where the energy came from (nothing is free in this world, says the first law!), it came from the constant-temperature bath.

Physically, if heat had not been added to the system, energy would have been lost and the temperature would have decreased as the gas expanded.

(Incidentally, you may be wondering how a gas can be expanded in such a way that the constant-volume free energy, F, remains the appropriate free energy to consider. One possibility is that the position of the piston is controlled by a screw that is turned very slowly—slowly enough for heat to be transferred to keep T constant. In this case, the volume is always strictly defined and does not fluctuate, in contrast to the case of constant pressure, which we consider below. Turning the screw requires work on the part of the person or machine turning it; however, when the gas pushes on the piston, it contributes some or all of the work according to $\int P\, dV$.)

7.5 *G* VS. *F*: OTHER FREE ENERGIES AND WHY THEY (SORT OF) MATTER

7.5.1 *G*, CONSTANT PRESSURE, FLUCTUATING VOLUME—A STATISTICAL VIEW

Real-life experiments typically occur under conditions of constant pressure, not constant volume. For instance, beakers in the lab are exposed to the air (pressure) and are not sealed in a fixed-size container. It turns out that this adds some complications to what we've learned, although it doesn't really change much in the way of new ideas. Certainly, it's worthwhile to learn the constant-pressure formulation, both to broaden our minds, and to understand solidly what it means when every chemistry and biochemistry book uses the Gibbs free energy G (instead of F).

To get started, let's recall the boring fact that a partition function is a Boltzmann-factor-weighted integral of all possible configurations of a system, and the free energy is proportional to the log of this. This same definition holds in the case when pressure is constant—and volume can fluctuate—except that there are now many more possible configurations. Also, the energy differences between states are a little different than before.

First, why are there more possible configurations? We can think again of a gas in box. We still integrate over all possible atomic positions, as before. However, now the system configuration also depends on the box's volume and we must also consider all possible volumes. We therefore want to consider a partition function of the form

$$e^{-G(P,T)/k_\mathrm{B}T} \equiv Z_\mathrm{p}(P,T) \propto \int\limits_{V=0}^{\infty} dV\, Z(T,V) e^{-E(V)/k_\mathrm{B}T}, \qquad (7.24)$$

where $Z(T,V) = e^{-F(T,V)/k_\mathrm{B}T}$ is the ordinary constant-volume partition function we have considered before. In other words, this is a Boltzmann-factor-weighted average of partition functions, with some mysterious energy $E(V)$. Mathematically, this partition function is equivalent to associating the energy $E(V)$ with every configuration in a volume V—that is, it's like setting $E_\mathrm{total} = U(x) + \mathrm{KE} + E(V)$. (The proportionality sign in Equation 7.24 is there to remind the experts that something has been left out—after all, the integral is not dimensionless, in contrast to the constant-volume partition function.)

A weighted average of partition functions! Is that meant to torture students? Perhaps a simpler way to think of Z_p is to consider it a regular partition function over configurations—except that now the configurations are "extended" in the sense that they not only include the usual configurational degrees of freedom $\mathbf{r}^N$, but also a coordinate describing the size of the container, namely, V. Since the volume can change and it affects the overall energy, we certainly should integrate over it. If we denote the extended coordinates by $\mathbf{r}^{ex} = (\mathbf{r}^N, V)$ with a corresponding extended potential energy $U^{ex}(\mathbf{r}^{ex}) = U(\mathbf{r}^N) + E(V)$, then the partition function looks more like what we saw before:

$$Z_p \propto \int d\mathbf{r}^{ex} \, d\mathbf{p}^{ex} \, \exp\{-\beta \, [KE(\mathbf{p}^{ex}) + U^{ex}(\mathbf{r}^{ex})]\}. \tag{7.25}$$

The only unusual aspect here is that you need to worry about the order of integration. Most simply, you can assume a particular container size V and then integrate over all coordinates consistent with it—that is, with no particles outside it! Of course, this is just what our original integral (7.24) was trying to "say."

It's time to clarify the energy term for the volume. The volume-dependent energy, in fact, is simply

$$E(V) = PV. \tag{7.26}$$

Why? First, note that the energy increases with volume: this makes sense because you have to push against the pressure to expand the volume, implying that larger volumes are higher in energy. Second, it is not hard to understand the simple proportionality to volume. Imagine expanding a cubic box of size L^3 in just one direction to a box of size $L^2(L + \delta L)$. The face of the box that moves is of size L^2 and this face pushes against the constant pressure p. Since pressure is just force per unit area, the total force on the moving face is PL^2, implying the work done is $PL^2\delta L$. Of course, $L^2\delta L = \Delta V$ is just the change in volume. In other words, the energy difference between the states with volumes V_1 and $V_2 > V_1$ is just $P(V_2 - V_1)$, implying the linear form is correct. (The result is not dependent on the cubic geometry, but is in fact fully general. You can show this by integrating over infinitesimal regions of the surface of the container, and accounting for the vectorial components of the force.)

But let's take a step back and consider the big picture. Essentially, we've proceeded in close analogy to the way we worked in the constant-volume picture. In that case, we had a Boltzmann-factor probability density $\rho(\mathbf{r}^N) \propto e^{-\beta U(\mathbf{r}^N)}$. By integrating over configurations constituting states, we saw that $e^{-\beta F_s}$ gave the probability of a state s. In the constant pressure case, we can proceed in the same way, except that our coordinate space is now extended by the fact that volume can change. We have $\rho(\mathbf{r}^{ex}) \propto e^{-\beta U^{ex}(\mathbf{r}^{ex})}$, and therefore the corresponding free energy G will give the probabilities of states. The key observable, of course, will be population ratios and the corresponding free energy differences ΔG.

Based on the statistical definition of G in Equation 7.24, you should expect that average quantities can be obtained by differentiation. In the constant-pressure ensemble, the volume fluctuates according to the probability $\text{prob}(V) \propto Z(T, V) \, e^{-E(V)/k_B T}$.

The average volume is given by

$$\langle V \rangle = \frac{\partial G}{\partial P},$$ (7.27)

which you will derive in the following problem.

PROBLEM 7.6

(a) Using Equations 7.24 and 7.26, show that $\langle V \rangle = \partial G/\partial P$. The partial derivative here refers to the fact that the other thermodynamic variables—T and N— are being held constant. Assume that the probability of V is proportional to $Z(T, V)\, e^{-E(V)/k_B T}$. (b) Calculate $\langle V \rangle$ for the ideal gas, starting from its constant-volume partition function. (c) Compare your result to Equation 7.21, and explain the limit in which the two results will become indistinguishable.

7.5.2 When Is It Important to Use *G* Instead of *F*?

We have developed our intuition using F, yet experimental conditions typically require G, so we must carefully explore when the difference matters. The short answer is that the difference is primarily important conceptually and when one is considering dilute systems. For solid and liquid phases, and for large molecules like proteins in solution, the difference is largely irrelevant. The reason is that volume fluctuations in experimental biological systems are tiny on a relative scale. Thus, if the experiments were somehow conducted at constant volume, the results would probably be indistinguishable from those obtained at constant pressure.

7.5.2.1 Volume Fluctuations Are Small in Macroscopic Systems

If you think about it, a gas will fluctuate in volume much more than a liquid, since the molecules are not being held together. Here we will examine the (relative) fluctuations in the volume of a gas under standard conditions to get a feel for the significance of volume fluctuations in macroscopic systems—that is, in systems on which experiments are likely to be performed.

Because the scale of fluctuations is set by the standard deviation σ, we will consider the fractional fluctuations of volume defined by

$$\hat{\kappa} \equiv \frac{\sigma_V}{\langle V \rangle} = \frac{\left[\langle V^2 \rangle - \langle V \rangle^2\right]^{1/2}}{\langle V \rangle}.$$ (7.28)

(For your information, our $\hat{\kappa}$ is a dimensionless version of the more standard isothermal compressibility κ_T.)

In the ideal gas, the necessary averages for Equation 7.28 are easy to calculate because the effective probability density of V is given by $Z(T, V)$ as is implicit in the integral (7.24) defining the Z_p partition function. That is, $\rho(V) \propto V^N e^{-\beta PV}$. If you consult a table of integrals (or a discussion of the gamma function), you will find

$\langle V \rangle = (N+1)k_B T/P$ and $\langle V^2 \rangle = (N+2)(N+1)(k_B T/P)^2$. Therefore, we have

$$\hat{\kappa}^{\text{idl}} = \frac{\sqrt{(N+1)(k_B T/P)^2}}{(N+1)(k_B T/P)} = \sqrt{\frac{1}{N+1}}. \tag{7.29}$$

It's just our old friend from the statistical "standard error," $1/\sqrt{N}$ (since $N+1 \simeq N$ if $N \gg 1$).

Now, we can answer the question—are the fluctuations big or small? Consider an ideal gas in a tiny volume of $1\,\text{mm}^3$. Recalling that 1 mol (6×10^{23} particles) of an ideal gas occupies $22.4\,\text{L} = 2.2 \times 10^{10}\,\text{mm}^3$, at $0°C$ and $P = 1$ atm, we can see that there are about 10^{13} molecules in a single cubic millimeter. That makes the relative volume fluctuations $\hat{\kappa}$ less than one part in a million! Not too many instruments could measure that and, of course, the volume fluctuations in a liquid will be a tiny fraction of those in a gas.

The bottom line is that volume fluctuations in systems of interest are negligible.

7.5.2.2 Specific Heat

The specific heat is a quantity that's relatively easy to measure and, interestingly, can reveal the difference between conditions of constant pressure and constant volume—even in the large N limit. We will therefore examine the specific heat in a few ideal situations to get some intuition about when the external condition really matters.

Let's start by considering a case where the difference between constant-volume and constant-pressure conditions is rather large: the ideal monatomic gas (i.e., the same one we've studied so far). Above, we derived that the specific heat at constant volume is $c_v = (3/2)Nk_B T$. What is it at constant pressure? The first law makes the question easy to answer. We know $c_p = \Delta Q/\Delta T = (\Delta E + P\Delta V)/\Delta T$. But the ideal gas equation of state tells us that $\Delta V = Nk_B \Delta T/P$, and so we find that

$$\text{Ideal gas:} \quad c_p = \left(\frac{5}{2}\right)Nk_B = c_v + Nk_B = \left(\frac{5}{3}\right)c_v. \tag{7.30}$$

This is 67% larger than in the constant-volume case.

Intuitively, it makes sense that the constant pressure specific heat is higher than that at constant volume. After all, the specific heat is proportional to the variance in the energy, and the energy in a constant-pressure system also fluctuates with the volume. To put it a different way, each ideal gas particle has x, y, and z degrees of freedom for motion ("translation")—each of which possesses $k_B T/2$ of energy—yielding the original $(3/2)Nk_B$ of c_v. Allowing the volume to fluctuate evidently adds two additional effective degrees of freedom per particle. To raise the temperature therefore requires energy to be put into more modes of motion.

Despite this dramatic effect of volume fluctuations in the monatomic ideal gas, the ratio c_p/c_v is expected to rapidly decrease to nearly 1 for more complex molecules. To see this, we can next consider an ideal gas of simplified diatomic molecules—that is, with each molecule consisting of two atoms, A and B, connected by an ideal spring.

This problem is also (essentially) exactly solvable, since the partition function can again be factorized for independent diatoms (see Equation 5.23).

The basic dependencies of Z for the diatomic system are not hard to determine based on Equation 5.23. Integration over $\mathbf{r}^A$ yields a factor of volume for each of N diatoms. Our main concern is with the molecular partition function q^{mol}, and we will make a quicker-and-dirtier calculation than we did previously in Section 5.2.3. Here, we simply observe that the energy is harmonic for an ideal spring, and we know from Chapter 3 that the configuration integral is proportional to $\sqrt{T}$ for each of the three degrees of freedom internal to the molecule. Thus, we obtain a factor of $T^{3/2}$ for the configurational coordinates. (We have relaxed the stiff-spring assumption of Section 5.2.3, which essentially eliminated one degree of freedom.) The three velocity degrees of freedom internal to the diatom contribute another $T^{3/2}$. Hence, we find $q^{\mathrm{mol}} \propto T^{6/2}$, and thus overall $Z \sim (T^{9/2}V)^N/N!$, accounting for the overall velocity terms normally implicit in λ.

You may be thinking, "So what?" But the fact that our simple molecule's ideal-gas partition function varies with a higher power of T, compared to the monatomic gas in Equation 7.17, is crucial. Basically, more degrees of freedom increase the power of temperature in Z. As the next problem shows, this results in a less significant difference between the constant-pressure and constant-volume ensembles. (We should note that our classical-spring-connected diatom is not a great model compared to experimental data for light diatomic molecules, but the effective power of T does indeed increase.)

PROBLEM 7.7

(a) Show that if $Z(T, V) \propto (T^y)^N = T^{Ny}$, then the constant-volume specific heat is given by $c_v = yNk_B$ regardless of the constant of proportionality in Z. Note that the factorizable form—that is, a factor raised to the Nth power—is indicative of a noninteracting "ideal" system. (b) Show that if $Z(T, V) \propto V^N$, then the average volume in a constant-pressure system is always given by the same ideal gas form. (c) Based on the two results above, and the first law of thermodynamics, show that the difference between the two specific heat values in an ideal system is always given by $c_p - c_v = Nk_B$. (d) Show that as y (the power of T) increases, the ratio c_p/c_v approaches one.

We can also consider a primitive model of a larger molecule, a "Gaussian chain," in which n atoms interact only via the ideal springs that connect them. In this case, we expect $(n - 1)(3/2)$ additional powers of T in $Z(T, V)$ based on the springs, plus additional kinetic powers. That is, the exponent y of Problem 7.7 will increase substantially, roughly in proportion to n. The relative difference between c_v and c_p becomes still less consequential.

7.5.3 ENTHALPY AND THE THERMODYNAMIC DEFINITION OF G

We saw above that under constant-pressure conditions, it is necessary to include the energy associated with changing the volume, $P\Delta V$. While this is indeed an energy, it is not the usual "internal" energy, and hence it (or, rather, its average) is given a new name, the enthalpy, and the symbol H. That is,

$$H = \langle \text{KE} \rangle + \langle U \rangle + P \langle V \rangle. \tag{7.31}$$

You may see this equation written as $H = U + PV$, but you should know enough by now to recognize which quantities are really averages.

Since H plays exactly the same role as average energy did in the case of constant volume, it makes good sense to define the associated Gibbs free energy as

$$G(T, P) = H(T, P) - T S(T, P) = F(T, \langle V \rangle) + P \langle V \rangle, \tag{7.32}$$

where all of these quantities (except T) represent statistical/thermodynamic averages (in the constant pressure ensemble corresponding to Z_p).

Unfortunately, the relation (7.32) between $G(T, P)$ and $F(T, V) = - k_B T \ln Z(T, V)$ cannot be derived in a simple way from the partition function Z_p of Equation 7.24. The reason is that Equation 7.32 is only appropriate in the thermodynamic limit—if we take the fundamental definition to be $G = -k_B T \ln Z_p$. (For experts, note that a Laplace transform is not the same as a Legendre transform! You can find more information in Callen's book.) We can, however, perform a very crude derivation in the thermodynamic limit ($N \to \infty$). In that case, volume fluctuations are negligible, and so we can approximate the integrand in Z_p by replacing V with $\langle V \rangle$. (It is also necessary to attribute some "width" to the integrand, which must cancel out the unspecified proportionality constant of Equation 7.24.) Of course, once $\langle V \rangle$ is assumed to dominate the integral, it's fairly clear that one will find the relation (7.32) among the free energies.

PROBLEM 7.8

Understanding G in the ideal gas case. (a) Using the ideal gas partition function, obtain $G(T, P)$—aside from an unknown additive constant—by integrating Equation 7.24. (b) Compare your result to the usual thermodynamic construction for G, namely, $F(T, \langle V \rangle) + P \langle V \rangle$, and show that they become equal in the thermodynamic limit—that is, for large V and N.

7.5.4 ANOTHER DERIVATIVE CONNECTION—GETTING P FROM F

When we calculated $\langle V \rangle$ from G using the partition function Z_p, in Equation 7.27, this was a straightforward use of the same trick we used for the average energy. The differentiation "brought down" the quantity of interest from a Boltzmann factor and yielded the exact average. We can actually do a sort of reverse trick with differentiation to obtain P from F.

To calculate the pressure from $F(T, V)$, we must take the physical point of view that the volume affects the pressure. In the ideal gas case, we know this because we know the explicit dependence of pressure on volume, namely, $P = N k_B T / V$. In the non-ideal case, there will be some other equation of state, but we expect that the volume will typically affect the pressure one way or another. In other words, we expect $V = V(P)$ and $\partial V / \partial P \neq 0$. Therefore, in differentiating a free energy with respect to P, we will need to use the rule (7.5).

The trick, which is not obvious until you are very comfortable with thermodynamics, is to differentiate G, and not F. The advantage of this is that we already know the answer (i.e., $\partial G / \partial P = \langle V \rangle$). Here, we will use the notation $\langle V \rangle = \overline{V} = \overline{V}(P)$, which makes it more convenient to show the P dependence. We find

$$\overline{V} = \frac{\partial G}{\partial P} = \frac{\partial}{\partial P} \left[F\left(T, \overline{V}(P)\right) + P\,\overline{V}(P) \right]$$

$$= \frac{\partial F}{\partial \overline{V}} \frac{\partial \overline{V}}{\partial P} + \overline{V} + P \frac{\partial \overline{V}}{\partial P}. \tag{7.33}$$

(Note that the partial derivatives here refer to the fact that T and N are not changing. However, it is not possible to change P at constant T and N without changing V—just think of the ideal gas law.)

Continuing with the math, since $\overline{V}$ appears on both the left- and right-hand sides of Equation 7.33, we can cancel these terms and divide by $\partial \overline{V} / \partial P$ to find

$$P = P(\overline{V}) = -\frac{\partial F}{\partial \overline{V}}. \tag{7.34}$$

It is mathematically equivalent to say that $P(V) = -\partial F / \partial V$, where F is defined in the usual way from the constant-volume partition function. Don't forget that in the thermodynamic limit, in which the relation (7.32) holds, there is no distinction between V and $\overline{V} = \langle V \rangle$.

7.5.5 SUMMING UP: G VS. F

Surely, G is the correct quantity to consider under constant-pressure conditions—but does it really matter? The short answer is that the difference between G and F is not really going to matter for the types of systems we're interested in. Basically, all of molecular-level biology and experimental biophysics occurs in the condensed phase and with plenty of molecules in every system. In condensed phases, molecules are strongly attractive and essentially fix their own volume. In such conditions, in other words, you can consider the volume constant.

From a theoretical point of view, the ability of volume to fluctuate, under constant pressure conditions, is quite interesting. It leads to predictions for the specific heat, which have indeed been measured in gases. It's particularly interesting that even in the large N limit when volume fluctuations are undetectable, the difference in specific heats (between conditions of constant pressure and volume) is readily measured. The key physical point is that the fluctuating volume acts like (two) additional degrees of freedom, and hence there are additional modes to take up energy. Nevertheless, once the system of interest gets much more complex than a monatomic ideal gas—like, say, a protein!—even the specific heat difference becomes negligible.

7.5.6 CHEMICAL POTENTIAL AND FLUCTUATING PARTICLE NUMBERS

Although it seems abstract at first, the chemical potential is very useful for understanding many aspects of chemistry. It also provides a useful way to understand biochemical binding equilibria.

In essence, the chemical potential is the free energy appropriate for chemistry. It is the free energy per molecule, not a whole-system quantity, and when the chemical potential is defined for different "conditions," it provides a natural means for determining equilibrium. For instance, if we define one chemical potential for unbound ligands of a certain type in a solution, and another for the same ligand bound to protein receptors, then setting the two potentials to be equal is a natural means of calculating the binding equilibrium. Molecules will move from high chemical potential to low until equilibrium is obtained and the chemical potential is constant throughout.

We'll start with the classic description of the chemical potential, which uses a direct analogy with constant-pressure and constant-temperature "baths" or reservoirs. Thus, we assume our system of interest is in contact with a very large particle reservoir, which can be characterized by a constant chemical potential, μ, which can be positive or negative. This is defined to mean that the energy cost of adding one molecule from our system to the bath is μ. (Said conversely, adding a bath particle to the system "costs" $-\mu$.) This cost describes the energy change of the bath only, and therefore the energy change to the combined system (system + bath) must also include the usual energy change to the system itself. Thus, the effective energy giving the probability in a Boltzmann factor for a system of N molecules is $KE + U - \mu N$, assuming constant volume and temperature.

Once the chemical potential has been defined, it is straightforward to write down the appropriate partition function, Z_μ. (In physics terminology, this is called the "grand canonical" partition function, often denoted by Ξ.) The partition function must not only consider all possible coordinates, but also different possible numbers of particles in the volume, and we therefore have

$$e^{-\Phi(T,V,\mu)/k_B T} \equiv Z_\mu(T,V,\mu) = \sum_{N=0}^{\infty} Z(T,V,N)e^{+\mu N/k_B T}, \qquad (7.35)$$

where the "+" sign in front of μN results simply from the definition of μ. The partition function $Z(T,V,N)$ is the usual constant-volume Z but with N emphasized as an argument. Note that our Φ notation is not standard, but it's not clear that there is a standard notation.

By now, you should know to expect some kind of derivative relation to pop out. Here it has to do with the average particle number $\langle N \rangle$. (N, of course, is no longer a constant.) In fact, from our past experience, it will be trivial for you to derive such a relation.

PROBLEM 7.9

Using the partition function (7.35), derive an expression for $\langle N \rangle$ based on a derivative of Φ.

7.5.6.1 Obtaining μ from *F* or *G*

To obtain μ from F or G, we can follow the example of Section 7.5.4. The details are left as a problem, but the result is

$$\mu = \frac{\partial F}{\partial N} = \frac{\partial G}{\partial N}. \tag{7.36}$$

Perhaps the key point here is that any free energy contains all the thermodynamic information. Any free energy can be derived from any other in the thermodynamic limit. The problems below demonstrate these ideas.

PROBLEM 7.10

Show that if the partition function (7.35) is dominated by a single term in the sum, then it is reasonable to write $\Phi(T, V, \mu) = F(T, V, \overline{N}) - \mu\overline{N}$. Why might this be approximately true in the thermodynamic limit?

PROBLEM 7.11

Derive the first equality of Equation 7.36 by differentiating both sides of $\Phi(T, V, \mu) = F(T, V, \overline{N}(\mu)) - \mu\overline{N}(\mu)$ with respect to μ. You are working in the thermodynamic limit, where we have assumed $N = \langle N \rangle = \overline{N}(\mu)$. You will also need the result of Problem 7.9, namely, $\langle N \rangle = -\partial\Phi/\partial\mu = \overline{N}(\mu)$.

PROBLEM 7.12

The following is a little tricky, since both P and N depend on μ, so you will ultimately require two groupings of partial derivatives. Derive the second equality of Equation 7.36 by differentiating $\Phi(T, V, \mu) = G(T, \overline{P}(\mu), \overline{N}(\mu)) - \overline{P}(\mu) V - \mu\overline{N}(\mu)$. You will need to make use of Equation 7.33, and be sure to consider V a constant.

More free energy relations can be derived using similar ideas. The interested reader is directed to the book by McQuarrie and Simon.

7.5.6.2 Chemical Equilibrium

The description of chemical equilibrium with chemical potentials is both very useful in practical terms, and also conceptually illuminating. We will only sketch the ideas and results here. A very good discussion of the details can be found in the book by McQuarrie and Simon.

We'll follow up on the idea that the chemical potential is the free energy per particle. From Equation 7.35, it's clear that μ is certainly some kind of energy per particle, and in the thermodynamic limit, this turns out to be the free energy.

We'll now imagine a scenario slightly different from that above. Instead of having our "system of interest" embedded in a much larger reservoir of identical molecules, imagine that each molecule can exist in one of two chemical states, such as bound or free for a ligand molecule, which can bind a protein receptor. In this case, the total number of ligands, N_L, will be partitioned among bound and free states so that $N_L^{free} + N_L^{bound} = N_L$. Hence, we can say that the free ligand molecules act as a reservoir for the bound molecules. Since the binding will be assumed reversible, the reverse is also true: the bound molecules act as a reservoir for the free. Transferring a molecule from the free to the bound state then involves the cost μ^{free}, which will be offset by the gain $-\mu^{bound}$. These chemical potentials are not constants but depend on the concentrations of all the molecules in the solution (ligand, receptor, solvent), perhaps in a complicated way. However, equilibrium will be achieved when the two chemical potentials are equalized by the transfer of an appropriate amount from one state to the other.

This equalization of chemical potentials is consistent with basic principles in two ways. First, if it is indeed true that μ is the free energy per particle, then equalizing chemical potentials is the same as equalizing probabilities. Second, we can consider the number of bound molecules, N_L^{bound}, to be an internal parameter that self-adjusts to minimize the free energy. (Free energy minimization will be discussed in detail below.) This is equivalent to equating of chemical potentials if we note that for any species i, $\mu_i = \partial G/\partial N_i$, in analogy with Equation 7.36.

PROBLEM 7.13

Using the species-specific chemical potential just defined (μ_i), show that if $G = G(T, P, N_a, N_b)$ and $N_a + N_b = N$, then minimization of G implies that $\mu_a = \mu_b$.

7.6   OVERVIEW OF FREE ENERGIES AND DERIVATIVES

7.6.1   THE PERTINENT FREE ENERGY DEPENDS ON THE CONDITIONS

First of all, the "conditions" that determine the appropriate free energy refer to what is held constant—like temperature, volume, pressure, or chemical potential. Once the conditions are set, the pertinent free energy is all of the following: (1) It is the logarithm of that partition function obeying the conditions, multiplied by $(-k_B T)$. (2) It quantifies the amount of work that can be done by the system under the conditions. (3) It tends to become minimized if the system is initially out of equilibrium, as we will see below.

The notation for free energies uses the conditions (the parameters held constant) as "natural" variables. Therefore, each free energy should always be written in a way that make the corresponding conditions obvious.

The free energies we have discussed are

- $F(T, V, N) = \langle E \rangle - TS$, the Helmholtz free energy
- $G(T, P, N) = F + P \langle V \rangle$, the Gibbs free energy
- $\Phi(T, V, \mu) = F - \langle N \rangle \mu$, the grand canonical potential

Note that the total energy is the sum of kinetic and potential contributions—that is, $E = KE + U$.

7.6.2 FREE ENERGIES ARE "STATE FUNCTIONS"

As we have seen more than enough times, a free energy is defined from a partition function. In turn, the partition function simply integrates to some number once the natural variables—for example, T, V, N in $Z(T, V, N)$—are fixed. Thus, the natural variables define the free energy: once you specify the variables, the free energy is some number.

A free energy is called a "state function" because it comes from a definite integral depending only on the equilibrium "state" of the system—that is, only on the natural variables. It does not depend on how that state was reached (maybe by a weird nonequilibrium process) because this doesn't change the equilibrium integral.

The notion of a state function is important in understanding free energy differences generally, and especially in understanding "thermodynamic cycles," which we shall study in the context of binding in Chapter 9.

7.6.3 FIRST DERIVATIVES OF FREE ENERGIES YIELD AVERAGES

One of the most important ideas of this chapter is that derivatives of free energies yield statistical averages. This is a mathematical consequence of the partition-function definitions of the free energies. However, the derivative relations typically make some intuitive sense—for example, the way a free energy changes with a volume change reflects the pressure.

The relations we discussed are these:

Constant T, V, N	Constant T, P, N
$\langle E \rangle = -T^2\, \partial(F/T)/\partial T$	$H = \langle E \rangle + P\langle V \rangle = -T^2\partial G/\partial T$
$S = -\partial F/\partial T$	$S = -\partial G/\partial T$
$P = -\partial F/\partial V$	$\langle V \rangle = \partial G/\partial P$
$\mu = \partial F/\partial N$	$\mu = \partial G/\partial N$
Constant T, V, μ: $\qquad \langle N \rangle = -\partial \Phi/\partial \mu$	

Recall that several of these are only valid in the thermodynamic limit. The partial derivatives mean that the other natural variables (other than the one being varied with the "∂" symbol) are held constant.

7.6.4 SECOND DERIVATIVES YIELD FLUCTUATIONS/SUSCEPTIBILITIES

Another consequence of the fundamental connection between thermodynamics and statistical mechanics is that second derivatives yield variances σ^2, in the traditional statistical sense. The relation of most interest to us arises in the specific heat

- $c_v = T\partial S/\partial T = \sigma_E^2/(k_B T^2)$

You could derive a similar relation for the constant pressure specific heat, and analogous relations for the fluctuations (variances) in the volume and number of particles. Of special physical significance, volume fluctuations are proportional to compressibility, which quantifies the way volume changes as pressure does. This latter result is easy to derive.

7.7 THE SECOND LAW AND (SOMETIMES) FREE ENERGY MINIMIZATION

The second law of thermodynamics is tremendously important conceptually, and also critical in terms of the way it places limitations on real-world machines—both big and small. At one level it's obvious, and tells you that heat flows from hot things to cold. On another, it's about entropy, and then it gets more subtle. When it comes to free energy, the second law is largely misunderstood, and so we'll be careful to say precisely what it means and what it does not. In particular, the second law does not apply to PMFs (potentials of mean force), although it's frequently claimed to do so (implicitly). For instance, the common statement, "A protein occupies the state of lowest free energy," is not fully correct, as we'll see.

7.7.1 A KINETIC VIEW IS HELPFUL

The second law often seems mysterious and abstract. However, a big part of it describes something we are already familiar with: relaxation back to equilibrium from a nonequilibrium state. This is the "self-healing" property of equilibrium and probability flows we discussed in Chapter 4. The same ideas have a thermodynamic expression, which reveals additional information about work and energy in large systems.

7.7.2 SPONTANEOUS HEAT FLOW AND ENTROPY

We'll start our discussion with the fundamental stuff.

You can learn a lot by reconsidering something you already know: the way heat will flow. We can consider a "thought experiment," adapted from the book by McQuarrie and Simon. If you consider two objects (or substances, as in Figure 7.4) in contact, initially at two different temperatures, it's obvious which way heat will flow. The hotter substance will lose heat to the colder one until their temperatures are equal.

The system illustrated in Figure 7.4 helps to turn this into more precise thermodynamics. The whole (compound) system is thermally isolated from the environment, so the net heat flow is zero ($Q_{tot} = 0$). Further the whole container, including the divider, is rigid so no work is done. If we assume that, initially, $T_1 < T_2$, then heat Q will flow to the colder system (increasing the energy $\langle E \rangle_1$) from the hotter one (decreasing $\langle E \rangle_2$). Of course, energy conservation—a.k.a., the first law—and the thermal isolation of the full system imply that the heat leaving one system enters the other $Q_{21} = \Delta \langle E \rangle_1 = -\Delta \langle E \rangle_2 = -Q_{12}$, where Q_{ij} is the heat flow from i to j.

It turns out that the information already given implies that the total entropy of the system, $S = S_1 + S_2$, has to increase. This can be shown mathematically, even for a large difference in temperatures, only by assuming $c_v > 0$. However, the calculation is complicated. (You can do it by writing the heat transferred in terms of the entropy, integrating by parts, and rearranging the terms to enable comparison of the total

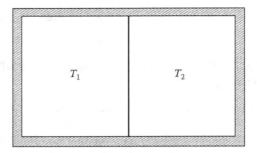

FIGURE 7.4 A thermally isolated container split into two parts by a rigid, heat-conducting divider. This setup can be used to demonstrate that entropy always increases in a spontaneous process for a thermally isolated system.

entropy before and after.) Of course, in our favorite system, the ideal gas, it's not hard to show that the entropy will increase during such an equilibration of temperatures. You can see this in the following problem.

PROBLEM 7.14

In Figure 7.4, assume that systems 1 and 2, initially at temperatures T_1 and T_2, are identical ideal monatomic gases: same atom type, $N_1 = N_2$, and $V_1 = V_2$. You can show that the entropy will increase when they are thermally connected in the following steps. (a) Sketch the function $\ln(x)$ vs. x to see that it is concave down—that is, the slope decreases as x increases. (b) Because of this concavity, show on your drawing why it is always true that $2\ln[(x_1 + x_2)/2] > \ln x_1 + \ln x_2$—that is, that the log of the average is greater than the average of the logs. (c) Because the specific heat is independent of temperature in an ideal gas, show that the final temperature of two identical ideal gases will be the average of the initial temperatures. (d) Using our earlier result for the entropy of the ideal gas, show that the entropy will increase when the gases are allowed to equilibrate.

7.7.2.1 The General Infinitesimal Case

For two systems brought into thermal contact as in Figure 7.4, we can always consider the case when the initial temperatures differ only infinitesimally—that is, when $T_2 - T_1 = dT > 0$ and $dT/T_1 \ll 1$. In this case, the incremental entropy changes for each system are readily computed from the incremental heat flows, dQ_{ij}, from the relation $c_v = dQ/dT = T\partial S/\partial T$, which implies

$$dQ_i = T_i\, dS_i. \tag{7.37}$$

We have used the fact that the temperature is essentially constant for this infinitesimal process. Because the first law, given the lack of work done, implies that $dQ_i = d\langle E\rangle_i$, we have that $d\langle E\rangle_i = T_i\, dS_i$. This implies the total entropy change of our infinitesimal process is

$$dS_{tot} = dS_1 + dS_2 = \frac{d\langle E \rangle_1}{T_1} + \frac{d\langle E \rangle_2}{T_2}. \qquad (7.38)$$

The fact that the energy changes are equal in magnitude but opposite in sign with $d\langle E \rangle_1 > 0$, means that the entropy change is positive

$$dS_{tot} = d\langle E \rangle_1 \left(\frac{1}{T_1} - \frac{1}{T_2} \right) > 0. \qquad (7.39)$$

Note that this demonstration was for two arbitrary—not necessarily ideal—systems as in Figure 7.4, assuming only that the change was infinitesimal.

The fact is, as we noted earlier, that the entropy will increase ($\Delta S > 0$) for any spontaneous process, whether infinitesimal or not, in a thermally isolated system. This is the second law of thermodynamics. But you already need to be careful. It's really the entropy of the universe that increases in general, whereas specific systems will only have this property if they are specially prepared to be thermally isolated—as in our demonstration above. The more pertinent consequences of the second law concern free energies—where even greater caution is required! That's coming soon.

Is it fair to say that because entropy is related to the widths of allowed states, and hence to probability (roughly, prob $\sim e^{S/k_B}$), that it's obvious that entropy should increase? Yes and no, although more "yes." First, on the "no" side, recall that spontaneous processes begin from nonequilibrium states, and it's not obvious how to define the entropy of a nonequilibrium state from the discussions in this book. On the other hand, if we think of such states as "constrained" to some nonequilibrium condition (e.g., a temperature difference as in Figure 7.4), then we can define a quasi-equilibrium system. Arguably, Problem 7.14 takes just this approach. Furthermore, it is obvious that a system will proceed to a state of higher probability from a less probable initial state. The classic example is a gas initially confined to half of a box. Once it is allowed to expand it will never spontaneously go back. After all, the probability to find all N gas molecules in one half of the container is $(1/2)^N$, which is essentially zero for real-world numbers of order 1 mol.

7.7.3 THE SECOND LAW FOR FREE ENERGIES—MINIMIZATION, SOMETIMES

Many of us are more familiar with the second law in the context of free energy. Here, we shall consider the constant-volume-and-temperature Helmholtz free energy $F(T, V)$ and show that it always decreases in spontaneous processes.

We will analyze the situation schematized in Figure 7.5, in which a system and its constant-temperature bath are contained within a thermally isolated container. Our starting point will be the necessity of entropy increase in the isolated "super-system" including the bath. We will determine the consequence for the system itself, shown at the center of Figure 7.5.

The total (positive) entropy change consists of the sum of the entropy changes to the bath and system. For the bath, the entropy change ΔS_{bath} results entirely from the transfer of heat Q at constant temperature T. Because Q is defined as the heat transferred to the system (although Q can be positive or negative), the entropy change to the bath is $-Q/T$. This simple form occurs only because of the constant temperature

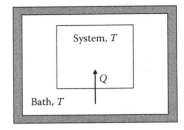

FIGURE 7.5 Another thermally isolated container, which contains a very large "bath" that maintains a system at constant temperature T. This setup is used to show that the free energy always decreases in a spontaneous process. Such a process (not shown) occurs inside the system only, not the bath. For example, a rigid partition within the system could be removed, which initially separates high- and low-density regions of a gas. The two parts of the gas would then mix and equilibrate spontaneously.

that results from the bath being much larger than the system. In the system, however, some unknown spontaneous process is occurring—which is what requires the input of heat in the first place. An example could be the mixing of two gases. The entropy change of this process is not known, but we call it ΔS_{sys}. Putting all this together, we have

$$0 < \Delta S_{tot} = \Delta S_{bath} + \Delta S_{sys} = \frac{-Q}{T} + \Delta S_{sys}. \tag{7.40}$$

However, we can carry this further. From the first law, we know that the heat transferred from the bath will be divided up into energy and work: $Q = \Delta \langle E \rangle_{sys} + W$, where the energy change and work are for the system, not bath. If we substitute this straightforward application of the first law into Equation 7.40 and multiply by T, we arrive at the result

$$\Delta \langle E \rangle_{sys} - T\Delta S_{sys} = \Delta F_{sys} < -W. \tag{7.41}$$

That is, if work is done, we see that it is bounded by the (negative) free energy change to the system: $W \leq -\Delta F_{sys}$.

If no work is done by the system ($W = 0$), we see that the free energy change is always negative.

$$\text{No work done:} \quad \Delta \langle E \rangle_{sys} - T\Delta S_{sys} = \Delta F_{sys} < 0. \tag{7.42}$$

This is the famous decrease of free energy for a spontaneous process at constant temperature. Again, all quantities here refer to the system only, excluding the bath.

The tendency of the free energy to decrease is a result that you will hear quoted (and misquoted) all the time. The key point is that the decrease in free energy refers to a change from nonequilibrium to equilibrium. This is what a spontaneous process is (in the present context).

There is another important restriction on the principle of free energy minimization. This has to do with the "natural variables"—that is, those parameters assumed constant, like T and V for F. Thus, F will decrease at constant T and V, but it is not correct to say that V will self-adjust so that F decreases. Your intuition should be that F is the right free energy to enable the comparison of different situations at constant T and V. This is the condition built into the partition function, in its fundamental probability role as a normalizing constant. The problem below will illustrate why minimizing F with respect to V doesn't make sense in condensed systems.

PROBLEM 7.15

(a) In the ideal gas, was the origin of the free energy term $Nk_B T \ln (V/V_0)$ from kinetic or potential energy? Answer by sketching the evaluation of the partition function. (b) Now consider a nonideal system, which will have free energy terms from both kinetic and potential energy. Assuming that the potential-energy-based terms in F do not become infinite as a function of V, what value of V leads to the minimum value of F? (c) Does such a minimization of F while varying V make sense for nonideal systems? Without calculation, think of a condensed-phase example to illustrate your point.

It is also extremely important to realize when the free energy does not become minimized, as we now discussed.

7.7.4 PMFs AND FREE ENERGY MINIMIZATION FOR PROTEINS—BE WARNED!

Based on my experience, many statements in the biophysics literature on the minimization of the free energy are not fully correct, and they are perhaps even somewhat misleading. The classic case involves protein structure, for which you will often read (or hear) that a protein's structure is the state of lowest free energy. Surely, if you have to pick just one state, it should be the one of lowest free energy. But what if there are three possible configurational states, with 45%, 30%, and 25% probability? Is it then reasonable to say that the protein even has a single structure?

This is not just idle speculation. For instance, a growing number of scientists are investigating what are called "intrinsically disordered" regions of proteins, which turn out to be extremely common. Furthermore, echoing the view presented in Chapter 1, if proteins are indeed machines, how can they work without moving, without changing configuration? Probably they cannot. Rather, it is reasonable to argue that most proteins have important conformational flexibility, and this view is almost a consensus among experimental and theoretical biophysicists.

Let's bring this discussion back to quantitative issues. First, recall that the tendency of the free energy to decrease (in accord with the second law of thermodynamics) refers to spontaneous changes from nonequilibrium to equilibrium states. However, many discussions of protein structure implicitly assume equilibrium based on the average conditions. Although equilibrium may not prevail for long times, it is indeed the most concrete basis for discussion. (Recall our "philosophical" discussion of equilibrium in Chapter 4.)

If a system is in equilibrium, what does it even mean to say the free energy is minimized? The intended meaning, even if it is not always explicit, is that the system

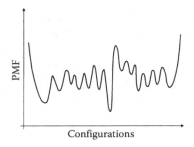

FIGURE 7.6 The complex landscape, revisited. We have seen this picture before, but now we make explicit reference to the protein landscape as a PMF with multiple minima. Each state occurs with probability determined by the Boltzmann factor of the PMF. In other words, multiple states "coexist" in an ensemble. The PMF is not minimized.

will occupy the state of lowest free energy as some (or all) system coordinates are varied. As statistical thermodynamicists, we know that an equilibrium free energy corresponds to integration over all possible system coordinates. On the other hand, a PMF represents an effective free energy for a set of coordinates that is not integrated over (see Figure 7.6). The Boltzmann factor of the PMF, as we saw, is defined to give the probability as a function of the chosen set of "important" coordinates.

But it is not correct to say that a PMF is minimized! Yes, it is true that the minimum value of the PMF represents the maximum single probability (of the chosen coordinate). However, the more fundamental fact about a PMF is that it is a description of the distribution of possible values of the important coordinates. The general expectation is that all possible values of the coordinates will occur. In fact, the most probable values of the coordinates do not necessarily reside in the *state* of maximum probability.

So to summarize... In common usage, the "free energy" often refers to a PMF. This is likely to be the case if the entire system is supposed to be in equilibrium. The minimum of the PMF represents only the single most probable value of the perhaps arbitrarily chosen "important coordinates." However, other values of the coordinates and other states may be important for the biological function of the protein.

In other words, be careful when you are tempted to say, "The free energy should be minimized."

7.7.4.1 Why Isn't Everything Minimized in the Thermodynamic Limit?

Earlier, we saw that the fluctuations in the volume of an ideal gas are tiny. In the thermodynamic limit, where $N \rightarrow \infty$, Equation 7.29 showed the fractional fluctuations are only of order $1/\sqrt{N}$. In fact, for many quantities, like the energy and almost anything expected to be proportional to system size (the so-called extensive quantities), the variability will be similarly small.

In sharp contrast, the number of important states available to proteins (those states of reasonable probability, say) does not tend to infinity, and certainly it has nothing to do with the number of proteins being studied! Thus, there is no reason to expect negligible fluctuations in the probabilities of protein states.

7.7.5   THE SECOND LAW FOR YOUR HOUSE: REFRIGERATORS ARE HEATERS

So here's a question. You're a poor graduate student and you can barely afford your rent. You thought you could survive summer without an air conditioner... but you do have an extra refrigerator. Should you leave the door open on one of your refrigerators in an attempt to cool the apartment?

The second law of thermodynamics, tells you that what refrigerates your food actually heats your house. The reason is that a refrigerator should be seen as a "heat pump." It takes heat from its interior and expels it to the exterior, that is, your kitchen. Now, the second law of thermodynamics tells us this can never happen spontaneously, since heat won't flow from cold to hot. Of course, your refrigerator requires electricity and performs work—that's why it makes noise.

What happens to the work coming from the electrical energy? This work is being put into our system, and it has to go somewhere. Nothing is lifted, and the refrigerator essentially stays in the same state all the time (in a time-averaged way). No mechanical or chemical work is done. Thus, the only choice is that the electrical work is converted to heat. And since refrigerator engineers are smart enough not to expel this heat into the inside of the refrigerator, which needs to be cold, the heat goes into your kitchen.

Now do the math. Some amount of heat, Q_0, is moved from inside the refrigerator to outside. Some amount of electrical energy is put into the process and must be released as an additional amount of heat Q_1. Basically, the second law ensures $Q_1 > 0$.

If you want to be smart about that second refrigerator, you can use it to heat your apartment in the winter! Of course, this probably won't be the most efficient way to heat a house, but that's a matter for a different book.

PROBLEM 7.16

Explain how a refrigerator can cool one selected room of an apartment, or perhaps a whole apartment (at some risk to passersby). Hint: This involves particular placements of the refrigerator, noting that heat is typically given off in the rear.

7.7.6   SUMMING UP: THE SECOND LAW

The second law of thermodynamics is confusing enough to warrant its own little summary. First of all, unlike the first law, it is truly thermodynamic in that it requires systems to be extremely large (e.g., have very many molecules). We will revisit this issue in Chapter 11, in discussing the Jarzynski relation.

One basic way of saying the second law involves entropy and free energy. Entropy will always increase in a spontaneous process in a thermally isolated, thermodynamically large system. For systems that are not isolated, it is the entropy of the whole universe (system + environment) that will increase in any given process. (Note that if the system is small, entropy can temporarily increase: for instance, in an ideal gas of just two particles, both will occupy the same half of the volume a quarter of the time. If we average over many processes in a small system, however, the entropy will always increase.)

For non-isolated systems, like those at constant temperature and/or pressure, the second law is stated in terms of our old friend, the free energy. The law for free energies, where S enters with a negative sign, is that the free energy will decrease in a spontaneous process, which takes a nonequilibrium state to equilibrium. Which free energy decreases depends on which variables are held constant.

Importantly, the PMF, which is like a free energy, does not necessarily tend toward its minimum. A protein may or may not predominantly occupy the configurational state of lowest free energy.

The second basic way of stating the second law involves heat flow. Heat must flow from hot to cold. Heat will not flow spontaneously from cold to hot (as in a refrigerator) without the addition of some work. These statements are also only (always) true for large systems. In a small system, you could imagine a strange set of collisions causing kinetic energy to flow to a hot region from a cold one—though on average, this will not happen.

7.8 CALORIMETRY: A KEY THERMODYNAMIC TECHNIQUE

Calorimetry describes a method that measures both the heat energy absorbed (or given off) by a system, as well as the temperature change. Sometimes, the heat will be added to the system to see what temperature change results, and sometimes the heat resulting from some spontaneous process will be measured. Different types of instruments are used for the different types of experiments, but either way, the specific heat is measured. Because most experiments are done at constant pressure, it is c_p that typically is measured (and not c_v).

For biophysics, there are two very important calorimetric strategies. One is a "differential" approach often used in studying protein folding/unfolding, and this is described below. The second, isothermal titration calorimetry, is commonly used to measure binding affinities—and this will be described in Chapter 9, on binding.

7.8.1 INTEGRATING THE SPECIFIC HEAT YIELDS BOTH ENTHALPY AND ENTROPY

From a microscopic point of view, the most interesting thing about the specific heat is that it quantifies energy fluctuations. However, from a thermodynamic perspective, the specific heat is even more important, because both enthalpy ($H = E + PV$) and entropy can be determined from it.

Recall that $c_p = dQ_p/dT$, where Q_p is the heat added in a constant-pressure process. The first law tells us that $dQ_p = dU + P dV$. But we also know that this is exactly the change in the enthalpy. (Note that there is no $V dP$ terms since P is constant.) In other words, the specific heat also gives the change in enthalpy

$$c_p = \frac{\partial H}{\partial T}, \tag{7.43}$$

which can be integrated to yield

$$\Delta H = H(T_2) - H(T_1) = \int_{T_1}^{T_2} dT \, c_p(T). \tag{7.44}$$

Since a calorimetry experiment will yield the heat added, this is exactly the enthalpy change. (No integration of experimental data is really required for the enthalpy since it is Q, which is measured directly.)

PROBLEM 7.17

Derive the relation analogous to (7.43) for a constant-volume system.

A similar integral relation for the entropy change can be derived. Since $c_p = T \partial S / \partial T$, we have

$$\Delta S = S(T_2) - S(T_1) = \int_{T_1}^{T_2} dT \, \frac{c_p(T)}{T}. \tag{7.45}$$

From the raw data measured in a calorimetry experiment, namely, Q_p values, it would be necessary to numerically approximate the c_p value (via numerical differentiation) and then perform numerical integration.

7.8.2 DIFFERENTIAL SCANNING CALORIMETRY FOR PROTEIN FOLDING

Differential scanning calorimetry (DSC) is a type of calorimetry that is specifically designed to determine the calorimetric properties of a solute—without the "background" effect of the solvent. In biophysics, the most important DSC studies have examined proteins in solution, but with solvent effects subtracted off.

The setup, not surprisingly, involves two separate samples. One contains solvent ("buffer") only, and the other contains the solute of interest in the identical solvent. The temperature is maintained to be uniform throughout the system at every time point, but over time, the temperature is gradually "scanned" over a range of values. A DSC machine has a clever feedback mechanism for maintaining temperature uniformity in the whole system, which includes separate heaters for each sample. The apparatus measures the difference in heat supplied to each of the two samples, and this reflects the Q_p for the solute alone.

7.9 THE BARE-BONES ESSENTIALS OF THERMODYNAMICS

1. Derivatives of thermodynamic functions often correspond to statistical averages.
2. Second derivatives can correspond to variances.
3. Thermodynamics applies in the thermodynamic limit ($N \to \infty$), where many fluctuations are tiny—of order $1/\sqrt{N}$.
4. Free energy means available energy, and some may come from the bath.
5. Different free energies are appropriate for different conditions, and the conditions define the free energy—making it a state function.
6. Your intuition from F generally will carry over to G.

7. The fact that G can be obtained from F and *vice versa* means any thermodynamic function has all the thermodynamic information about a system.
8. The first law of thermodynamics is just energy conservation and the second law requires entropy increase in the whole universe (system + bath).
9. The second law also tells us that free energies decrease in nonequilibrium spontaneous processes, but PMFs do not as they represent distributions in an equilibrium system.
10. Because the specific heat can be measured in calorimetry, thermodynamic functions can also be measured.

7.10 KEY TOPICS OMITTED FROM THIS CHAPTER

The interested student is direct to a more complete statistical mechanics book to learn about the following.

Extensive and intensive properties. Many thermodynamic functions are expected to be proportional to the system size. These "extensive" quantities include the free energies themselves, the (average) energy, and the entropy. If a function is extensive, you can always write it as N times the average per-molecule value, and this has important consequences in many thermodynamic calculations. Other properties should be independent of system size—such as temperature, pressure, and chemical potential—and these are called "intensive."

Third law of thermodynamics. This law states that the entropy must be equal to zero at $T = 0$, providing an absolute reference point. The law makes intuitive sense, because from a statistical perspective, the Boltzmann factor suggests a single configuration (or a small subset) will dominate as $T \to 0$. Of course, quantum mechanics takes over at low temperature, with the implication that a system has only a single state at $T = 0$.

Phase transitions. Phase transitions are physically and mathematically complex phenomena beyond the scope of this book. Even though true phase transitions may not occur in biological systems, many of the same ideas are useful in discussions of protein folding and lipid bilayers.

FURTHER READING

Callen, H.B., *Thermodynamics and an Introduction to Thermostatistics*, Wiley, New York, 1985.
McQuarrie, D.A. and Simon, J.D., *Physical Chemistry*, University Science Books, Sausalito, CA, 1997.
Phillips, R., Kondev, J., and Theriot, J., *Physical Biology of the Cell*, Garland Science, New York, 2009.
Reif, F., *Fundamentals of Statistical and Thermal Physics*, McGraw-Hill, New York, 1965.

8 The Most Important Molecule: Electro-Statistics of Water

You already know that water is essential to life, and here we will explain some of the molecular-level behavior underlying that role.

Water is not just essential, it's ubiquitous. To a good approximation, biochemistry is chemistry in an aqueous environment. Water plays a key "structural" role in both the general folding of proteins and in specific details of protein structures themselves. Besides its role in driving protein folding and stability, the hydrophobic effect also provides the cells with stable membranes to compartmentalize various functions. Water's remarkable properties arise from its electrostatic character in a fluctuating thermal environment—from "electro-statistics."

The main biophysical topics we will study are: (1) a microscopic/statistical explanation for the hydrophobic effect; (2) the origins of dielectric phenomena in pure water; and (3) screening that occurs as a result of the presence of ions in water. All these phenomena are electro-statistical.

8.1 BASICS OF WATER STRUCTURE

8.1.1 WATER IS TETRAHEDRAL BECAUSE OF ITS ELECTRON ORBITALS

The most important thing to know about the water molecule is that it is essentially tetrahedral. You should not think of it as oxygen bonded to two hydrogens, but rather as an oxygen molecule with four tetrahedral orbitals (electron clouds). Two orbitals are bonded to hydrogens and overall, including the hydrogens, are fairly positive in charge. The other two orbitals do not participate in covalent bonds, so these "lone pairs" remain negative (see Figure 8.1).

8.1.2 HYDROGEN BONDS

Hydrogen bonds ("H-bonds") are critical to the structure and biological behavior of water. A hydrogen bond forms between a positively charged hydrogen atom and a negatively charged atom—generally N, O, or F—that are not covalently bonded. The hydrogen atom is called the "donor" and the other molecule's negative atom is called the "acceptor." The donor and acceptor may be part of the same molecule or different molecules. Because H-bonds form between molecular groups that are charge-neutral overall, any charged atom on an H-bonding molecule necessarily is compensated by opposite charge somewhere else on the molecule: hence a dipole is formed. In water,

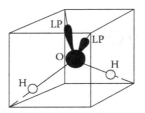

FIGURE 8.1 A water molecule is best described as containing five elements: the central oxygen nucleus, two positively charged hydrogen atoms, and two negatively charged "lone pairs" (unpaired electrons, denoted LP). The lone pairs and hydrogens are tetrahedrally oriented around the oxygen. The figure shows that tetrahedral symmetry is readily visualized using nonadjacent corners of a cube. (Adapted from Ben-Naim, A.Y., *Statistical Thermodynamics for Chemists and Biochemists*, Springer, Berlin, Germany, 1992.)

for instance, the partial charges on the hydrogens are compensated by the negative charges on oxygen's lone pairs.

Hydrogen bonds are directional, because the dipoles on each molecule participate. That is, the strength of a hydrogen bond depends strongly on the relative orientations of the participating dipoles. There are two reasons for this. First, the dipoles tend to align along a single axis (see ice structure in Figure 8.2) to maximize electrostatic interactions. But further, because the electron density of one dipole becomes further polarized by the presence of the other molecule's dipole, the hydrogen bond is strengthened and additionally straightened. This latter quantum mechanical effect explains why H-bonds are considered to have a "quasi-covalent" character. Based on directionality, it has been estimated that H-bonds can vary in effect from 1 to 10 kcal/mol.

It is not possible, however, to assign a fixed (free) energy value to a hydrogen bond, and not only because of the directional sensitivity just described. The free energy—that is, net stability—of a particular H-bond can be greatly decreased if other polar groups are present that can "compete" for H-bonding partners. This is indeed the case in a thermally fluctuating fluid environment like water.

8.1.3 ICE

As shown in Figure 8.2, water will form a fairly open structure when its energy becomes minimized at low temperatures—that is, when it freezes. The energy is optimized simply by ensuring that all hydrogen bonds are satisfied (four per molecule) so the structure of ice is dictated by the tetrahedral geometry of the water molecules. The open structure is less densely packed than can be achieved by a more random placement of neighbors around a given water. In other words, ice is less dense than liquid water. When pressure is applied to the ordered ice structure, it can only become more dense by becoming less ordered (i.e., more liquid-like). This contrasts sharply with most substances, in which the solid phase is the most dense.

There are some other interesting facts related to ice. First, there are several crystal forms of ice, with different types of order, as described in van Holde's book. Second, on a biological note, some fish that live in low-temperature conditions synthesize

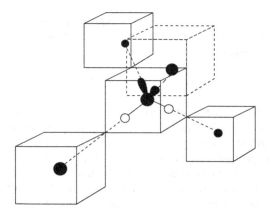

FIGURE 8.2 The open, tetrahedral structure of ice as exemplified by a single tetrahedral cluster of five water molecules. Each water "cube" from Figure 8.1 forms four hydrogen bonds with neighboring water molecules. The H-bonds are shown with dotted lines connected to the neighbors' oxygen nuclei. Because every molecule is fully hydrogen bonded, the structure is quite open. Of the 26 neighboring cubes around a central water, only 4 of the corner cubes are occupied.

antifreeze peptides to prevent freezing of their blood serum, as described in Daune's book. (Meanwhile the ocean does not freeze because of its high salt concentration.)

8.1.4 FLUCTUATING H-BONDS IN WATER

You can think of the structure of water as "ice plus fluctuations." That is, the hydrogen-bonding forces try to keep the structure of neighbors around any water as ice-like as possible—an enthalpic effect that competes with the thermal motions and entropy. It is thought that molecules in liquid water maintain at least three out of four hydrogen bonds in a time-averaged way. The presence of thermal fluctuations tell us, however, that the identity of the three bonds is changing rapidly in time: a given water may change from having two H-bonds to four to three, and then to a different set of three.

In addition to the orientational fluctuations that will cause changes in the hydrogen-bond network, water molecules also fluctuate in position—that is, they diffuse around. For reference, the self-diffusion constant of water is $D \simeq 0.2\,\text{Å}^2/\text{ps}$—recall Chapter 4.

8.1.5 HYDRONIUM IONS, PROTONS, AND QUANTUM FLUCTUATIONS

You may know that protons are important in biological systems, and that there are special proton "pumps" (i.e., specialized membrane proteins) that have the purpose of maintaining the right concentration of protons in the cell. A proton is often described as a hydrogen atom with a missing electron, that is, H^+. However, chemists do not believe that protons exist that way in water.

When water dissociates into positive and negative parts, what really happens is that a proton is added to another water molecule, effectively bonded to one of its lone pairs. The corresponding reaction is therefore

$$2H_2O \; \rightleftharpoons \; H_3O^+ + OH^-. \tag{8.1}$$

You can think of this as an ordinary binding/unbinding equilibrium characterized by a dissociation constant—a description we shall take up in detail in Chapter 9.

The fact that protons really reside on hydronium ions (H_3O^+) is of great biophysical importance. If protons existed in the form H^+, then for a proton to diffuse in a cell (or in a narrow channel) would require the fairly slow process of waiting for water molecules to move out of the way. On the other hand, when the extra proton is transferred from a hydronium ion to the neighboring water (changing that water into a hydronium ion and *vice versa*) the diffusional motion can be much faster. That is, protons move through water by repeated occurrences of the reaction,

$$H_3O^+ + H_2O \; \rightleftharpoons \; H_2O + H_3O^+, \tag{8.2}$$

which allows a "proton" to hop around from water to water using the magic of quantum mechanics. It means that neighboring water molecules can act as a wire for protons.

8.2 WATER MOLECULES ARE STRUCTURAL ELEMENTS IN MANY CRYSTAL STRUCTURES

When protein structures are determined by crystallization, the atoms of the protein are in nearly identical locations in every repeating "unit cell" of the crystal. But protein crystals contain more than proteins. The crystals are often described as "wet" because they contain a substantial amount of water. Indeed, proteins could not maintain their structural stability without water. Nevertheless, most of the water molecules do not seem to be in the same positions in every repeating cell of the protein crystal. Yet some "structural" waters are "frozen" in place, and the locations of these molecules is determined along with the protein's atoms.

The existence of such structural waters means that, in addition to nonspecific effects that water has—like the hydrophobic effect described below—water also takes part in specific interactions with protein atoms. In a sense, these waters are a part of the protein structure, typically because the water molecules form hydrogen bonds with the protein which are very strong. For this reason, the aqueous environment of a protein must not be thought of as unstructured. Rather, some water molecules may be exerting highly specific forces that can be as important for the detailed structure of the protein as forces exerted by the protein's own atoms.

8.3 THE pH OF WATER AND ACID–BASE IDEAS

Acid–base chemistry is important biologically for many reasons, but our focus will be on the underlying (un)binding equilibrium and the electrostatics. These general phenomena operate against the backdrop of two important facts. First, although pure water does indeed dissociate according to Equation 8.1, the effect is quite minor, and therefore a small amount of acidic or basic solute can substantially change the amount of reactive hydronium or hydroxyl ions: more on this below. Second, in the realm of protein structure, a number of amino acids have "titratable groups" that will undergo association or dissociation depending on the local electrostatics and ion concentrations. For instance, the histidine and cysteine side chains can both lose a proton (H^+) under the right conditions.

The association and dissociation processes in water—or in other molecules—will tend to reach an equilibrium governed by statistical mechanics under fixed conditions. This equilibrium is influenced by several factors, but especially by the local electrostatic environment and the concentrations of the dissociating species. For instance, a low pH value (and correspondingly high concentration of H_3O^+ ions) will tend to disfavor the dissociation of H^+ from a cysteine or from the NH_3^+ group at the N terminus of a protein. A nearby negative charge on the protein will have a similar effect.

Water always undergoes the unbinding and rebinding described by Equation 8.1. In the pure state, you probably recall that the pH of water is 7.0, which means the molar concentration of hydronium is $[H_3O^+] = 10^{-7}\,M$. This is based on the well-known definition of pH,

$$pH = -\log_{10}[H_3O^+], \tag{8.3}$$

when the concentration is in molar units. (And since there is one hydroxyl ion for every hydronium in pure water, we also know that $[OH^-] = 10^{-7}\,M$.) This is not much dissociation. In fractional terms, because the concentration of pure water is about 55 M, fewer than one in 10^8 water molecules is dissociated.

With the case of pure water as a baseline, you can begin to appreciate the huge impact that acids can have. Although most biological processes take place near the neutral pH of 7, there are many exceptions, such as in the digestive system. In analogy to Equation 8.1, acids dissociate according to

$$HA + H_2O \rightleftharpoons H_3O^+ + A^-, \tag{8.4}$$

where H is hydrogen and "A" represents the rest of the acidic compound.

You may be familiar with pH measurements significantly smaller than 7, where every unit decrease in pH means an order of magnitude increase in hydronium concentration. Nevertheless, the overall change in pH reflects not just the "strength" of the acid—how much it typically dissociates—but also the amount of acid that is present. After all, the increase in hydronium concentration cannot exceed the acid concentration.

We will quantify these ideas in the chapter on binding.

PROBLEM 8.1

(a) Explain how the acid equilibrium equation (8.4) can be used to describe a titratable group on a protein. (b) Make a sketch to illustrate the different factors that can influence the equilibrium between the left and right sides of the equation for the case of a protein.

8.4 HYDROPHOBIC EFFECT

8.4.1 HYDROPHOBICITY IN PROTEIN AND MEMBRANE STRUCTURE

Hydrophobicity may be the single-most important phenomenon driving the formation of protein structure. Certainly, on a global level, one finds that the core of a typical soluble (non-membrane) protein is hydrophobic. In discussions of the dynamics of protein folding, you will frequently see the initial compactification described as "hydrophobic collapse" in which hydrophobic (nonpolar) residues are nonspecifically attracted to each other. Amino acid side chains can also be hydrophilic if they have polar or charged groups.

Aside from proteins, there is no doubt that hydrophobicity is the principal determinant of the bilayer structure of membranes. The polar (hydrophilic) headgroups of lipids arrange themselves to shield the nonpolar hydrocarbon (hydrophobic) interior of the membrane. While many details of specific lipids and their interactions with other membrane components (e.g., cholesterol, proteins) are biologically critical, the basic function of the membrane—to separate different environments—is possible only because of the dual hydrophobicity and hydrophilicity of lipids.

PROBLEM 8.2

Consider a transmembrane protein consisting of two helices that span the membrane. Explain how the hydrophobic nature of the protein side chains could be sufficient to determine the relative orientation of the helices in the membrane.

8.4.2 STATISTICAL/ENTROPIC EXPLANATION OF THE HYDROPHOBIC EFFECT

The hydrophobic effect is special, distinct from other forces. By way of contrast, consider "elementary" forces like van der Waals attraction, electrostatics, and hydrogen bonding. These forces will be exerted among the atoms of a biomolecule, or any molecule or set of molecules, regardless of the environment (i.e., solvent) in which the molecule is placed. Hydrophobicity is fundamentally different because hydrophobic interactions depend completely on the solvent. Hydrophobic "attractions" are intrinsically statistical. Mathematically, they come from the derivative of a free energy (i.e., potential of mean force [PMF]) rather than the potential energy.

The hydrophobic effect is largely entropic. To see why, look at Figure 8.3. Around a hydrophobic solute, water tends to become structured in order to maintain a reasonable number of hydrogen bonds—approximately three per molecule, as in bulk water. This behavior of water at the interface between a solute and the bulk is sometimes called "caging." When two hydrophobic solutes come into contact, the

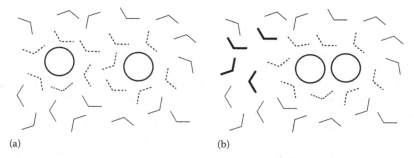

(a)   (b)

FIGURE 8.3 Entropy in the hydrophobic effect. (a) The hydrophobic solutes (circles) are always surrounded by structured water (dashed angles) that fluctuate much less than the bulk water (solid angles). When the two hydrophobes are next to each other, (b), there are fewer structured waters at the solute/solvent interface. The thicker bulk waters represent formerly interfacial waters that have substantially increased in entropy by the transfer to bulk.

number of interfacial water molecules decreases because water is excluded from the interface. These excluded waters enter the bulk, where each can maintain hydrogen bonding without giving up entropic fluctuations. Thus, although hydrogen-bonding energy may differ little for a water molecule at the interface compared to bulk, the entropy difference is significant.

In summary, the attraction of two hydrophobic solutes really reflects the fact that water molecules "prefer" to be surrounded by other waters, and they can maximize the fraction of waters in bulk by minimizing the exposed surface area of solutes. Even though there may be quite weak attractions between nonpolar solutes in the absence of water, the probability to find such solutes next to one another is high reflecting the low free energy (due to high entropy) of that state.

Our discussion has assumed that solutes are large enough to disrupt the H-bonding network of water. Very small molecules (e.g., O_2) can fit in the naturally occurring gaps in water, and only minimally disrupt H-bonding. A more detailed discussion of the physical chemistry of hydrogen bonding can be found in the book by Dill and Bromberg.

PROBLEM 8.3

Toy model of hydrogen bonding, illustrated in Figure 8.4. (a) Omitting kinetic energy, calculate the partition function, average energy, and entropy of each of the two states in the figure. (b) Which state is more probable, and is the favorability due primarily to the entropy or the energy difference? (c) Discuss the origin of an apparent force between two objects (the hydrophobes) that do not directly interact. Use the following rules: (1) hydrophobes do not interact with one another or with water; (2) hydrogen bonds formed by head-to-head arrows each lead to an energy of $-\epsilon$; (3) other water configurations have zero energy; (4) the temperature is such that $e^{+\epsilon/k_B T} = 5$; and (5) the waters are assumed to be able to rotate among only two orientations, vertical and horizontal, with no other movement allowed.

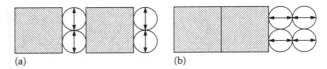

(a)   (b)

FIGURE 8.4 A simple model illustrating the entropic origin of the hydrophobic effect. Square hydrophobic solutes are in contact with circular water molecules, which can hydrogen bond when two "dipole" arrows touch head to head. Each water dipole has two possible orientations—vertical or horizontal—so there are 2^4 configurations for each state of the square hydrophobes. Panel (a) illustrates the separated state and (b) the together state. Only two hydrogen bonds can be formed in both states, but the together state can form this maximal of H-bonds in more ways.

PROBLEM 8.4

"Lipo-phobic effect." Consider two rigid "protein" cylinders embedded in a pure lipid bilayer, and assume there are no direct interactions between the proteins. Given that lipid chains are flexible, explain what net interaction you expect between the proteins. Apply reasoning analogous to that of Section 8.4.2.

8.5   WATER IS A STRONG DIELECTRIC

We have all learned that water is a strong dielectric, and that dielectrics reduce electrostatic forces and energies. In fact, the dielectric constant of water is about $\epsilon \simeq 80$, which means that electrostatic interactions are reduced by a factor of 80! Recall that the force law between two charges q_1 and q_2 separated by a distance r in a dielectric is given by

$$f = k \frac{q_1 q_2}{\epsilon \, r^2}, \tag{8.5}$$

where k is a constant depending on the system of units (e.g., $k = 1/4\pi\epsilon_0$ in MKS units) and $\epsilon = 1$ if the two charges are in vacuum.

Two questions come to mind: (1) What is the physical origin of dielectric behavior? (2) How does the high dielectric constant of water affect the interactions among charges on a protein—given that a protein has a much lower dielectric constant? We must understand the physical origins of dielectric behavior in order to answer the second question.

Dielectric behavior should be distinguished from "screening" that occurs due to ions in solvent, which will be discussed below. The physical origins of the two phenomena are not quite the same, and the mathematical consequences are strikingly different.

8.5.1 Basics of Dielectric Behavior

Dielectric behavior results from two things: First, dielectric phenomena reflect a choice to focus on a subset of charges of interest—for example, those in a protein—while ignoring other charges—for example, those of the water molecules. Thus, one makes a somewhat artificial separation between the charges of interest $(q_1, q_2, \ldots)$ and those in the rest of the system that is now considered a background "medium." Second, and here there is some real physics, the medium should be polarizable. That is, the medium should be able to change the arrangement of its charges—its charge distribution—in reaction to the charges of interest $q_1, q_2, \ldots$. Typically, a medium will consist of molecules that individually can change their dipole vectors.

The first point above may sound like semantics, but it is fundamental. It is very possible to represent the electrostatics of charges in a dielectric medium without using a dielectric constant. One does so simply by summing over all forces, including those from charges in the medium $q'_1, q'_2, \ldots$—which could be the partial charges on water molecules. If there are just two charges of interest, q_1 and q_2 and an additional M charges in the medium, then the net force acting on q_1 has two parts given by

$$\vec{f_1} = k\frac{q_1 q_2 \hat{e}_{12}}{r^2} + \sum_{i=1}^{M} k\frac{q_1 q'_i \hat{e}_{1i}}{r_{1i}^2}, \tag{8.6}$$

where r_{1i} is the distance between q_1 and q'_i and $\hat{e}_{1i}$ is the unit vector from charge i to charge 1. By comparison to Equation 8.5, we can see that $\epsilon = 1$ here, but there should be no mystery about this. The basic rule of physics is that electrostatics is completely described by the Poisson force law with $\epsilon = 1$ so long as all charges are included. Dielectric theory approximates all of the effects of the sum in Equation 8.6 into a parameter $\epsilon > 1$ in Equation 8.5.

8.5.1.1 One or the Other

We have an either/or situation: you must use one description or the other, but do not mix them. That is, either you focus on a subset of charges of interest and represent the effects of all other charges by the single dielectric constant ϵ—as in Equation 8.5—or you sum over all electrostatic forces to get the exact, complicated answer—as in Equation 8.6 (see Figure 8.5). Do not sum over all charges and use a dielectric constant other than $\epsilon = 1$.

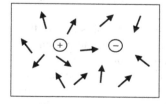 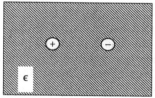

FIGURE 8.5 The two mutually exclusive descriptions of a dielectric system. The dielectric constant approximates the effects of all the dipoles (arrows) in the explicit system at left.

8.5.1.2 Dipole Review

Because dipoles will be so important to our discussion of water's dielectric behavior, we should be sure to understand one key point. This is another "either/or" situation. That is, you either describe the electrostatic field of water as due to partial charges on the different atoms (and perhaps lone pairs)—or you can say that the field is due to the dipole moment of water. A dipole, after all, is nothing more than two nearby charges of opposite sign. More generally, any charge distribution can either be represented by summing over the effects of all individual charges, somewhat like Equation 8.6, or by a set of "multipole moments"—namely, the dipole, quadrupole,.... Since water is generally considered to have three or four charge sites, it has a dipole moment and also higher moments, but we typically only discuss the dipole moment. Again, the bottom line is that it's either the partial charges or the multipole moments (approximated by the dipole alone) but not both.

8.5.1.3 PMF Analogy

It is time to think again of our old friend, the PMF. Recall that the PMF for a given coordinate (think of the distance r) reflects the effective energy (free energy) that results from averaging over all other coordinates (positions of other charges/molecules). The PMF is that energy that yields the probability distribution along the selected coordinate. Although we did not discuss it earlier, the PMF can be differentiated in exactly the same way as can a regular potential energy to yield a statistically averaged force. Thus, we can rewrite Equation 8.5 as an approximate derivative:

$$f = k\frac{q_1 q_2}{\epsilon\, r^2} \simeq -\frac{d}{dr}\text{PMF}(r). \tag{8.7}$$

We are not going to attempt to perform the averaging necessary to compute a PMF, but one needs to imagine it in order to understand the meaning of the dielectric approximation.

8.5.2 Dielectric Behavior Results from Polarizability

Polarizability is the ability of a medium to change its charge distribution in response to an electric field, and this is what causes the dielectric effect. In our simple case of studying two charges, q_1 and q_2, in such a medium, we will consider the electric field of the second charge on the first. Since the force on the first charge can generally be written as $q_1 E_1$, where E_1 is the electric field at q_1, we know the field due to the second charge is $E_1 = kq_2/\epsilon r^2$.

We have said that, in general, the dielectric constant is greater than one ($\epsilon > 1$), which means the field experienced by q_1 is weaker in the presence of a dielectric. Why? The answer is polarizability—the ability for charges (or partial charges) to rearrange themselves and/or move. Since the charges in the medium can move, those that are attracted to q_2 will move toward it. For instance if q_2 is positive, negative charges will move toward it. From the point of view of the first charge, it will appear that the magnitude of q_2 is reduced by the presence of the polarized charges—of opposite sign—that surround it. After all, the electric field experienced by q_1 is due

FIGURE 8.6   An illustration of fully aligned dipoles symmetrically surrounding two point charges.

to all charges—recall Equation 8.6. Needless to say, the second charge will see the same thing happen to the first. To put it another way, electrostatic forces will always cause the charges in the polarizable medium to rearrange themselves so as to shield or cloak the charges of interest with oppositely signed partners. Thus, the net interaction is reduced by some factor $\epsilon > 1$. The reduction in the electric field at "a charge of interest" is illustrated in Problem 8.5.

A fluid "medium" (into which our charges are inserted) can be polarized by two different means, and water exhibits both behaviors. These are (1) the reorientation of molecules with dipole moments and (2) the shift of charges (i.e., electron density) within individual molecules. Both of these mechanisms will lead to the shielding charge redistribution described above. Note further that both of these effects are accompanied by energy changes that will indeed be governed by statistical mechanics.

PROBLEM 8.5

Consider the two point charges of Figure 8.6. Show that the net electric field acting on the positive charge is reduced due to the presence of the aligned, symmetrically placed dipoles. Assume that the dipoles are identical and made up of two point charges. Further, assume that the two dipoles neighboring each point charge are exactly equidistant from the central charge. You do need to know the specific distances to perform this exercise.

8.5.3   WATER POLARIZES PRIMARILY DUE TO REORIENTATION

Water molecules are polarizable both due to shifting electron density and due to reorientation—but reorientation is the dominant effect. A single water molecule in isolation will have a large dipole moment, due to oxygen's high electronegativity, but a water in aqueous solution will have an even larger moment since it will be polarized by the electric fields of its neighbors. When our "charges of interest" are inserted in water, they can further polarize individual molecules a little (by shifting the electron density) but they can cause significant reorientation. The ability of water molecules to reorient is the primary cause of the high dielectric constant.

8.5.3.1   Thermal Effects

When a charge in placed in water, the water dipoles reorient in response to electrostatic forces—so by definition the system is lowering its potential energy. This ordering of the dipoles naturally will be opposed by thermal fluctuations to higher energy states. Thus, although it is not the main point of our discussion of dielectrics, you should be aware that statistical mechanics indeed enters into dielectric phenomena. For instance, as temperature increases, one expects ε to decrease because large thermal fluctuations will strongly resist reorientation of the dipoles.

8.5.4 CHARGES PREFER WATER SOLVATION TO A NONPOLAR ENVIRONMENT

Because of water's high dielectric constant, it is energetically favorable for a charge to be in water as compared to an environment with a low dielectric constant. In biological systems, protein interiors and lipid bilayers are fairly nonpolar. Charged groups tend to be found on the surfaces of proteins.

The reason why charges prefer a high-dielectric/polar environment is not hard to see. Looking again at Figure 8.6, for example, it is clear that charges interact favorably with the dipoles of the dielectric.

The magnitude of a single charge's interactions with a dielectric medium can be quantified using the Born equation. To transfer a charge q of radius a from vacuum to a dielectric characterized by ε leads to a decrease in energy

$$\Delta G^{\text{Born}} = k\frac{q^2}{2a}\left(\frac{1}{\epsilon} - 1\right) < 0. \tag{8.8}$$

Although this is a true free energy (because dielectric effects result from thermal averaging), the derivation uses ordinary electrostatics and will not be described here. Details can be found in the book by Dill and Bromberg.

For every rule, there's a caveat, of course. Here we note that some (low-dielectric) proteins do indeed have charged groups in their interiors. As explained by Warshel, Russell, and Churg, these charges tend to be "solvated" by nearby polar groups—that is, by locally favorable electrostatics.

8.5.5 CHARGES ON PROTEIN IN WATER = COMPLICATED!

There is no simple and correct way to treat charges on a protein in solution using dielectric constants. That is, even if you believe it is correct to assign a dielectric constant (perhaps $\epsilon \simeq 2$) to a protein, and the charges of interest are located on the protein, it is still difficult to describe the interaction energy and force between the charges. (There is significant controversy over the best value of ε to assign to a protein, with suggestions ranging from about 2 to more than 10. However, it is important to realize that ε is fundamentally a bulk quantity so it is not likely that an effective value can be assigned that would be correct for arbitrary charge locations in the protein. Nevertheless, we want to explore the consequences of assuming some value of ε is appropriate for a protein.)

The key point can be appreciated from Figure 8.7. As we have emphasized, it is always true that the force on a given charge can be calculated by summing over all charges present, as in Equation 8.6. To compute the force on, say, q_1 in Figure 8.7, we must sum over the charges both on the protein and in the water. (Recall that each dipole is composed of positive and negative charges.) Therefore, it is clear that the dielectric properties of water embodied in its oriented dipoles will affect the force on q_1, even though that charge is in the protein. It is incorrect, in other words, to use only the protein's dielectric constant in Equation 8.5 to describe the interaction between q_1 and q_2.

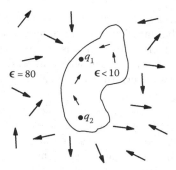

FIGURE 8.7 Dielectric effects are complicated when two media are present. In this schematic drawing, both water and a protein "medium" are present, with the low-dielectric protein containing the charges q_1 and q_2. It is not simple to average the two dielectric constants, $\epsilon = 80$ for water and a much smaller value for the protein, shown schematically by the large and small dipole arrows. Each dipole can be thought of as a pair of opposite charges. The basic physics embodied in Equation 8.6 suggests the correct procedure: all charge interactions must be summed over, and this includes not just the charges between q_1 and q_2. Thus, the effective ϵ is not determined solely by the dielectric constant of the medium in which the charges of interest are placed.

8.6   CHARGES IN WATER + SALT = SCREENING

We saw that a uniform, polarizable medium like water will reduce the effect of charges in a fairly simple way (ignoring the protein complications). According to Equation 8.5, charges feel a net electric field, which is simply reduced by a factor of ϵ. In this standard description of dielectrics, there is no distance dependence to the effect of the medium—that is, the original $1/r^2$ behavior remains. By contrast, when salt is added to water, the force law becomes more complex and the distance dependence changes. As you probably know, salt is always present under physiological conditions.

Physically, when a salt (e.g., NaCl) is added to water, it dissociates into mobile ions (e.g., Na^+ and Cl^-), which are then free to respond independently to electric fields. Thus, if a positive charge is placed in an ionic solution (water + salt), the negative ions will tend to migrate toward the charge and positive ions away. The negative ions surrounding the positive charge are said to "screen" it, since the net electric field due to the charge will be reduced by the fields of the oppositely charged ions. From this basic point of view, screening is no different from the polarization that led to the dielectric effect. Yet we have already said the effects are dramatically different in terms of the governing equations. Why is this?

The primary difference between the mobility of ions and the ability of water molecules to reorient is essentially one of magnitude. Both effects redistribute charge based on basic electrostatic forces. However, when considering the reorientation of water dipoles, it is clear that the maximum effect—which occurs when all waters are fully oriented—is limited by the fact that dipoles are causing the effect. After all, in a dipole, where there is a charge of a certain sign, an opposite charge will be close by. Dissociated salt ions have no such limitation. As many oppositely charged ions as necessary can screen a charge. Of course, there will be a limit due to the repulsion among like-charged ions, but this is taken care of automatically by the electrostatic forces in combination with thermal fluctuations.

As in the dielectric case, thermal effects will be important for screening. However, we will now be in a position to perform a classical statistical mechanics analysis to understand basic screening behavior.

The Poisson–Boltzmann (PB) equation and Debye–Hückel (DH) theory given below describe the electrostatic potential—and hence the electric field and forces—due to a charge in a solution with mobile ions. The solvent is accounted for by a simple dielectric constant ϵ, but the ionic charges and densities are incorporated explicitly. The ability of the ions to redistribute themselves based on electric fields is treated in a statistical mechanics approximation. We will start with the PB description and see how DH is derived from it.

PROBLEM 8.6

Explain whether you expect the speed of diffusion for ions in water to be greater or less than that of protons. [Hint: Consider the "hopping"/water-wire mechanism above.]

8.6.1 STATISTICAL MECHANICS OF ELECTROSTATIC SYSTEMS (TECHNICAL)

Our basic interest is again to understand the electrostatic field/potential at a charge of interest—which we'll call q_0 to distinguish this discussion from what came before. We will assume this charge is fixed at the origin of coordinates, and that ions are free to rearrange as necessary, all the time with a "background" dielectric of ϵ. For simplicity, we will assume there are equal numbers of two types of monovalent ions with charges $\pm e$, with e the charge of an electron (although this assumption is not necessary). Let the concentrations ("number densities") of these ions be $n = N_+/V = N_-/V$. To make the problem mathematically tractable, however, we will need to assume that our charge of interest is immersed in a solution (as opposed to being on a protein) so that we expect the environment around it to be spherically symmetric. In other words, average ion concentrations can only vary with the radial distance r from q_0.

Our specific goals are to determine the average radial distribution of the ions about q_0 and the corresponding electrostatic potential around q_0, both as functions of r. In other words, recalling Chapter 6, our goal is essentially to compute the radial distribution functions $g_+(r)$ and $g_-(r)$ given the condition that the charge q_0 is at the origin. Our calculation will use the overall average radial distribution of charge about q_0, excluding the central charge itself, which is

$$\langle \rho_0(r) \rangle = n e \left[g_+(r) - g_-(r) \right]. \tag{8.9}$$

This is an exact equation, given our assumptions and recalling that g is defined to be dimensionless. (Note that ρ_0 is not a probability density here, as it has units of charge and is normalized to the net charge in a system.) All the complicated behavior resulting from electrostatics and van der Waals forces is incorporated into the g functions (which we do not yet know). We can express these conveniently in terms of PMFs, as described in Chapter 6, by writing

$$g_+(r) \equiv e^{-\text{PMF}_+(r)/k_B T} \quad g_-(r) \equiv e^{-\text{PMF}_-(r)/k_B T}, \tag{8.10}$$

but we still do not know the distributions or, equivalently, the PMFs. In other words, Equations 8.9 and 8.10 are exact but they really contain only definitions and not physical information.

To proceed, we need to shift our focus to the electrostatic side, and we need to be careful with some of the calculus details. Recall from elementary physics that the electrostatic potential due to a charge q at a distance r in a dielectric ϵ is simply $kq/\epsilon r$. It is not so elementary, but also not so surprising, that one can determine the electrostatic potential ϕ at any point in space simply by integrating over the potential due to all charge density—which is essentially what we discussed about dielectrics with Equation 8.6. But it is also possible, in essence, to differentiate the integral relation and write the charge density ρ as a second derivative of the potential ϕ. This famous relation of advanced electrostatics is the Poisson equation, namely,

$$\nabla^2 \phi(\mathbf{r}) = -\left(\frac{4\pi k}{\epsilon}\right) \rho(\mathbf{r}), \tag{8.11}$$

where k is the usual electrostatic constant from Equations 8.5 and 8.6. Equation 8.11 applies to any particular distribution of charge ρ. Those not familiar with this equation may wish to complete Problem 8.9 to see that the Poisson equation embodies all the familiar electrostatics. Note that $\mathbf{r} = (x, y, z)$ is a regular position vector in three-dimensional space because ϕ and ρ describe "real space"—as opposed to configuration space.

We can adapt the Poisson equation to our situation in several steps. We first focus on our fixed charge q_0 in a sea of ions by considering $\phi_0(\mathbf{r})$, which is the potential at any point $\mathbf{r}$ due to the distribution $\rho_0(\mathbf{r})$ for the ions. The Poisson equation (8.11) applies to the distribution ρ_0 and the potential ϕ_0 since it applies to any distribution.

Next, we can average both sides of the Poisson equation—that is, average over all Boltzmann-factor-weighted configurations of ions consistent with the fixed charge q_0 at the origin. We denote the average spatially varying electrostatic potential by the letter psi, so that $\psi_0(r) = \langle \phi_0(\mathbf{r}) \rangle$, where we have used the fact that the average potential will only depend on the radial distance r from the origin. Incorporating the PMFs from Equation 8.10, which represent averages by definition, we obtain

$$\nabla^2 \psi_0(r) = \frac{1}{r^2} \frac{d}{dr}\left[r^2 \frac{d\psi_0}{dr}\right] = -\left(\frac{4\pi k}{\epsilon}\right) ne\left[e^{-\text{PMF}_+(r)/k_B T} - e^{-\text{PMF}_-(r)/k_B T}\right], \tag{8.12}$$

which applies to all space excluding the ion itself and has assumed spherical symmetry and just two ion types. We have written the explicit Laplacian (∇^2) for our case where there is only r dependence. Nevertheless, under our assumption that water can be represented by a dielectric constant ϵ, Equation 8.12 is an exact combination of the Poisson equation and the laws of statistical mechanics. Unfortunately, on the left-hand side, we have the average electrostatic potential ψ_0, whereas on the right we have the PMFs, which include not just electrostatics but all other forces, such as van der Waals'.

To make progress with the equations, we must make approximations.

8.6.2 First Approximation: The Poisson–Boltzmann Equation

The first approximation we make is based on our primary focus on electrostatics. We ignore all other forces (i.e., van der Waals) and equate the two PMF functions with the electrostatic potential energy. Mathematically, this means we set

$$\text{PMF}_+(r) \simeq +e\,\psi_0(r) \quad \text{and} \quad \text{PMF}_-(r) \simeq -e\,\psi_0(r). \tag{8.13}$$

This approximation will not account for steric interactions among ions in the "sea of ions" ($r > 0$), and hence may be unphysical when the ion concentration tends to become high in any region such as near q_0.

Substituting the approximations (8.13) into Equation 8.12, we finally obtain the famous—but approximate—PB equation for the case with two monovalent ions (charges of $\pm e$):

$$\frac{1}{r^2}\frac{d}{dr}\left[r^2\frac{d\psi_0}{dr}\right] \simeq -\left(\frac{4\pi k}{\epsilon}\right)ne\left[e^{-e\psi_0(r)/k_B T} - e^{+e\psi_0(r)/k_B T}\right], \tag{8.14}$$

where we recall that there is a concentration of ions of each charge. Our approximation has led to an equation that at least has the same basic function (ψ_0) on both sides. Nevertheless, Equation 8.14 is nonlinear and therefore is often further simplified.

8.6.3 Second Approximation: Debye–Hückel Theory

As we have seen many times, solving a simple problem exactly often provides important physical insights. This is the case here. In order to progress mathematically with (the already approximate) Equation 8.14, we consider the limit of (relatively) high temperatures—that is, when the electrostatic energy $e\psi_0$ is small compared to $k_B T$. We therefore approximate the exponential terms by setting $e^x \simeq 1 + x$, which leads to the linearized PB equation, also known as the DH equation:

$$\frac{1}{r^2}\frac{d}{dr}\left[r^2\frac{d\psi_0}{dr}\right] \simeq \kappa^2\psi_0(r), \tag{8.15}$$

where

$$\kappa^2 = \frac{8\pi kne^2}{\epsilon k_B T} \tag{8.16}$$

for the monovalent solutions we have been considering.

In retrospect, if we examine the definition of κ^2, which is the only place temperature enters the DH equation, we can see that our assumption of high temperature is equivalent to a condition of low concentration, n. It is shown in McQuarrie's book, in fact, that the DH equation becomes exact in the limit of low salt concentration, regardless of temperature (assuming a simple dielectric medium fully characterized by ϵ).

We will now (partly) solve the DH equation—enough to learn the key physical lessons. Although the radial Laplacian on the left-hand side of Equation 8.15 appears complicated, we can make quick progress using our physical and mathematical intuition. First, we expect that the statistically averaged electrostatic potential ψ_0 should be a modification of the normal Coulomb potential, which is proportional to r^{-1}. Second, we see that writing ψ_0 in a way to account for modified Coulomb effects leads to mathematical simplification. Specifically, if we define a function f by $\psi_0(r) \equiv f(r)/r$, then Equation 8.15 reduces to $(1/r)f''(r) = (\kappa^2/r)f(r)$. This equation has simple growing and shrinking exponential solutions, $f(r) \propto \exp(\pm\kappa r)$. Because the potential ψ_0 due to a single charge must vanish for large r, we obtain the solution

$$\psi_0(r) = C\frac{e^{-\kappa r}}{r}, \qquad (8.17)$$

where C is a constant determined by the boundary conditions (see the books of Daune or McQuarrie for details).

PROBLEM 8.7

(a) Calculate the screening length κ^{-1} in 100 mM NaCl aqueous solution at $T = 300$ K. (b) Compare this length to the size of a typical protein. (c) Explain how you think this might affect interactions among charged atoms on a single protein, and also interactions between proteins.

PROBLEM 8.8

Generalizing the PB and DH equations for arbitrary ion species. Consider a charge q_0 surrounded by a sea of ions with charges $\{z_1e, z_2e, \ldots\}$ and concentrations $\{n_1, n_2, \ldots\}$, with zero overall charge. For this general case, accordingly revise Equations 8.9, 8.10, 8.12 through 8.14, and 8.16. Note that the form (8.15) remains unchanged, so long as the definition of κ is generalized.

8.6.3.1 The Key (Bio)physical Consequences of Screening by Ions

The simple mathematical form (8.17), combined with the definition (8.16), embodies physical phenomena that are critical for biology. (1) The Coulomb interaction is not simply weakened by a constant factor as in dielectrics, but in fact takes on a different mathematical form, which can decay rapidly if κ is significant. In this important sense, screening differs from dielectric behavior. (2) Since the argument of an exponential function must be dimensionless, we see that κ^{-1} is a key physical lengthscale, which is known as the Debye screening length. Roughly speaking, electrostatics effects will be unimportant at distances beyond the screening length. (3) Electrostatic interactions are screened more—and hence less important—at higher salt concentrations. This has allowed many biophysical experiments to "salt out" electrostatic effects and separately study them from other effects such as hydrophobicity. (4) Finally, although perhaps less biologically important, we see that screening becomes weaker at higher temperature, when powerful thermal motions destroy the correlations set up by Coulombic forces.

8.6.4 COUNTERION CONDENSATION ON DNA

Double-stranded DNA is a fairly stiff polymer with a highly negatively charged backbone, due to the presence of phosphate (PO_4) groups. As you would expect from our previous discussion, the presence of negative charges attracts, on average, positive ions ("counterions"). Although we could analyze the statistical behavior of the ions around DNA in analogy to our treatment of the screening of a single charge, the discussion becomes a bit technical without a sufficient payoff in terms of physical insight.

We turn instead to a simpler, more static analysis because of the special geometry and extreme charge density of DNA. The linear nature of a DNA molecule means that its electrostatic field will decay much more slowly than $1/r^2$—in fact, as $1/r$, as you can check in a text on electricity and magnetism. Positive counterions thus are strongly attracted to DNA. These forces will attempt to create a "mechanical" equilibrium—that is, to induce motion until the forces are reduced. There will be some thermal agitation opposing the ions' attraction to DNA (which will prevent all phosphate groups from being paired with counterions) but most phosphate groups will tend to be "covered" by a positive ion. This is an elementary discussion of the condensation of counterions on DNA.

From a statistical mechanics point of view, the basic DH picture has been turned on its head. In the linearized DH analysis, the potential and fields are relatively weak and permit a Taylor expansion of the Boltzmann factor. In that case, the electrostatics may be considered a small perturbation disrupting the uniformity toward which thermal disorder tends. By contrast, in DNA, the typical ionic distribution is far from uniform. Perhaps it would be better to say that thermal agitation causes a perturbation to neutralized DNA—which otherwise would tend to be fully charge-neutralized by counterions.

For a more detailed discussion of this important biophysics problem see the books by Daune and by Phillips et al.

8.7 A BRIEF WORD ON SOLUBILITY

Solubility reflects the degree to which a molecule tends to stay in solution, surrounded by water molecules, rather than "precipitating" out to aggregate with other solute molecules. Every molecule is soluble to some degree, even hydrophobic molecules, but in our discussion of the hydrophobic effect we saw that molecules that are not "liked" by water will tend to be attracted to one another. Arguably, you could say that the hydrophobic core of a protein consists of molecular groups that have aggregated—even though the protein itself may be fully solvated (i.e., surrounded by water).

Our discussion of the hydrophobic effect also highlighted the importance of the way water structures itself around a solute. The layer of waters in contact with a solute is called a "hydration layer" or "solvation shell" and can be considered to increase the effective size of a molecule for many diffusional effects. Thinking ahead to the phenomenon of binding, when two molecules bind, they must first shed their hydration layers of partially structured waters.

In terms of drug design and the analysis of signaling, the relative solubility of a given molecule in water, compared to the membrane, is important. Some molecules are small enough and hydrophobic enough to diffuse through the membrane—and then emerge to the other side. This ability is important for signaling between cells by small molecules, and also for the design of drugs. Many modern drugs are thought to act by permeating the membrane to enter the cell.

Finally, we note that the relative solubility between aqueous and hydrophobic phases is quantified by the "partition coefficient" P, which measures the ratio of a molecule's concentration in octanol and water when the solute molecule is equilibrated between the two substances. It is often reported as $\log P$.

8.8 SUMMARY

There is a lot to know about water. We reviewed structural aspects of water due to hydrogen bonding and its nearly tetrahedral symmetry. The open structure of ice only melts partially in liquid water, which tends to maintain most of its hydrogen bonds. Hydrogen bonding and fluctuations, in turn, lead to the entropic nature of the hydrophobic effect. We also reviewed key solvation and electrostatic phenomena: the dielectric effect and screening by ions, both of which include thermal effects. Both of these "phenomena" stem from a choice to focus on certain charges of interest (e.g., on a protein) while considering everything else (water, ions) to be background.

8.9 ADDITIONAL PROBLEM: UNDERSTANDING DIFFERENTIAL ELECTROSTATICS

PROBLEM 8.9

Understanding, or at least believing, the "differential form" of the Poisson equation (8.11). Our strategy will not be to understand the fundamental math behind the Poisson equation, but rather to see whether we believe its most basic predictions. In other words, what does the equation tell us? We will show that the differential Poisson equation indeed leads to the usual electrostatic potential $\phi(r) \propto 1/r$ and Coulomb force law.

First of all, the Laplacian in Equation 8.11 simply stands for a sum of second derivatives: $\nabla^2 = (\partial/\partial x^2) + (\partial/\partial y^2) + (\partial/\partial z^2)$. For better or worse, our interest is in determining the radial (r) dependence of the potential and field, and hence we want to use spherical polar coordinates—which in turn require a Jacobian as in the integrals of Chapter 5. If we use the perfect spherical symmetry of our problem—that is, the fact that the potential and field around a point charge depend only on the distance from it—the Laplacian becomes purely radial: $\nabla^2 = r^{-2}(\partial/\partial r)(r^2 \, \partial/\partial r)$.

We are almost ready to attack the Poisson equation, but we need to reckon with the charge density. For a point charge, the charge density is given by $\rho(r) = q \, \delta(r)$. The Dirac delta function has the nice property that it is zero away from the charge itself (i.e., for $r > 0$), as discussed in Chapter 2. In other words, for our simple point charge case, the Poisson equation away from the charge simplifies to: $\nabla^2\phi = r^{-2}(\partial/\partial r)(r^2\,\partial\phi(r)/\partial r) = 0$.

(a) Show by differentiation according to the radial Laplacian that $\phi(r) = kq/(\epsilon r)$ satisfies $\nabla^2\phi = 0$. (b) Also show that any other power of r would not satisfy the equation. (c) Using the relations between electric potential, field, and force, show that the potential from (a) leads to the usual Coulomb force law (8.5).

FURTHER READING

Ben-Naim, A.Y., *Statistical Thermodynamics for Chemists and Biochemists*, Springer, Berlin, Germany, 1992.

Daune, M., *Molecular Biophysics*, Oxford University Press, Oxford, U.K., 1999.

Dill, K.A. and Bromberg, S., *Molecular Driving Forces*, Garland Science, New York, 2003.

McQuarrie, D.A., *Statistical Mechanics*, University Science Books, Sausalito, CA, 2000.

Phillips, R., Kondev, J., and Theriot, J., *Physical Biology of the Cell*, Garland Science, New York, 2009.

van Holde, K.E., Johnson, W.C., and Ho, P.S., *Principles of Physical Biochemistry*, Pearson, Upper Saddle River, NJ, 2006.

Warhsel, A., Russell, S.T., and Churg, A.K., *Proceedings of the National Academy of Sciences, USA*, 81:4785–4789, 1984.

9 Basics of Binding and Allostery

Arguably, binding is *the* fundamental molecular process of biology. Almost all communication among cells and within cells (signaling) results from binding or molecular recognition. Those arrows you may see in cell-biology figures—connecting proteins in a network or even connecting genes—result from binding. After all, genes are regulated largely by binding events, such as transcription factors to DNA (plus many other examples involving RNA, siRNA, histones, etc.). Even catalysis, which might be considered a distinct process, relies on binding of substrate to enzyme before it can even get started.

From a theoretical point of view, the study of binding is an excellent opportunity to integrate the various perspectives we have been studying: microscopic and macroscopic, statistical and thermodynamical, dynamical and equilibrium.

On the biological side, we will apply our analysis of binding to key phenomena: the origin of ATP's free energy, allostery/cooperativity, catalysis, and pH/pK_a phenomenology.

9.1 A DYNAMICAL VIEW OF BINDING: ON- AND OFF-RATES

This book emphasizes that equilibrium can be understood as a balance of dynamics, an averaging of many time-dependent processes. In fact, although our studies began with simple equilibrium concepts in probability theory, we might well have started from a dynamical point of view, which is perhaps more physical. After all, as we've discussed, nature doesn't calculate partition functions. Nature generates forces, which lead to dynamics. Let's start there to understand binding.

Imagine you could sit on a single molecule in a test tube and watch all of its doings. Assume further that the tube also contains another type of molecule (circle vs. square in Figure 9.1a) to which your molecule can bind. What would you see? Probably something like Figure 9.1b.

In other words, if you were sitting on a protein receptor (R, perhaps the circles) you would see a free ligand (L, square) bind and stay bound for a time, followed by unbinding, followed by rebinding. Of course, once a particular ligand unbinds, it may well diffuse away, so the rebinding event may be with a different L molecule. A similar sequence of events would occur for a ligand. (Note that the ligand could be a small molecule or ion, or it could equally well be another protein. The identity of L makes no difference in our discussion.) These are the dynamics underlying the equilibrium implied in

$$R + L \underset{k_{\text{off}}}{\overset{k_{\text{on}}}{\rightleftharpoons}} RL, \qquad (9.1)$$

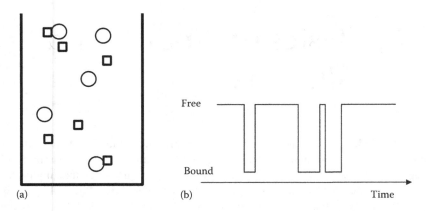

FIGURE 9.1 (a) A test tube filled with a ligand and a receptor that can bind and unbind reversibly. (b) A trajectory of binding and unbinding.

where RL represents the bound complex. It should be understood that the rightward binding process is characterized by the rate k_{on} and unbinding process by the rate k_{off}. These rates will be discussed in more detail below.

Let's quantify this dynamical or trajectory-based picture, starting with unbinding, which is fairly straightforward. If we are focusing on a particular RL complex, then the usual (reasonable) assumption is that there is fixed probability to unbind per unit time. In other words, there is a microscopic unbinding rate, more commonly called the off-rate, k_{off}. This off-rate corresponds exactly to the (inverse of the) average amount of time one has to wait in a trajectory for unbinding to occur in the prevailing conditions. Thus, k_{off} has units of s^{-1}. If we want to describe the behavior of all the complexes in the test tube, then the expected number of unbinding events per unit volume (and per unit time) is just $k_{off} \cdot [RL]$, where [X] denotes the concentration or number density of any molecule X.

The frequency of binding events is a bit more difficult to quantify. To see this, imagine you are following a particular unbound receptor in our test tube (Figure 9.1). We can divide the binding process into two steps: (1) collision and (2) complexation. That is, the receptor and ligand must first come into proximity ("collide") and then they have a chance to form a complex. Although the second step, complexation, is much more complicated in principle and depends on structural details of the system, we will be satisfied with saying that once a collision has occurred, there is some fixed probability for complexation. Structural details will affect this probability, but we are still correct in saying there is some overall average probability.

The overall binding rate is going to be the product of the collision frequency and the probability for complexation, but the collision frequency depends on the whole sample. For instance, if there is only one ligand molecule in the whole test tube, the collision frequency will be very low. In fact, the collision frequency for our single receptor will be proportional to the concentration of ligands, [L], so long as [L] is not too high. Of course, this will be true for every receptor, so the overall collision frequency per unit volume will be proportional to both concentrations—that is, to

[R][L]. Note that these factors represent the unbound molecules only and not the total concentrations, which are the sum of bound and unbound:

$$[R]_{tot} = [R] + [RL] \quad \text{and} \quad [L]_{tot} = [L] + [RL]. \tag{9.2}$$

Note further that we would have determined the same overall proportionality (to [R][L]) if we had started by considering a single ligand molecule—there's a symmetry here.

But what of the on-rate (k_{on}) itself? Recall that we always want rate constants to represent intrinsic physical properties of the molecules (and unavoidably, external conditions like temperature, pH, and salt concentration). In particular, we do not want the on-rate to depend on the concentrations of the binding molecules themselves. Therefore, the on-rate is defined so that the total number of binding events per unit volume and per unit time is given by $k_{on}[R][L]$. This makes perfect sense, but we now have the consequence—strange at first sight—that the on-rate has units of inverse seconds multiplied by inverse concentration ($s^{-1} M^{-1}$, where M is molarity). Of course, these apparently weird units also make sense, since binding frequency, unlike for unbinding, is proportional to concentration.

Thinking back to our trajectory sketched in Figure 9.1, those stretches of time in the unbound state reflect a particular concentration of the other species ([L] if we are watching R, or [R] if we are watching L). Those waiting times for binding to occur are shorter at higher concentration and *vice versa*.

We have now completed the connection between the microscopic dynamics—that is, individual binding trajectories—and the average dynamics embodied in rate constants. The next step is to incorporate our understanding into a time-dependent picture.

Before moving on, in the following problem we note a caveat that is largely unimportant for biochemistry, but is nonetheless of physical and conceptual importance.

PROBLEM 9.1

The assumed proportionality of binding events to [R][L] is very reasonable at small and perhaps moderate concentrations. It is called a "mass action" picture. Give a qualitative explanation of why the proportionality should be expected to break down at very high concentrations of either R or L. A sketch may aid your understanding.

9.1.1 Time-Dependent Binding: The Basic Differential Equation

We want to write down the differential equation governing the flow of population (probability), analogous to what we did when studying flux balance in Chapter 4. Compared to describing the isomerization of a molecule—or motion from one part of a physical space to another—the dependence on concentrations is just one step more complicated.

We have already developed expressions for the overall binding and unbinding frequencies (per unit volume), and we can therefore write an equation for the way

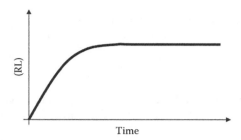

FIGURE 9.2 Binding over time when an experiment is initialized with no complexes.

the population of bound complexes will change:

$$\frac{d[\text{RL}]}{dt} = [\text{R}][\text{L}]k_{\text{on}} - [\text{RL}]k_{\text{off}}. \tag{9.3}$$

Each term should make sense, including the units, based on our previous discussion.

It is worthwhile to consider one particular time-dependent case in more detail. We discussed in Chapter 4 that equilibrium is self-correcting—so that if the system exhibits nonequilibrium populations at one point in time, they will relax toward the equilibrium values. Here we will consider what happens when a solution containing only receptors has ligands added to it at time $t = 0$. This means that at $t = 0$, there are no complexes ($[\text{RL}] = 0$). Of course, complexes will form as time proceeds, but at short times, the first term on the right-hand side of Equation 9.3 will dominate the behavior because $[\text{RL}]$ will be small. Thus, we expect linear behavior in time, at short times: $[\text{RL}] \simeq k_{\text{on}}[\text{R}]_{\text{tot}}[\text{L}]_{\text{tot}}\, t$. In physical terms, since so few complexes will have formed, an even smaller—and negligible—fraction will have had the chance to unbind. Equivalently, we can approximate $[\text{R}] \simeq [\text{R}]_{\text{tot}}$ and $[\text{L}] \simeq [\text{L}]_{\text{tot}}$ at short times for the right-hand side of the equation. Eventually, a sufficient number of complexes form so that there is substantial unbinding (second term on the right), leading to a turnover from linear behavior and ultimately unbinding exactly balances with binding in equilibrium—at the longest times. This should correspond with Figure 9.2

PROBLEM 9.2

Write down a differential equation analogous to Equation 9.3 for the rate of change of $[\text{RL}]$ when the complex RL can undergo a chemical reaction to a product-complex, RP. That is, you will consider the reactions $\text{R} + \text{L} \rightleftharpoons \text{RL} \rightleftharpoons \text{RP}$, with rates for the $\text{RL} \rightleftharpoons \text{RP}$ reaction denoted k_+ (forward) and k_- (backward).

9.2 MACROSCOPIC EQUILIBRIUM AND THE BINDING CONSTANT

We can easily compute the key equilibrium relation by setting to zero the time derivative in Equation 9.3. This will lead to a ratio of rates, which in turn defines an association constant. It turns out that in biochemistry it is more common to use the reciprocal of the association constant, which is simply the dissociation constant, K_d. In terms of the equations, we find that in equilibrium

$$\frac{[R][L]}{[RL]} = \frac{k_{off}}{k_{on}} \equiv K_d. \tag{9.4}$$

The three-line equals sign indicates that this is a definition of the dissociation constant. You can always remember which concentrations go in the numerator, because dissociation means the un-complexed molecules are the focus of the equation: it is the off-rate that appears in the numerator.

We can rewrite Equation 9.4 in a way that is very useful for data interpretation. Recall first that the total concentrations for reaction (9.1) must consist solely of bound and unbound parts, so that $[R]_{tot} = [R] + [RL]$. This means that when exactly half of the receptors are bound, $[R] = [RL]$. If we substitute this 50%-condition into Equation 9.4, we find

$$K_d = [L]_{50}. \tag{9.5}$$

where the subscript "50" reminds us that this is the particular ligand concentration required to bind exactly 50% of the receptors. We will see shortly that this simple equation allows direct graphical determination of K_d from simple titration experiments.

Something about Equation 9.4—and especially Equation 9.5—should seem strange to you. We have emphasized how important it is to have rate constants reflect intrinsic physical quantities, which are independent of receptor and ligand concentrations. The same criteria apply to the dissociation constant, which after all is just a ratio of two intrinsic, physical rates. So why does K_d appear to depend on concentrations?

In fact, K_d is just as independent of concentrations as the rates are! The amazing thing is that the ratio of concentrations in Equation 9.4 and the 50%-value $[L]_{50}$ are themselves independent of the concentrations used in any particular experiment. That is, these particular quantities self-adjust to constant values (for the given external conditions). This has to be true if the rate constants themselves (off and on) are concentration independent, as we have claimed.

9.2.1   INTERPRETING K_d

Because the dissociation constant is so important in biophysics and biochemistry, it is worthwhile to familiarize ourselves with the meaning of K_d values. The first point is to note from Equation 9.4, that K_d is not dimensionless—it is not a simple ratio—but in fact has units of concentration (M). This makes it somewhat easier to discuss.

The most important thing is that stronger binding is described by smaller K_d. Equation 9.5 makes this clear: when a smaller ligand concentration is required to bind half the receptors, it must correspond to stronger binding. Similarly, in Equation 9.4, it is clear that a higher on-rate or larger fraction of bound receptor will make K_d smaller.

Specific values corresponding to biochemical phenomena of interest are important. The strongest binding molecules are said to have nanomolar affinity, which is used a bit casually to mean $1\,\text{nM} < K_d < 1\,\mu\text{M}$—basically anything less than micromolar. Of course, there is a big difference between 10 and 900 nM—so always ask for a specific value if someone just says nanomolar. In the field of drug design, for a compound to be promising it usually will have nanomolar affinity (i.e., something less than micromolar). Drugs, after all, will only be present in low concentrations in the cell, and they often have to compete with endogenous binding molecules—that is, with those molecules naturally present in the cell.

9.2.2 The Free Energy of Binding ΔG_0^{bind} Is Based on a Reference State

Finally, the free energy! The free energy difference between the bound and unbound states is defined by

$$\Delta G_0^{\text{bind}} \equiv +k_B T \ln\left(\frac{K_d}{1\,\text{M}}\right). \tag{9.6}$$

The use of a plus sign instead of the minus you may have expected is the simple consequence of using the dissociation constant instead of its reciprocal, the association constant.

Most importantly, the presence of the 1 M reference concentration makes the free energy (difference) of binding quite a bit different from other free energies we have encountered. The "0" subscript is there to remind us of that. The difference is that ΔG_0^{bind} does not directly correspond to the ratio of populations we have come to expect. Instead, ΔG_0^{bind} effectively has an arbitrary constant built into it. To see this, note that a different value of ΔG_0^{bind} would result, for instance, if 1 mM were used as the reference concentration in the definition (9.6). Because of this, ΔG_0^{bind} does not correspond to the actual energy released due to binding. Fortunately, differences in binding free energies, $\Delta\Delta G^{\text{bind}}$ values, are observable and will be discussed in detail below.

PROBLEM 9.3

The concrete meaning of molarity. Imagine that one type of protein is arrayed on a perfect cubic grid in three dimensions. (a) Calculate the center-to-center distance corresponding to a 1 M concentration. (b) Compare this with the typical size of a protein.

You may also have been surprised that this book would define a free energy difference without making reference to partition functions. Don't worry, we'll get to the partition functions later—the delay is because they turn out to be more complicated than usual in the case of binding. The basic reason for all these complications is that binding is a "second-order process" in which two molecules join to become one object, as opposed to something like a molecular isomerization in which the number of objects stays constant. The second-order nature of binding is also the reason why the on- and off-rates have different physical units.

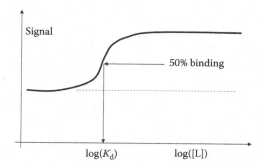

FIGURE 9.3   Determining the binding constant from an experiment.

9.2.3   MEASURING K_d BY A "GENERIC" TITRATION EXPERIMENT

Titration experiments are very useful to discuss because they are so simple in design that even theorists can understand them! In a titration experiment, the concentration of one species, say L, is gradually increased and a signal is measured. We assume the concentration of free ligand [L] can be measured.

The signal used must satisfy two criteria. First, it must distinguish between the bound and free states. Second, its intensity must be proportional to the number of molecules involved. While this may sound restrictive at first, in fact, there are several types of measurements that could work. First, one can monitor the intensity of fluorescence, which is proportional to the number of photons released—which in turn will be proportional to the number of molecules. Fluorescence of tryptophan side chains, for instance, often will be sensitive to binding if they are near a binding site, or if a conformational change is involved. Circular dichroism may also fill the role if a conformational change results from binding.

Titration curves, on a log scale, always have a classic sigmoidal shape. Even without knowing the details of the interactions underlying the measured signal, we can readily determine the point at which 50% of the receptors are bound, as in Figure 9.3. K_d is then determined via Equation 9.5.

Other equilibrium techniques for determining K_d are described in the books by van Holde and by Phillips et al.

9.2.4   MEASURING K_d FROM ISOTHERMAL TITRATION CALORIMETRY

Because isothermal titration calorimetry (ITC) measurements for K_d are very common, it is worthwhile to understand the underlying theory. Your first thought might be that calorimetry, being a measurement of heat, might somehow measure ΔG_0^{bind} directly. However, as we discussed in Section 9.2.2, the binding free energy is not a directly observable quantity since it is based on an artificial reference state.

Surprisingly, ITC uses the heat measured simply to count the number of complexes formed. That is, the key (reasonable) assumption is that the heat evolved during binding is proportional to the number of complexes formed—specifically,

$$Q = [RL] \cdot \Delta H \cdot V. \tag{9.7}$$

Here ΔH is the molar enthalpy change of binding and V is the volume of the experimental system. The volume is known, and ΔH is treated as a fitting parameter.

The ITC procedure measures heat as a function of total ligand concentration $[L]_{tot}$ at some fixed total receptor concentration $[R]_{tot}$. (Both $[L]_{tot}$ and $[R]_{tot}$ are known because they are added to the sample by the experimenter.) Therefore, in order to use Equation 9.7, what's needed is the dependence of $[RL]$ on $[L]_{tot}$. Fortunately, the dependence on total ligand concentration is straightforward to determine. The three equations in (9.2) and (9.4) contain three unknowns ($[L]$, $[R]$, $[RL]$), which can be solved for in terms of K_d, $[L]_{tot}$ and $[R]_{tot}$.

We are almost ready to determine K_d from ITC data. To proceed, one can substitute in Equation 9.7 for $[RL]$ in terms of K_d, $[L]_{tot}$, and $[R]_{tot}$. After measuring Q for a range of ligand concentrations $[L]_{tot}$, one simply fits the two "parameters" ΔH and K_d. The fitting is usually done to the differential form:

$$\frac{dQ}{d[L]_{tot}} = \Delta H \left(\frac{1}{2} + \frac{1 - X_R - r}{2\sqrt{(X_R + r + 1)^2 - 4X_R}} \right), \tag{9.8}$$

where $X_R = [L]_{tot}/[R]_{tot}$ and $r = K_d/[R]_{tot}$.

Full details can be found in the paper by Indyk and Fisher. While our discussion was specific to monovalent binding, the same approach can be generalized to multisite binding.

9.2.5 MEASURING K_d BY MEASURING RATES

Some experiments can directly measure on- and off-rates, thus yielding K_d via the ratio (Equation 9.4). Such techniques include various timed mixing approaches and surface plasmon resonance. See the book by Fersht for further details.

9.3 A STRUCTURAL-THERMODYNAMIC VIEW OF BINDING

The description of binding in terms of on- and off-rates is both correct and simple. For better or worse, however, a lot of physics and chemistry goes into the rates. To really understand binding, we must dig a little deeper and look at this complex array of phenomena that occur. The good news is that all the behavior can be understood in terms of the principles we have already been investigating. There is nothing more than entropy and enthalpy. The "trick" here is to apply our thermo-statistical knowledge to specific molecules and their interactions.

9.3.1 PICTURES OF BINDING: "LOCK AND KEY" VS. "INDUCED FIT"

While we will not dwell on the history of binding theory, it has been studied for over a century. The earliest view, which was prevalent for decades, held that two molecules that bind fit together like a lock and key, which was understood to mean that there was perfect complementarity of shapes that remained static before, during, and after binding. Later, the notion that the receptor would be changed by the ligand—that is, "induced" to bind—became dominant. Considering the microscopic physics, it is hard to see how a ligand could bind to a receptor (and each exert forces on the other) without changes occurring to both.

In the current literature, one sees a discussion of induced fit as being different from an "ensemble picture" in which bound and unbound conformations are assumed to coexist prior to binding. The alleged discrepancy is this: induced-fit theory sees the ligand as changing the receptor to a completely new conformational state, whereas the ensemble picture holds that the "induced" state has always been present in the unbound ensemble but just decreased in free energy upon binding. From the author's point of view, statistical mechanics tells us that all conformations will occur (under any condition—e.g., unbound), although perhaps with low probability. Thus, there is no such thing as a new state. Of course, a state could have such a small probability as to be practically unobservable—and this would not be surprising given that probability is exponentially sensitive to (free) energy as we know from the Boltzmann factor. However, at bottom, it is hard to say that the induced-fit viewpoint is fundamentally different from the ensemble picture. The difference would seem to depend on whether something that is not easily observed (the bound conformation in the unbound state) really occurs—and this book will not make scientific distinctions on philosophical grounds.

9.3.2 MANY FACTORS AFFECT BINDING

We want to understand fully the "before" vs. "after" description of binding sketched in Figure 9.4. There are three important components here—receptor, ligand, and solvent (R, L, and S)—implying three pairs of interactions. Each interacting pair will undergo an enthalpy change—for example, for the receptor-solvent pair, we have ΔH_{RS}. Further, each component, individually, will undergo both an entropy change ΔS and an enthalpy change ΔH. Thus, there are nine types of "thermo-statistical changes" to account for. (Don't forget that the designations "receptor" and "ligand" are somewhat arbitrary and can refer to any two molecules. Also, "solvent" may be primarily water, but typically also includes salts and possibly other molecules. Finally, for "enthalpy" you should think simply of energy, as the pressure component is typically irrelevant.)

In other words, we are dividing up the binding free energy change as follows:

$$\Delta G_0^{\text{bind}} = \Delta H_{RL} + \Delta H_{RS} + \Delta H_{LS} + \Delta H_R + \Delta H_L + \Delta H_S$$

$$- T\left(\Delta S_R + \Delta S_L + \Delta S_S\right). \tag{9.9}$$

While there are indeed entropy changes that result from interactions among pairs of components, we will assign these changes to the individual components; you may want to refer back to our discussion of two and three component systems in Chapter 5. A key example is the translational entropy lost by receptor and ligand due to complexation.

The net binding affinity depends on the balance of all the terms in Equation 9.9. While we can group certain interactions together (e.g., all entropies or all

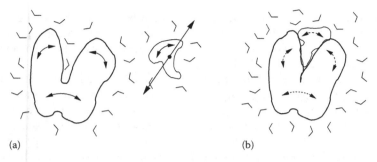

(a)   (b)

FIGURE 9.4 Binding: before and after. The figure shows a receptor and ligand (really, two arbitrary molecules) with solvent represented by the many angled lines, both (a) before and (b) after a binding event. Many factors affect binding—based on the receptor, ligand, solvent, their interactions, and their entropies. The most obvious interaction driving binding is the enthalpy decrease expected from the ligand and receptor coming into contact. For the complexation to occur, however, solvent interactions with both ligand and receptor will be decreased, typically leading to a gain in solvent entropy (due to fewer surface-associated waters)—but solvent-associated enthalpy may also change. The internal entropy of both the ligand and the receptor may change, as represented schematically by curved double-ended arrows that differ in the "before" (solid) and "after" (dashed) pictures. The translational entropy will always decrease upon binding, because the centers of mass of the receptor and ligand can no longer move separately: this is shown schematically by the straight double-sided arrow that occurs solely before binding, assuming the receptor's center of mass is fixed in both pictures. Finally, the internal enthalpy of each molecule may change if binding induces conformational changes.

receptor-ligand interactions), no group by itself is the determinant of binding. It is valuable to consider each term separately:

- Receptor–ligand enthalpy change, ΔH_{RL}. This is the easy one—usually. We expect that if R and L indeed bind, then the enthalpy of the ligand–receptor pair decreases significantly upon complexation. Note that this may be only weakly true in the case of purely hydrophobic solutes in water, when primarily the solvent drives the observed "binding."
- Ligand–solvent enthalpy change, ΔH_{LS}. Before binding, the ligand is completely exposed to solvent, while after complexation some (or all) of the ligand's surface must be buried in the receptor. In general, the loss of contacts between ligand and solvent implies an unfavorable enthalpy change (i.e., increase) even if the lost interactions were only van der Waals contacts.
- Receptor–solvent enthalpy change, ΔH_{RS}. The considerations here are similar to those for ΔH_{LS}. Although there is no strict rule as to whether a receptor-binding pocket will be solvent-exposed or buried, one typically expects that at least some water will reside in the binding site of an unbound receptor. Therefore, one also anticipates a somewhat unfavorable $\Delta H_{RS} > 0$.
- Receptor enthalpy change, ΔH_R. This enthalpy refers to interactions among atoms of the receptor, which are not expected to change too much upon

binding. Also, the details and sign of the change may change from system to system.

- Ligand enthalpy change, ΔH_L. A discussion analogous to that for ΔH_R applies here.

- Solvent enthalpy change, ΔH_S. What about interactions "internal" to the solvent? Although water molecules may be stripped from the surfaces of receptor and ligand, such surface waters tend to maximize their hydrogen bonds, as we discussed in the context of the hydrophobic effect (Chapter 8). Therefore, a minimal change is expected in the enthalpy of water. On the other hand, the solvent also contains salt, which might undergo more significant changes as the result of the possible burying of surface charges or polar groups upon complexation. Because such changes will be highly system specific and complicated, we will not discuss them here.

- Receptor entropy change, ΔS_R. We really need to speak of "internal" and "external" entropy. Internal entropy refers here to the change in the fluctuations of the receptor atoms—that is, the entropy change if the receptor were considered the whole system. Typically, one expects complexation with a ligand to make the receptor somewhat stiffer and decrease the internal entropy—but this is not required in principle. External entropy refers to the overall translational entropy of the receptor molecule, which decreases by definition upon complexation. Two molecules moving separately have more entropy than the same two moving together (see Section 5.2.3). The change in external entropy clearly is shared by the ligand and so part of the entropy loss can be assigned to each.

- Ligand entropy change, ΔS_L. Ligands typically exhibit some flexibility in solution (due to rotatable bonds) that will be suppressed on binding. Thus, ΔS_L generally will oppose binding. This has interesting consequences for drug design. A ligand that is designed to be as "stiff" as possible in solution (few rotatable bonds) will suffer a minimal ΔS_L penalty on binding. Nevertheless, the ligand always loses some translational entropy.

- Solvent entropy change, ΔS_S. Because water molecules will be released from the surfaces that get buried in the complex, one can expect the entropy of those solvent molecules to increase. As we discussed in the context of the hydrophobic effect, the entropy of a water molecule in bulk solution is expected to be substantially higher than a surface water.

9.3.3   ENTROPY–ENTHALPY COMPENSATION

The idea that entropy and enthalpy "compensate" each other—to yield an overall modest binding affinity—is quite important in biological binding. In a living cell, after all, not only must binding occur, but so too must unbinding. In signaling processes, for example, there are often feedback loops to slow down a process after it has produced the desired effect. If binding were not readily reversible, it would be hard to regulate cellular processes precisely.

The overall picture of binding, as we have seen, is complex. Further, the entropy–enthalpy compensation should be expected to apply to the net or overall binding process—rather than to individual terms in Equation 9.9. For instance, solvent entropy is expected to increase, favorably, on binding (since more water molecules will go to the bulk) but solvent enthalpy is not expected to change much. On the other hand, translational entropy of the ligand and receptor will decrease on binding (unfavorably) while their enthalpy will decrease (favorably). The net binding free energy is a balance among all these terms, but one generally expects that entropy lost will be made up with enthalpy.

In other words, entropy–enthalpy compensation is a useful qualitative picture to bear in mind, but it is not a rule of thermodynamics. In fact, the association of two hydrophobic solutes would seem to be a counterexample, since it is primarily the increased entropy of solvent that compensates the lost translational entropy due to complexation.

9.4 UNDERSTANDING RELATIVE AFFINITIES: $\Delta\Delta G$ AND THERMODYNAMIC CYCLES

9.4.1 The Sign of $\Delta\Delta G$ Has Physical Meaning

In many situations of biophysical and biochemical interest, there is a need to measure or estimate relative binding affinities. For instance, imagine you are trying to design a drug that will interfere with a process already occurring—perhaps the binding of a receptor R to a ligand L_a. Your drug candidate molecule will be called L_b. If you are to be effective at stopping or slowing down the L_a binding, the absolute affinity $\Delta G_0^{\text{bind}}(b)$ doesn't really matter. All that matters is that your drug candidate bind with strength comparable to L_a. If possible, you would like the relative binding affinity to favor L_b binding—that is, $\Delta\Delta G_{ab}^{\text{bind}} = \Delta G_0^{\text{bind}}(b) - \Delta G_0^{\text{bind}}(a) < 0$. What's important here is not the convention used to order the subscripts, but that we specify $\Delta G_0^{\text{bind}}(b)$ be smaller than $\Delta G_0^{\text{bind}}(a)$ so that L_b binds more strongly to receptor R.

You will find $\Delta\Delta G$'s everywhere in the experimental and computational literature, and it is essential to understand them. In fact, as we will detail further below, it is often easier to measure a relative binding affinity in an experimental assay. In general, it is much easier to estimate relative affinities computationally, rather than absolute affinities.

Thermodynamic cycles provide a pictorial way of understanding $\Delta\Delta G$. Figure 9.5 shows a thermodynamic cycle for relative binding affinities. The species on the left side are not bound to each other, while those on the right are bound. As we will see, such cycles have a nice and simple mathematical property due to the fact that each corner represents a well-defined state—once the concentrations and other conditions of each state have been specified. You may want to review our discussion of "state functions" in Section 7.6.2. (The reference concentration is usually taken to be 1 M, which seems a logical choice until you realize it is physically impossible for most proteins—see Problem 9.3.)

One essential point about thermodynamic cycles is that they are very general. For instance, even though the cycle of Figure 9.5 apparently describes the binding of a "receptor" to a ligand, it could equally well describe the binding of two

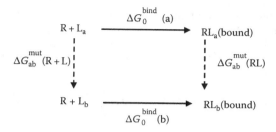

FIGURE 9.5 A thermodynamic cycle that explains relative binding affinities ($\Delta\Delta G^{bind}$) of two ligands L_a and L_b to one receptor R. Each corner of the cycle corresponds to a well-defined thermodynamic (equilibrium) state. For instance, the upper-left corner indicates the state composed of all unbound configurations of the receptor R and the ligand L_a at fixed reference concentrations. Other corners are defined similarly. Note that the horizontal arrows connote measurable binding affinities ΔG_0^{bind}. This cycle was employed by Tembe and McCammon for the calculation of relative binding affinities using "alchemical" methods.

proteins—that is, if the "ligands" are themselves proteins. Indeed R, L_a, and L_b could be three arbitrary molecules, as will be shown in the problems below.

PROBLEM 9.4

Draw a thermodynamic cycle embodying the binding of a single ligand to both a "wild-type" receptor and to a mutated receptor.

PROBLEM 9.5

Draw a thermodynamic cycle representing the relative solvation free energy of two different molecules. Solvation refers to the difference between a molecule in water and in vacuum ("gas phase").

The key quantitative point of a thermodynamic cycle is that once the conditions are specified, the absolute free energy G of each state (corner of the cycle) is fully determined. This is the state function idea of Section 7.6.2. Because G depends only on the state, the free energy differences ΔG must sum to zero going around the cycle. This sum will depend on the way the ΔG values are defined—that is, by the direction of the arrow between two states. The convention is that the arrow points from the initial state to the final state, so the free energy difference is that of the final state less that of the initial: $\Delta G(\text{state X to state Y}) = G(Y) - G(X)$. Although the vertical arrows correspond to mutations (ΔG^{mut}) that could not be enacted in an experiment, such "alchemical" processes will turn out to be important in computations. Regardless of this, the free energy differences ΔG^{mut} are well defined—based on simple subtraction—since the absolute free energy is defined for each corner.

For the cycle of Figure 9.5, we therefore have

$$\Delta G_0^{bind}(a) + \Delta G_{ab}^{mut}(\text{bound}) - \Delta G_0^{bind}(b) - \Delta G_{ab}^{mut}(\text{free}) = 0, \qquad (9.10)$$

which is equivalent to an exact relation between the relative binding affinity and the difference in mutation free energies:

$$\Delta\Delta G_{ab}^{bind} = \Delta G_{ab}^{mut}(bound) - \Delta G_{ab}^{mut}(free) \equiv \Delta\Delta G^{mut}(free \to bound).$$

(9.11)

Below, we will see how this apparently trivial relation can be of practical importance in computations.

9.4.2 COMPETITIVE BINDING EXPERIMENTS

Relative affinities can be measured from competitive binding experiments. In competitive binding, a ligand (L_a) with a measurable signal (see Figure 9.3) is first added to a solution containing the receptor until saturation occurs—binding of all receptors. Then a second ligand (L_b) that also binds the receptor is titrated into the solution and the (now decreasing) signal is measured as a function of the increasing concentration of L_b. Based on the data collected, the relative affinity of the second ligand can be measured. From an experimental point of view, this is very convenient, since it is only necessary to measure the signal from one ligand.

In the problem below, you will show that the data just described is sufficient for determining the relative affinity or the second dissociation constant, if the first is known. Note that the ability to measure an ordinary binding curve for L_a implies that the dissociation constant for L_a should be considered known.

PROBLEM 9.6

(a) Write down the five equations governing competitive binding between two ligands. [Hint: See Section 9.2.4.] (b) Show that these equations contain enough information to determine the second dissociation constant from a competitive binding curve.

9.4.3 "ALCHEMICAL" COMPUTATIONS OF RELATIVE AFFINITIES

It turns out to be fairly simple to design a calculation, which, for the "mutational" free energy differences (ΔG^{mut}) of Equation 9.10, can be combined to yield the physically important relative affinity $\Delta\Delta G^{bind}$. We will finally call on our statistical mechanics to make this possible.

The key point is to realize that—to a computer program—the only thing that differentiates two ligands (e.g., L_a, and L_b) is the potential energy function (forcefield) used to describe them (see Section 5.3.3). In our case, we will use the notation U_a and U_b for the two different forcefields, which will include ligand, receptor, and solvent. Configurations of the whole system will be denoted $\mathbf{r}^N$. For now, we will assume that both ligands have the same number of atoms, so that the forcefields differ only in their parameters (force constants)—recall Chapter 5. In fact, ligands with different numbers of atoms can also be treated, as described in Problem 9.7.

Consider the calculation of the bound-state mutation free energy difference, $\Delta G_{ab}^{mut}(bound)$, which is the right vertical leg in Figure 9.5. The configuration integral for the state with ligand L_a (upper right corner) is $\hat{Z}_A = \int d\mathbf{r}^N \, e^{-\beta \, U_a(\mathbf{r}^N)}$, and there is an analogous expression for L_b. The free energy difference is therefore given by the ratio of integrals:

$$\exp\left[-\Delta G_{ab}^{mut}(\text{bound})/k_B T\right] = \frac{\int_{bound} d\mathbf{r}^N \, \exp\left[-\beta \, U_b(\mathbf{r}^N)\right]}{\int_{bound} d\mathbf{r}^N \, \exp\left[-\beta \, U_a(\mathbf{r}^N)\right]}$$

$$= \frac{\int_{bound} d\mathbf{r}^N \, \exp\left[-\beta \, U_a(\mathbf{r}^N)\right] \exp\left[-\beta \, \Delta U(\mathbf{r}^N)\right]}{\int_{bound} d\mathbf{r}^N \, \exp\left[-\beta \, U_a(\mathbf{r}^N)\right]},$$

(9.12)

where we have used the notation $\Delta U = U_b - U_a$. The integrals are to be performed only over configurations defined as bound (perhaps via the RMSD "distance" to a known experimental structure of the complex; RMSD is defined in Equation 12.11).

The somewhat daunting ratio of integrals in Equation 9.12 can be rewritten, however, using a standard "trick." Notice that the rightmost ratio is exactly in the form of a canonical average (for the U_a ensemble) of the Boltzmann factor of ΔU. This means that the free energy difference can be written in a very simple form amenable to computation, namely,

$$\exp\left[-\Delta G_{ab}^{mut}(\text{bound})/k_B T\right] = \left\langle \exp\left[-\beta \, \Delta U(\mathbf{r}^N)\right]\right\rangle_{a,bound},$$

(9.13)

where the canonical average $\langle \cdots \rangle_{a,bound}$ must be taken with respect to the Boltzmann factor of U_a and include only bound configurations. The unbound free energy difference, $\Delta G_{ab}^{mut}(\text{unbound})$, is given by the same expression as (9.13), except integrated over those configurations are not deemed to be bound.

The computational procedure for evaluating Equation 9.13 is very straightforward. One first generates an equilibrium ensemble for the potential U_a—that is, a set of configurations $\mathbf{r}^N$ distributed according to $e^{-\beta \, U_a(\mathbf{r}^N)}$. Then one computes the average of the Boltzmann factors $e^{-\beta \, \Delta U(\mathbf{r}^N)}$ over all the sampled configurations. That's it! (At least, that's it, in principle. In practice, if the two forcefields U_b and U_a are too different, then the calculation is numerically difficult—since the configurations in the U_a ensemble are not important in the U_b ensemble. See Chapter 12 for more information.)

At the end of the day, one returns to Equation 9.11 to calculate the physically important quantity $\Delta\Delta G^{bind}$ using the bound and unbound estimates for ΔG^{mut}.

PROBLEM 9.7

If the ligands L_a and L_b have different numbers of atoms, then the associated configuration integrals in the mutation steps of a thermodynamic cycle will not form a dimensionless ratio. Show that the addition of noninteracting "dummy atoms," which are statistically independent from the rest of the molecule, (a) can lead to a dimensionless ratio and (b) do not affect the final value of $\Delta\Delta G$.

9.5 ENERGY STORAGE IN "FUELS" LIKE ATP

"ATP is the fuel of the cell," we often hear. Although this is one of the most important facts in biochemistry, energy storage in ATP is described quite differently in different places. The explanations tend to be confusing, at best, and misleading at worst.

In fact, free energy storage in ATP (and other molecules) can be understood quite directly from our discussion of binding. At first, this seems a strange idea, but every chemical reaction can be analyzed in terms of binding concepts. Here, we're concerned with the synthesis/hydrolysis of ATP, described as

$$ADP + P_i \rightleftharpoons ATP, \tag{9.14}$$

for which there is a standard dissociation constant,

$$K_d = \frac{[ADP][P_i]}{[ATP]}. \tag{9.15}$$

Although the reaction (9.14) is almost always catalyzed in both directions in the cell, that does not change the fundamental equilibrium between the species. Equilibrium depends only on the free energies of the species (in the given conditions). To put it more directly, given the conditions at hand (e.g., pH, salt concentration, temperature, etc., of a living cell), the ratio of concentrations defining K_d is determined—at equilibrium!

If we want to be a bit more careful, we can imagine that Equation 9.15 is general and simply defines K_d even out of equilibrium, while a special equation holds at equilibrium and defines ΔG_0^{bind}, namely,

$$K_d^{equil} = \frac{[ADP]^{equil}[P_i]^{equil}}{[ATP]^{equil}} = 1\,M\,\exp\left(\frac{\Delta G_0^{bind}}{k_B T}\right). \tag{9.16}$$

Then, following up on our discussion of Section 9.2.2, we can calculate a physical free energy $\Delta\Delta G$ using the relation

$$e^{\Delta\Delta G/k_B T} \equiv \frac{K_d}{K_d^{equil}}. \tag{9.17}$$

Equation 9.17 holds the secret to chemical fuel. If the K_d ratio of concentrations is at its equilibrium value, then there is no free energy available. However, the cell works hard (burning food!) to make sure that K_d is far from equilibrium, such that $\Delta\Delta G \simeq -20k_B T$. In other words, the cell shifts the reaction (9.14) far to the right, so there is always a superabundance of ATP. Splitting ATP (into ADP and P_i) releases some of the free energy stored in the far-from-equilibrium concentrations.

Equation 9.17 immediately implies that the energy stored in ATP depends directly on the concentrations of species. Therefore, ATP is emphatically *not* a fuel in the vernacular sense: the energy available from "burning" one ATP molecule fully depends on the environment—that is, the nonequilibrium concentrations, as well as the conditions embodied in K_d^{equil}.

You might object that if all this were true, then ATP could never power biochemical activity at the single-molecule level. That is, the concentrations in the cell (or test tube) would not seem to affect a single protein's use of ATP. Well, yes and no. An enzyme that has an ATP bound to it will indeed perform the required hydrolysis, almost regardless of the environment. At least, this enzyme will perform the catalysis that one time. But to repeat the performance, the protein will need to release the products ADP and P_i, and that's where the catch is. If we call the enzyme E and ignore P_i for the moment, it is critical to realize that the E-ADP complex will be in a dynamical equilibrium with the surrounding environment. When there is sufficient ADP in the environment, ADP will tend to occupy the binding site and prevent ATP from binding.

Really, this is quite a beautiful story, and simple concepts explain it. You can read more about this in the books by Nicholls and by Phillips et al.

9.6   DIRECT STATISTICAL MECHANICS DESCRIPTION OF BINDING

Although Equation 9.13 is used indirectly to calculate binding free energies, it is less than satisfying as a statistical mechanics explanation of the binding affinity. Here, we will express ΔG_0^{bind} in terms of a ratio of partition functions.

9.6.1   WHAT ARE THE RIGHT PARTITION FUNCTIONS?

With the goal of writing down the statistical mechanics derivation of ΔG_0^{bind}, we might first try writing down a partition function for all the molecules involved and then work from there. Indeed, such an approach is possible (see the articles by Gilson et al. and by Woo and Roux), but here we will take a simpler path. Our analysis will apply most directly to the case of relatively high affinity binding.

Recall that the reciprocal Boltzmann factor of ΔG_0^{bind}—that is, K_d—is equal to the free ligand concentration in molar units when exactly half the receptors are bound (see Equation 9.5). Fortunately for us, the dimensionless fraction 1/2 comes into play. A ratio of partition functions is exactly the right tool for estimating a fractional probability.

To construct our partition-function ratio, we will consider the case when receptor and ligand are present in equal overall concentrations—that is, $[L]_{tot} = [R]_{tot}$. (The generality of this apparently special case is explained below.) With equal concentrations, we can then study just a single ligand and receptor in a fixed volume, since interactions with other ligands and receptors should be negligible in the concentration ranges of interest (see below). The basic idea is that for R and L in a very large volume—at very low concentrations—only a tiny fraction of the ensemble will be bound, while at small volumes, when the ligand and receptor are forced to be nearby one another, the bound fraction will exceed 1/2. Therefore, if the partition function is considered over a range of volumes V, a special value V^* can be found, which leads to exactly 50% binding. The volume V^* directly determines the free ligand concentration at half-binding: since the ligand is free half the time, $[L]_{50} = 1/2V^*$.

In principle, then, one solves for the volume V^*, which makes the ratio of bound and unbound partition functions equal to 1:

$$1 = \frac{\int_{V^*,\text{bound}} d\mathbf{r}^N \, e^{-\beta \, U(\mathbf{r}^N)}}{\int_{V^*,\text{unbound}} d\mathbf{r}^N \, e^{-\beta \, U(\mathbf{r}^N)}}. \tag{9.18}$$

One then has $K_d = [L]_{50} = 1/2V^*$.

The point here is to appreciate the relation between the microscopic statistical mechanics and the thermodynamic quantity ΔG_0^{bind}. In practice, using an equation like (9.18) to estimate ΔG_0^{bind} computationally would not be easy.

9.6.1.1 Technical Point: The Assumption of Equal Overall Concentrations

Our analysis is based on a seemingly strange assumption of equal total concentrations of ligand and receptor: $[L]_{\text{tot}} = [R]_{\text{tot}}$. However, in the case of sufficiently strong binding, this turns out to be fully general. To see why, consider first the limit of very low (equal total) concentrations of R and L. Because the ligands and receptors are so dilute, they will rarely find each other in the sample cell: equivalently, the on-rate will be tiny and the translational entropy of the unbound state will be dominant. As we increase the concentration to the point where R and L become close together, at least half the receptors will become bound if the affinity is reasonably strong. The fraction of bound receptors indeed increases with growing overall concentrations, as can be seen from $[RL]/[R] = [L]/K_d$, since $[L]$ will increase as $[L]_{\text{tot}}$ increases.

The only remaining question is then whether the case of equal total concentrations is sufficient to lead to binding of half the receptors for typical K_d values. We need to determine whether the affinity is strong enough (whether K_d is low enough) to permit 50% binding before the R and L molecules crowd into one another. For $K_d < 1$ mM, which would seem to include most biologically relevant binding, the affinity should be strong enough. The reason is that $K_d = 1$ mM means that the free ligand concentration is $[L] = 1$ mM, which in turn implies each free ligand occupies a volume of about $(120 \text{ Å})^3$ (see Problem 9.3). With equal ligand and receptor concentrations at half-binding, 120 Å is also the typical spacing between receptors and complexes. This will allow sufficient space to prevent crowding for relatively small receptors and thus permit 50% binding. Of course, if binding is stronger (K_d lower), then the spacing will be even bigger and our assumptions will certainly be valid. Most biological interactions reported to date have $K_d \ll 1$ mM.

9.7 ALLOSTERY AND COOPERATIVITY

9.7.1 BASIC IDEAS OF ALLOSTERY

Allostery is another one of those biophysics topics with a very intimidating sound to it. The basics of allostery are not hard to understand, however, if we use the binding concepts already developed.

Allostery can occur when a molecule (or molecular complex) can bind more than one ligand. As always, these ligands might be small molecules, or they might be proteins. They might be different or the same. Allostery means that the binding of one ligand affects the binding (affinity) of the other. Hemoglobin is the classic example, where there are four binding sites that usually take up the same ligand at any given time (molecular oxygen or carbon monoxide). Allostery is most interesting when the binding sites are well separated on the protein, as is the case with hemoglobin and many other examples.

The phenomenon of binding at one site affecting binding at another site is also termed cooperativity because the sites are said to work together. However, the friendly sound of the word should not mislead you about the phenomenon. There can be positive or negative cooperativity. Positive cooperativity between two sites implies that the binding of the first ligand increases the affinity of the second site from what it would be in isolation, while negative cooperativity implies the affinity of the second site is decreased.

Although allostery generally refers to cooperativity in binding (either positive or negative), the word now has a somewhat broader meaning. Essentially any time a binding event has an effect elsewhere on the protein, this is considered allostery. Such an effect might be a dramatic conformational change or simply a change in the fluctuations.

From a structural point of view, allostery is thought to be somewhat nonintuitive. Indeed, the binding sites for different ligands are often well separated since most allostery takes place in multi-meric proteins so that ligand separations are about the size of the individual protein monomers. From this perspective, since the binding of one ligand affects another in an allosteric system, the effects of binding "propagate" to distant parts of the protein or complex. Such action-at-a-distance, historically, has been the unexpected part of allostery.

From a statistical mechanics point of view, one could argue that allostery is almost obvious and should generally be expected (see the article by Gunasekaran et al.). After all, compared to the ensemble representing the unbound state, one expects the ensemble of a bound state to be different. Since a free energy difference like the binding affinity for a second ligand reflects the balance between ensembles (singly bound vs. doubly bound), it must be affected in principle by anything that modifies either of the ensembles.

Note that cooperative binding is not simply ordered binding. That is, the situation is not that the highest-affinity site takes up the first ligand, and the next highest takes the second, and so on. If this were true, then all apparent cooperativity would be negative. As we said, both positive and negative cases occur.

Figure 9.6 illustrates the effects of idealized positive cooperativity on traditional binding curves. The curve for $n = 1$ is a standard binding curve simply based on the definition for K_d, Equation 9.4, where we have used simple algebra to solve for $[RL]/[R]_{tot}$ using $[R]_{tot} = [R] + [RL]$. The other curves use the Hill model for n ligands described below, which assumes that all ligands after the first bind extremely strongly. The key point is that when there is cooperativity, it will be visible in binding measurements. This notion is quantified further in our discussion of the Hill constant below.

Figure 9.6 also suggests why hemoglobin, which has $n > 1$, would provide superior oxygen exchange compared to an $n = 1$ oxygen carrier. The lungs have a much

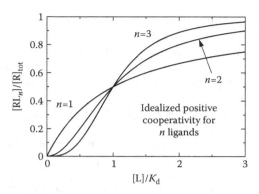

FIGURE 9.6 Idealized, positively cooperative binding curves. Receptors binding different numbers (n) of ligands are contrasted. Because the binding is favorably cooperative for $n > 1$, at high ligand concentrations, there is greater saturation of the receptors. When there is lower free ligand concentration [L], however, there is less saturation for $n > 1$ because n ligands are required for any binding in this model. These curves help to illustrate the importance of hemoglobin for oxygen transport, as explained in the text. The units of the horizontal axis are somewhat arbitrary for $n > 1$, where "K_d" is shorthand for $\hat{K}_d^{1/n}$ based on Equation 9.20.

higher concentration of the free ligand oxygen—that is, higher $[L]/K_d$—compared to the tissues that receive oxygen from hemoglobin. Positively cooperative binding, which is sketched in the figure for $n > 1$, permits both a greater saturation of hemoglobin at the lungs and also a greater release of the ligands when the tissues are reached. This is a more efficient delivery of ligand than could be achieved by a noncooperative binder, which would release fewer oxygen molecules for each switch from lungs to tissue.

PROBLEM 9.8

(a) Explain which of the free energy terms discussed in Section 9.3 reflect the modification of the receptor ensemble discussed in the context of allostery. (b) Of course, the perturbation to the receptor ensemble is mirrored by changes to the ligand ensemble. Which of the free energy terms reflect such changes?

9.7.2 QUANTIFYING COOPERATIVITY WITH THE HILL CONSTANT

How do we begin to quantify our ideas about allostery? As usual, we will be well served by looking at the simplest case first and trying to understand that as well as possible.

Let's assume that our receptor R can bind n identical ligands. In general, this would lead to a series of equilibria among many species:

$$R + L \rightleftharpoons RL_1 \quad K_d^{(1)}$$
$$RL_1 + L \rightleftharpoons RL_2 \quad K_d^{(2)}$$

$$RL_2 + L \rightleftharpoons RL_3 \quad K_d^{(3)}$$

$$\vdots$$

$$RL_{n-1} + L \rightleftharpoons RL_n \quad K_d^{(n)}, \tag{9.19}$$

with a separate equilibrium constant $K_d^{(j)}$ for each reaction. If the system described by the reactions (9.19) is allosteric (cooperative), then by definition the equilibrium constants will differ. Nevertheless, a quantitative analysis of these coupled reactions is fairly involved. It can be simplified somewhat by assuming that each binding site can occupy only two states, as discussed in a famous paper by Monod et al.

We will focus on a far simpler analysis due to Hill, which captures much of the essential behavior. The key idea is to assume initially that the intermediates $RL_1, RL_2, \ldots, RL_{n-1}$ do not occur—that is, their populations are so small as to be irrelevant. (Later we will see that the analysis can discern intermediates.) Instead of the coupled equilibria of Equation 9.19, we therefore write only the single equation for n ligands binding a single receptor:

$$R + nL \rightleftharpoons RL_n \quad \hat{K}_d, \tag{9.20}$$

where we assume a single effective equilibrium constant defined by

$$\text{Hill analysis:} \quad \hat{K}_d \equiv \frac{[R][L]^n}{[RL_n]}. \tag{9.21}$$

Note from dimensional considerations that $\hat{K}_d$ has the same units as $(K_d)^n$.

If the model embodied in Equation 9.20 were exact, note that we would expect binding curves precisely as shown in Figure 9.6. These curves are derived from Equation 9.21 using the substitution $[R] = [R]_{tot} - [RL_n]$ and a little algebra.

PROBLEM 9.9

(a) Write the time-dependent differential equation corresponding to Equation 9.20. (b) Explain why the binding constant $\hat{K}_d$ (Equation 9.21) is indeed constant and appropriate given the reaction and the assumption of no intermediates.

Our analysis really has two goals: (i) to estimate the assumed constant n based on an experimentally measurable parameter and (ii) to see if experimental behavior follows the assumptions built into our simple model. As before, we will assume that we can measure some kind of spectroscopic signal of the bound fraction of receptors, which here corresponds to

$$y \equiv \frac{[RL_n]}{[R]_{tot}}, \tag{9.22}$$

where $[R]_{tot} = [R] + [RL_n]$. By solving for $[RL_n]/[R]$ in terms of y, substituting in Equation 9.21 and taking the logarithm, one finds that

$$\log \left(\frac{y}{1-y} \right) = n \log [L] - \log \hat{K}_d. \tag{9.23}$$

Equation 9.23 provides a very powerful analysis of allosteric binding data—that is, of pairs of y and $[L]$ values. If indeed $\hat{K}_d$ and n are constants, then they can be determined by the intercept and slope of a plot of $y/(1-y)$ vs. $[L]$, which would be linear. The reality is somewhat more complicated in typical cases, but also much richer.

Let's analyze a schematic Hill plot in detail (Figure 9.7). First, we see that the behavior is not purely linear, but almost looks like a concatenation of three linear regimes. However, we can still extract quantitative information about cooperativity. Indeed, the degree of cooperativity is generally defined by the slope n of Equation 9.23 at half-saturation—that is, at $y = 1/2$ so that $\log(y/(1-y)) = 0$. This particular slope is named the Hill constant, h. It is approximately three for hemoglobin—slightly less than the maximum value of four based on the four binding sites.

We can gain more insight into Figure 9.7 by considering the intermediates that are apparently present. (If there were no intermediates, we would have had a straight line.) The easiest way to do this is by examining the extreme limits of low and high concentrations. In each of these cases, a ligand really has only one choice of which site to bind: at high concentrations, there will be only one site remaining on each receptor, while at low concentrations the "first site" (the site with highest affinity in the receptor's unbound state) will get the ligand. Even if there are two or more identical sites, at low-enough ligand concentrations only one of these will bind a ligand. In both limiting concentrations, then, a Hill plot can be expected to display an effective slope of $n = 1$ in Equation 9.23. Furthermore, extrapolating these limiting linear behaviors back to $\log[L] = 0$ should provide insight into the apparent

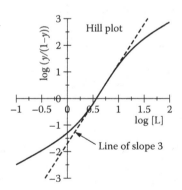

FIGURE 9.7 A schematic Hill plot for oxygen binding in hemoglobin, which has four binding sites. Although a simple "all or none" model—lacking intermediates—suggests the data (solid line) should follow a straight line with slope $n = 4$, the observed behavior tends to be much more complicated. Because the slope is not constant, we can conclude that intermediates do play a role. The maximum slope, $n \simeq 3$, indicates strong—though not total—cooperativity. More generally, we can observe that the maximum slope in a Hill plot should depend on the difference between the limiting behaviors at low and high ligand concentrations.

equilibrium constants for the first and last ligands. The larger the separation between the two limiting straight lines, the larger the difference between the effective binding constants, and the larger the cooperativity.

Our discussion and Figure 9.7 have focused on the case of positive cooperativity, which means that the effective binding affinity increases ($\hat{K}_d$ decreases) as more ligands bind. However, the opposite case of negative cooperativity is also possible.

PROBLEM 9.10

Sketch what a Hill plot would look like for negative cooperativity.

9.7.3 FURTHER ANALYSIS OF ALLOSTERY: MWC AND KNF MODELS

Although the Hill plot is extremely informative, it does not really tell us about possible microscopic origins of the cooperativity it reveals. We will discuss two models that attempt to provide such details, although the analysis remains at a fairly abstract level—attempting to discern how many discrete conformational states best describe an allosteric process. In other words, we will not learn specific structural/geometric details. Nevertheless, estimating the number of important states is a major step in understanding the underlying structural biology.

In both approaches we briefly look at, the basic plan is (i) to assume that every "binding condition," specified by the number of ligands, is described by a small number of conformational states and (ii) to assign a binding constant to each conformational state and binding condition. Thus, in addition to the equilibrium between binding and unbinding, there is also a balance between the different conformational states.

9.7.3.1 MWC Model

Monod, Wyman, and Changeux (MWC) developed a simple picture based on two conformational states for the whole receptor. To be concrete, we will consider a receptor that, like hemoglobin, can bind four identical ligands, L. (The generalization to fewer or more ligands is simple.) Assume that the receptor can occupy two states, A and B, which can interconvert when no ligands are bound. Also assume that states A and B can bind any number of ligands, leading to the complexes denoted by AL_j and BL_j with $j = 1, 2, 3, 4$. This reaction scheme is given by

$$A \rightleftharpoons B$$
$$A + L \rightleftharpoons AL \qquad B + L \rightleftharpoons BL$$
$$AL + L \rightleftharpoons AL_2 \qquad BL + L \rightleftharpoons BL_2$$
$$AL_2 + L \rightleftharpoons AL_3 \qquad BL_2 + L \rightleftharpoons BL_3$$
$$AL_3 + L \rightleftharpoons AL_4 \qquad BL_3 + L \rightleftharpoons BL_4 \qquad (9.24)$$

Where does the allostery come in? Well, assume that state A is dominantly populated in the unbound state ($[A] \gg [B]$), but that the ligand has much greater affinity for B than A. In the absence of any ligand, the (total) concentration of A will be much greater than B. As soon as ligand is added, however, the overall equilibrium of A vs. B starts to shift to B. When this happens, the observed (average) affinity increases because a larger fraction of the receptors occupy the high-affinity B state.

While the MWC picture nicely explains positive cooperativity, the discussion just presented shows why the MWC model is unable to explain negative cooperativity. With increased presence of ligand, the stronger affinity state will always win out in the balance of conformations, and the average affinity will therefore increase.

PROBLEM 9.11

By using only the first two lines of Equation 9.24—the first three reactions—show that even with just a single binding site of higher affinity for B, the overall concentration of B increases upon addition of ligand. That is, if the unbound conformational equilibrium is governed by $K^{eq} = [B]/[A]$ and the two binding equilibria are governed by $K_d(A) > K_d(B)$, show that $([BL]/[AL]) > K^{eq}$. This implies, in turn, that binding has increased the overall proportion of the B state—that is, $[B]_{tot}/[A]_{tot} > K^{eq} = [B]/[A]$. Additional sites will increase the effect.

There are more details to the MWC model, which can be found in their original paper. For instance, the model assumes that the "microscopic" binding constants are not affected by multiple ligand binding. Nevertheless, the following problem draws your attention to a technical consideration arising from even this simple assumption.

PROBLEM 9.12

When there are four identical binding sites on a receptor, a ligand has a chance to bind to each. (a) Taking this fact into account, write down the differential equation governing the time dependence of the concentration of [AL], based solely on the reaction $A + L \rightleftharpoons AL$. (b) Show that the observed equilibrium constant will therefore differ from the single-site constant by a factor of 4. Similar, but more complicated, considerations apply for the binding of the second ligand.

9.7.3.2   KNF Model

The KNF model—due to Koshland, Nemethy and Filmer—while also simplistic, contains important generalizations over the MWC model. Most importantly, as each ligand binds in the KNF model, the binding constants for subsequent ligands can change. This allows for negative cooperativity, as the set of binding constants are free parameters in the model. However, the KNF scheme becomes more like a fitting procedure, rather than a predictive model, with this increased number of parameters. Note that the KNF model is largely equivalent to earlier work by Adair.

9.7.3.3   Comprehensive Descriptions of Allostery

The allostery discussion of the last few pages has attempted to present the key ideas, but it hardly described the many possible analyses of allostery. The reader interested in this topic should consult the discussions in the books by Cantor and Schimmel, and by van Holde, Johnson, and Ho.

9.8 ELEMENTARY ENZYMATIC CATALYSIS

Binding is a fundamental part of biological catalysis, the speeding up of chemical reactions. As we discussed in Section 4.3.2, catalysts can lower barriers to reactions, although not the intrinsic (free) energies of the reactants and products. The reactant and product free energies are state functions.

Enzymes are biology's protein catalysts, which are capable of speeding reactions by factors of many orders of magnitude! Catalytic processes may work on one or more ligands, called substrate molecules, which gets transformed into product(s) after the catalyzed reaction. For example, when the energy-storage molecule ATP is used, it is enzymatically cut into two parts, ADP and a phosphate. The dissociation of ATP into two parts, which involves breaking a covalent bond (unlike the unbinding we have been discussing above), is a chemical reaction that would almost never occur without catalysis. Most covalent bond breakage and formation in biomolecules occurs via catalysis—and these events are critical in the life of the cell. Aside from the ATP use just described, additional examples include the synthesis of essential molecules, such as ATP itself, as well as the covalent attachment/detachment of phosphate groups to proteins that are critical in signaling processes.

Every biochemistry book, and many biophysics books too, gives a description of catalysis by enzymes within the Michaelis–Menten model. Our goal here is not to repeat what has been said many times, but to ensure that our understanding of the basic assumptions is very solid. We do this by treating catalysis directly within the context of our binding discussions. The goal of this brief discussion is to enable you to approach other catalysis descriptions with confidence.

We would like to describe catalysis in analogy to the expressions (9.1) and (9.3) for simple binding. We will describe enzymatic catalysis in the accepted way, as consisting of two steps. First the enzyme (E) and the substrate (S) bind together to form an uncatalyzed complex (ES), followed by the chemical transformation of substrate to product (P). These steps are equivalent to the scheme

$$E + S \underset{k_-}{\overset{k_+}{\rightleftharpoons}} ES \overset{k_{cat}}{\rightarrow} E + P. \qquad (9.25)$$

The rates for each step are given above and below the arrows. However, note that the inverse of the chemical transformation step is assumed not to occur at all (i.e., no reverse arrow). In reality, the reverse transformation would occur, but so infrequently as to be negligible.

While Equation 9.25 is indeed the standard formulation of catalysis, there is something quite weird about it, from an equilibrium point of view. While the leftmost binding of E and S to form ES will try to maintain an equilibrium according to the ratio of the rates (k_-/k_+), the essentially irreversible transformation of S to P on the right will constantly deplete the complex ES. Thus, as time goes on, eventually there will be no substrate left at all unless it is added to the system.

9.8.1 THE STEADY-STATE CONCEPT

Michaelis and Menten realized that biological cells don't just perform a "task" like catalysis and then stop working. Rather, most cellular processes operate continually. In the case of enzymatic catalysis, this means that at the same time that substrate is being transformed into product, more substrate is being formed. In fact, Michaelis and Menten modeled catalysis by assuming that exactly as much substrate is added to the system as is catalyzed into product. In such a situation, the whole system described by Equation 9.25 is not in equilibrium since substrate is continually being turned into product. It is sometimes said that large substrate concentration leads to a steady state, but this is only approximately true, as we will see.

Instead of equilibrium, for a fixed total concentration of enzyme ($[E] + [ES] =$ const.), one can have what is called a "steady state." The amount of product produced per unit time is constant, as are all other concentrations. Indeed, if product is steadily produced, then one also expects a constant (in time) ES concentration since product production is proportional to [ES]. We analyze the model by first noting the key differential equations for the scheme (Equation 9.25), namely,

$$\frac{d[P]}{dt} = k_{cat}[ES], \qquad (9.26)$$

$$\frac{d[ES]}{dt} = k_+[E][S] - k_-[ES] - k_{cat}[ES], \qquad (9.27)$$

If we make the steady-state assumption ($d[ES]/dt = 0$), then we find that

$$\frac{[E][S]}{[ES]} = \frac{k_- + k_{cat}}{k_+} \equiv K_M, \qquad (9.28)$$

which defines the equilibrium-like constant, K_M.

(As an aside, note that Equation 9.28 is, in fact, a condition on the various concentrations for a steady state to hold. In principle, given some $[E]_{tot}$, one can adjust the rate at which substrate is added until the condition holds over time. From this point of view, we can see that simply having large [S] is not sufficient for a true steady state—although we will see later that there is an approximate steady state in that limit. Further, one can have a steady state at any [S] value—so long as new substrate is added at a rate that keeps [E][S]/[ES] constant in time.)

Since the ratio of rates in Equation 9.28 is assumed constant, as are [E] and [ES], the steady state requires [S] to be constant, although not necessarily large. In words, as product is created, substrate must be added to balance the loss—and indeed this seems a reasonable description of a continually functioning process in a living cell.

The reason why large substrate concentration leads approximately to a steady state is embedded in Equation 9.28. With large enough [S], essentially all the enzyme will be bound ($[ES] \gg [E]$). Further, whenever product is released, additional substrate will immediately be available to bind, implying fairly constant values of [ES] and [E]. Equation 9.28 shows that near-total complexation will occur when $[S] \gg K_M$.

9.8.2 THE MICHAELIS–MENTEN "VELOCITY"

To make contact with the standard presentations of the Michaelis–Menten analysis, note that one can define the "velocity" $v = d[P]/dt = k_{cat}[ES]$ of product formation. Using the equations already derived, v can be determined as a function of substate concentration to produce the classical Michaelis–Menten plot. You can pursue this further in any biochemistry text or in the book by Fersht.

By continuing our steady-state analysis, we can see that the product formation rate is $v = (k_{cat}/K_M)[E][S]$, which means that the ratio k_{cat}/K_M is the effective rate constant for catalysis. Physically, K_M^{-1} measures the tendency for enzyme-substrate binding to occur while k_{cat} is the probability per time for catalysis once binding has occurred. It is a remarkable fact that many enzymes work at a rate (k_{cat}/K_M) nearly equal to the collision frequency for enzyme and substrate expected from diffusion alone. In other words, every time E and S come into contact in solution—in the highly efficient cases—catalysis occurs. This is called "diffusion-limited" catalysis.

Finally, it is worth noting that our analysis has only considered single-substrate enzymes, whereas many enzymes also require an additional energy-providing ligand (ATP) to drive the catalysis. The analysis of this case is beyond the scope of the book.

9.9 pH AND pK_a

Because pH and pK_a are fundamental to the chemistry of proteins and other biomolecules—and also because pH and pK_a describe equilibrium-binding phenomena—we will discuss them briefly.

Recall our Chapter 8 discussion of water, where we noted that because of the way quantum mechanics governs the chemistry, water auto-dissociates into hydroxyl and hydronium ions (and not simple protons, H^+). Thus, the basic reaction is

$$2H_2O \underset{k_{on}}{\overset{k_{off}}{\rightleftharpoons}} H_3O^+ + OH^-. \tag{9.29}$$

Because the dissociation rate k_{off} is not zero, we can expect that an equilibrium occurs.

The equilibrium of the process (9.29) can be described in the way we have seen before—despite the quantum mechanical origins of the process. That is, although we may not fully understand the details underlying the association and dissociation events, we can still describe the equilibrium. If you think about it, the same was true for our study of non-covalent binding events: our analysis did not require knowledge of the details.

Continuing with an equilibrium analysis, we write

$$K_d = \frac{[H_3O^+][OH^-]}{[H_2O]^2} \tag{9.30}$$

$$= \frac{[H_3O^+]^2}{[H_2O]^2} = \frac{[OH^-]^2}{[H_2O]^2} \quad \text{(pure water)}, \tag{9.31}$$

where the second line uses the 1:1 ratio that must hold for hydronium and hydroxide ions in pure water. In general, the presence of other species (acids or bases) can change the concentration of either H_3O^+ or OH^-. To connect with some of the conventions for water, we can define $K_w = K_d[H_2O]^2$, which is equal to 10^{-14} at $T = 25°C$. Note that in pure water and all but the most dense aqueous solutions, the concentration of water essentially is constant at $[H_2O] = 55.4\,M$.

9.9.1 pH

Once we see that water and its ions can be treated via the usual equilibrium analysis, it is clear that the definition of pH is a simple convention, namely,

$$pH = -\log_{10}[H_3O^+]. \qquad (9.32)$$

(In general, $pX = -\log_{10}[X]$, with "H" being a shorthand for H_3O^+ in the case of pH.) In pure water, the pH is 7. In the presence of an acid, however, there will be more H_3O^+ and the pH will decrease due to the minus sign in Equation 9.32. On the other hand, the pH can increase if there are positive (basic) species present that can bond to H_3O^+—since pH refers only to the concentration of free hydronium.

9.9.2 pK_a

Elementary acid/base ideas can help us understand the chemistry of protonation in proteins. A number of amino acids can gain or lose protons depending on the local environment—that is, on the proton concentration in accessible water and the nearby electrostatic charges. For instance, the protonation state of histidine is known to be very sensitive to the specific environment in which it is found. Similarly, the oxidative environment can affect the protonation state of cysteine—and hence its ability to form disulfide bonds.

We will see that the "same old" statistical ideas also provide insight into protonation state. Let's take a look at one of the nitrogens in the histidine ring. It can undergo the reaction

$$\text{(deprotonated)} \quad -N + H_3O^+ \rightleftharpoons -NH^+ + H_2O \quad \text{(protonated)}, \qquad (9.33)$$

where "–N" indicates that this N is part of larger compound. Like any reaction, the equilibrium point of (9.33) can be pushed in a certain direction (e.g., to the right) by increasing the concentration of one of the species (e.g., H_3O^+). That is, in a more acidic environment, it is more likely for histidine to be protonated. The state can also be affected by nearby charges: for instance, if a negative charge were located nearby, that would also increase the electrostatic favorability for the protonated state. From a dynamics perspective, the protonation state of any individual histidine should fluctuate in time with average behavior determined by the environment.

We can quantify the average/equilibrium for histidine or other ionizable groups in the usual way. We'll consider the equilibrium between a general acid, HA, and its dissociated state:

$$H_2O + HA \rightleftharpoons H_3O^+ + A^-. \qquad (9.34)$$

The equilibrium constant is $K_a = [H_3O^+][A^-]/[HA]$, so that a larger K_a means a stronger, more dissociated acid. By taking the log of the definition of K_a and also using the definition of pH, we find

$$pH = pK_a + \log_{10}\left(\frac{[A^-]}{[HA]}\right), \qquad (9.35)$$

where pK_a is based on the usual pX convention. Equation 9.35 is the Henderson–Hasselbalch equation. As we said above, the level of dissociation for a given group— for example, an aspartic acid or histidine residue—depends on the local electrostatic environment, in addition to the pH. That is, the pK_a depends not only on the identity of a chemical group but also on its location if it is in a structured macromolecule like a protein. Effects particular to the local environment are embodied in the $[A^-]/[HA]$ ratio which actually occurs.

There is one key lesson from Equation 9.35, namely, the way in which the protonation state is related to pH. When $pH = pK_a$, then a given group (at a given site) will be protonated exactly half of the time. Similarly, if the pH is less than the pK_a, the excess of "protons" will push the equilibrium to the protonated state, HA.

9.10   SUMMARY

If there is a common theme to the many classical biophysics phenomena described here, it is this: essential equilibrium and steady-state analyses can be performed without knowing the structural details of the component biomolecules. That is, the use of rates and equilibrium constants enables the understanding of a broad array of biophysical phenomena—binding, allostery, catalysis, protonation, and even the source of ATP's free energy. A few simple ideas can go a long way.

We also related the rates and K_d values to a molecular picture whenever possible and, further, to the underlying statistical mechanics—free energy, entropy, and enthalpy. Indeed, the rates themselves—which constitute the basic building block of the whole chapter—can only be understood from a probabilistic/statistical point of view.

FURTHER READING

Berg, J.M., Tymoczko, J.L., and Stryer, L., *Biochemistry*, 5th edition, W.H. Freeman, New York, 2002.

Cantor, C.R. and Schimmel, P.R., *Biophysical Chemistry, Part III*, W.H. Freeman, New York, 1980.

Fersht, A., *Structure and Mechanism in Protein Science*, W.H. Freeman, New York, 1999.

Gilson, M.K., Given, J.A., Bush, B.L., and McCammon, J.A., *Biophysical Journal*, 72: 1047–1069, 1997.

Gunasekaran, K., Ma, B., and Nussinov, R., *Proteins*, 57:433–443, 2004.

Indyk, L. and Fisher, H.F., *Methods in Enzymology*, 295:350–364, 1998.

Koshland, D.E., Nemethy, G., and Filmer, D., *Biochemistry*, 5:365–385, 1966.

Monod, J., Wyman, J., and Changeux, J.-P., *Journal of Molecular Biology*, 12:88–118, 1965.

Nicholls, D.G. and Ferguson, S.J., *Bioenergetics*, 3rd edition, Academic Press, Amsterdam, the Netherlands, 2002.

Phillips, R., Kondev, J., and Theriot, J., *Physical Biology of the Cell*, Garland Science, New York, 2009.

Tembe, B.L. and McCammon, J.A., *Computers and Chemistry*, 8:281–283, 1984.

van Holde, K.E., Johnson, W.C., and Ho, P.S., *Principles of Physical Biochemistry*, Pearson, Upper Saddle River, NJ, 2006.

Woo, H.J. and Roux, B., *Proceedings of the National Academy of Sciences, USA*, 102:6825–6830, 2005.

10 Kinetics of Conformational Change and Protein Folding

10.1 INTRODUCTION: BASINS, SUBSTATES, AND STATES

Conformational transitions are fundamental in structural biology and biochemistry. In almost every case, the motion of a protein either *is* the function or is a necessary part of the function. As examples, consider motor proteins whose job is to generate motion or enzymes, which typically rearrange conformation to perform catalysis and/or release their products. "Signaling"—the transmission of information from one part of the cell to another—is typically accomplished by allosteric binding events (see Figure 10.1). In allosteric signaling, protein X binds to Y, causing conformational changes in Y that lead it to bind Z, and so on. (The fact that Y didn't bind Z before binding X makes it allostery.) Of course, protein folding is a change of conformation, by definition.

The "kinetic" picture—based on average dynamics embodied in rate constants—is extremely useful for describing the array of conformational transitions: those occurring in allostery, those occurring in equilibrium, and folding. The kinetic approach divides a system's conformation space into distinct states among which transitions occur according to rate constants. Although the kinetic approach is the simplest description of dynamics, the kinetic characterization of a complex process such as allostery or folding is not trivial! We will build on our previous discussions of simple dynamics and kinetics (Chapters 4 and 9), as well as on our earlier introduction to equilibrium aspects of binding (Chapter 9) and folding (Chapter 3).

10.1.1 SEPARATING TIMESCALES TO DEFINE KINETIC MODELS

Most generally, as sketched in Figure 10.2, a kinetic picture can be said to "coarse grain" the energy landscape. (From a statistical mechanics point of view, many dynamical degrees of freedom are integrated out, such as variations among individual configurations and among individual transitioning trajectories.) Such coarse graining requires a separation of timescales (i.e., of rates) into groups—usually fast and slow. For instance, if there are rapid interconversion rates among a set of basins—and only slow conversions from these basins to all others—then the set of basins can be categorized as a physical state (see Figure 10.2). The rates for converting among basins within a state will—by definition—be much higher than the rates among states. In the kinetic picture, one only needs to know the identities of the states and the rates between them. This hierarchical description of the protein landscape has

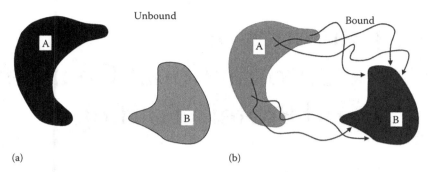

FIGURE 10.1 Conformational change resulting from allosteric binding. A cartoon of configuration space shows a protein's two conformational states, A and B, with relative populations indicated by the shading. (a) Before binding, state A is predominantly occupied, while (b) binding induces a population shift to state B. The ensemble must shift along dynamical paths (arrows) to state B.

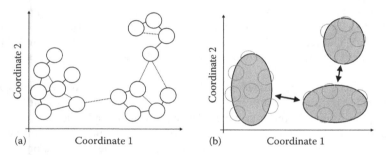

FIGURE 10.2 From basins to a kinetic description. The energy landscapes of large molecules consist of many basins (small circles). Transitions among some basins may be rapid (solid lines), significantly slower among others (dotted lines), and essentially nonexistent for the rest. Groups of basins that can rapidly equilibrate via such transitions can be considered as physical states (large ovals), while transitions between states are much slower by comparison (double-sided arrows). A kinetic description thus vastly simplifies the complexity of dynamics by considering only a small number of physical states and transitions among them. Such a hierarchical classification of dynamics will not be possible if groups of rates are not cleanly separated but rather span a spectrum. The coordinates used are arbitrary, and simply are assumed to distinguish separate basins or substates.

a long history and was particularly emphasized by Frauenfelder et al. (see also the article by Chodera et al.).

In our discussion of kinetics, we will build directly on the concepts introduced in Chapter 4. Equilibrium always "hides" continual but balanced dynamical transitions. If a system is out of equilibrium—perhaps because it was just unfolded, or because a ligand just bound to it—then dynamics will return it to equilibrium. The rate constants, of course, will determine how quickly these events occur.

One caveat is important. Not all systems can be described well by a small number of states and rate constants. If there are many basins, but no set can be grouped together based on a separation of timescales, then a kinetic description typically will not be very helpful. Such a system can be described as diffusive because transitions among states will be fast and somewhat randomly directed. Nevertheless, with advanced computational approaches, such as those described in the work of David Wales, a surprisingly large number of basins can be treated.

10.1.1.1 Underneath the Kinetics, Trajectories

It's worth a moment to paraphrase from Chapter 4 and point out that nature doesn't calculate rate constants (or identify physical states); she only performs dynamics. In other words, the kinetic behavior we discuss results from averaging over many trajectories, each driven by forces in the usual way.

It is therefore useful to think first about the trajectories underlying kinetic descriptions. Take a look at Figure 10.3, which shows a schematic trajectory in the same landscape as Figure 10.2. All information we might want is contained in the ensemble of such trajectories. The rate from one state to another can be determined based on the average waiting time in the first state before a transition to the second. States and intermediates can be identified as regions of configuration space where trajectories tend to dwell. Try to carry this picture with you. There must be straightforward dynamical processes underlying all kinetic (and equilibrium) phenomena. The statistical mechanics formalism of "trajectory ensembles"—in analogy to equilibrium ensembles of configurations—will be introduced in Chapter 11.

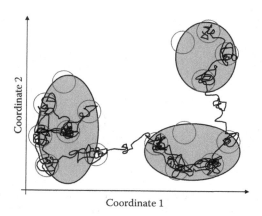

Coordinate 2

Coordinate 1

FIGURE 10.3 A schematic dynamical trajectory, which would form the underlying basis for a kinetic description. The rate constants to be used in this chapter implicitly result from statistical mechanics averages of many trajectories. As shown, the separation of timescales here is not very large: there are only a few transitions among substates or basins before transitions among states. With better separation, there would be more intrastate transitions (within states) before each interstate transition.

10.1.1.2 If You're Not a Protein. . .

It is worth emphasizing that the kinetic formulation presented here applies not only to proteins but to any "functioning" molecule—for example, to RNA and DNA systems, which are far from being simple sequences of bases. Structure-based functional roles for both RNA and DNA—that is, catalysis and gene regulation—are well known, and our appreciation of these roles is growing rapidly. These are not "passive" molecules!

10.2 KINETIC ANALYSIS OF MULTISTATE SYSTEMS

10.2.1 REVISITING THE TWO-STATE SYSTEM

First things first. Let's revisit—more carefully—the kinetic description of the very simple two-state system. We'll consider states A and B, with transitions governed by the forward and reverse rates k_{AB} and k_{BA}.

We imagine initiating a large number of trajectories in one state (A) at time $t = 0$ and examining the fraction (the "population") that arrives in another state, B. In the case of a double-well potential, state A would be the left well and B the right. For the molecular example of butane, state A could be the *trans* state near $\phi = 180°$ and state B could be *gauche*+ near $\phi = 300°$.

Figure 10.4 shows the evolution of the fractional or absolute population of state B. Initially, at $t = 0$, all probability is in state A, so the population of B is 0. At the earliest times $t > 0$, there is transient period of duration t_b reflecting the finite time necessary for the first transitions to occur. Thereafter, a fixed fraction of the population makes the transition per unit time, leading to the linear behavior with time. (Think of a Poisson process.) Since the rate k_{AB} is the conditional probability per unit time to make a transition, the slope of the graph is the rate itself. Once

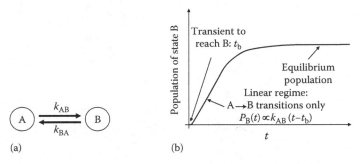

(a)   (b)

FIGURE 10.4 The evolution of the population of the "final" state B after initiating many trajectories in state A at $t = 0$. First, there is a transient period of duration t_b, which is the travel time for the arrival of the first transitions. Then, a constant fraction per unit time makes the transition to B, leading to linear behavior for short times. The fraction making the transition in a given time interval Δt is given by $k_{AB}\Delta t$, based on the definition of the rate k_{AB}. Once an appreciable fraction of the population has reached state B, reverse transitions will occur, leading to the turnover in the population curve. At very long times ($t \to \infty$), the populations will equilibrate, reflecting a balance of flux in both directions.

enough probability has crossed over to state B, reverse transitions back to A occur, which prevent the continued linear growth in the population of state B. However, the populations of state B continues to grow with time, and hence so too do the reverse transitions. Eventually, the flux will balance out, as discussed in Chapter 4, leading to equilibrium described by $P_A(t \to \infty) \cdot k_{AB} = P_B(t \to \infty) \cdot k_{BA}$.

From this point forward in this section, we will make a simplifying assumption about the transient behavior: namely, that it is irrelevant. In particular, we will assume the barrier-crossing "travel time" (excluding the waiting time) is much smaller than the waiting time, so that we can simply set it to zero ($t_b = 0$). In complex systems like proteins, however, t_b could be substantial—and will depend sensitively on the definitions of the states A and B.

We can be more precise about the kinetic process by using the same type of phenomenological rate theory employed to describe the dynamics of binding (see Equation 9.3). In fact, the A–B isomerization within a single molecule is even simpler. Using the definition of the rate, we see that the amount of probability flow from state i to j (either of which can be A or B) is $P_i(t) \cdot k_{ij}$. These flows directly imply the differential equation description of two-state kinetics to be

$$\frac{dP_A}{dt} = k_{BA}P_B(t) - k_{AB}P_A(t),$$

$$\frac{dP_B}{dt} = k_{AB}P_A(t) - k_{BA}P_B(t), \tag{10.1}$$

which means that the population of a state changes according to the difference between the inflow and the outflow of probability. The equations become more complicated if there are additional states.

These equations can be solved exactly with fairly simple exponentials. However, a simple linear approximation will be quite good at short times when the barrier is large, as the following problem shows.

PROBLEM 10.1

Perturbative solution of the kinetic equations. Consider discretizing the kinetic equations, so that $\delta P_i(t) \equiv P_i(t + \delta t) - P_i(t) = [k_{ji}P_j(t) - k_{ij}P_i(t)] \cdot \delta t$, where δt is a short time interval.

  a. Write down the probability of state B at times $t = 0$, δt, and $2\delta t$ according to the discretization. Be sure all inflow and outflow of probability is included, and check your units.
  b. Explain the condition on δt necessary for P_B to increase in an approximately linear way. Your answer should also describe the timescale on which P_B turns over significantly.

The exact solution to Equations 10.1 is not difficult to determine, either. We first examine a special case, when the equilibrium populations of A and B are equal.

PROBLEM 10.2

Exact solution of the differential equations (10.1) for a symmetric system. Given that equilibration must occur for $t \to \infty$, a reasonable guess for the form of the solution in a symmetric system—where $k_{AB} = k_{BA}$ is $P_B(t) = (1/2)(1 - e^{-\lambda t})$.

    a. Explain why the guess is physically reasonable.
    b. Check that the guess works and solve for λ.

The fully general solution is also not hard to determine, especially if a few standard "tricks" are used.

PROBLEM 10.3

Fully general solution of differential equations (10.1).

    a. To eliminate the second equation, substitute $P_B(t) = 1 - P_A(t)$ in the first equation. After you solve for $P_A(t)$, you can get P_B by simple subtraction.
    b. We know by now that $P_A(t)$ will decay exponentially to its equilibrium value. Therefore, substitute $P_A(t) = P_A^{eq} + ae^{-\lambda t}$ into the remaining equation, where a and λ are constants to be determined. Explain why the constant a need not be solved for—that is, explain its role in defining the arbitrary initial condition at $t = 0$.
    c. By equating the time-dependent terms on both sides of the equation, show that $\lambda = k_{AB} + k_{BA}$.
    d. Based on your fully general solution, sketch the behavior of $P_B(t)$ for two special cases: $P_A(0) = P_A^{eq}$ and $P_A(0) = 1$.

10.2.1.1 What Can Be Observed?

Kinetic descriptions, such as those presented here, are widely used in the interpretation of experimental data. It is very important to consider which rate quantities can be measured, in principle.

    Although we have stressed the linear regime that may occur when all population is initialized in a single state, this regime may not always be observable in an experiment. Typical laboratory studies have a "dead time" due to limitations of the measuring equipment—while the linear regime will only persist for an amount of time of the order of an inverse rate.

    However, the exponential relaxation lasts indefinitely and can often be observed in experiments. For this reason, the inverse of the relaxation time is sometimes called the observed rate. If you have not worked Problem 10.3, you should do so now to appreciate why, in a two-state system, the "observed rate" is given by the sum of the two rates:

$$\lambda \to k_{obs} = k_{AB} + k_{BA}, \tag{10.2}$$

where the "$\rightarrow$" indicates we are renaming the symbol. This relation reflects the fact that the system as a whole can relax by two processes, both of which contribute to the speed of relaxation.

10.2.1.2 Caveats on the Phenomenological Kinetic Equations

We emphasize, again, that there are two caveats to the phenomenological description embodied in the differential equations (10.1). Neither is very important for our discussion, but it is useful to recognize the limitations of any model. First, as we discussed earlier, the rates are just constants, which means they are assumed not to depend on any internal dynamics within the states. All dynamic processes besides the transition are thus assumed to occur on timescales much shorter than the inverse rates, implying a good separation of timescales. Second, the transitions are assumed to be instantaneous, so that as soon as probability leaves state A, it enters state B (i.e., $t_b = 0$) (see Figure 10.6). In principle, states can always be defined so that this is true based on a suitable dividing surface in configuration space; for a complicated system, however, it will be difficult to determine a surface that maintains the necessary separation of timescales.

10.2.1.3 A Useful Cartoon View

Figure 10.5 is just another view of the same picture of the evolution of the population of state B. However, the use of the state "bubbles" and arrows will prove very useful in schematizing more complex systems with intermediates. The arrow widths in the figures represent the total probability flowing in a given direction at a few different times. The rates, as always, are constant in time—so a given arrow width results from the product of the rate and the probability of the state (at the indicated time) where the arrow originates.

10.2.1.4 A Simple Ensemble Description of the Two-State System

Where is the ensemble in a kinetic description of a two-state system? Here, we give a simple answer to the question, reserving a more complete treatment for Chapter 11. In

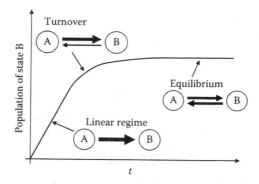

FIGURE 10.5 Probability evolution of state B, with cartoons. The arrow widths indicate the total probability fluxes at various times—and not the rates, which are physical constants.

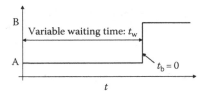

FIGURE 10.6 Another view of two-state kinetics. In the simplest kinetic model of A→B transitions, the rate k_{AB} can be said to dictate the distribution of "waiting times" t_w for trajectories started in A to make the transition to B. The simple Poisson process is described by an exponential distribution of waiting times (Equation 10.3).

the case of transitions directly from state A to B (i.e., no intermediates), the situation is schematized in Figure 10.6, where the waiting time t_w reflects a Poisson process—that is, a constant probability per unit time for a transition. There is an ensemble—or distribution—of waiting times. As we saw in Chapter 2, the distribution of waiting times, t_w, prior to a transition will simply be exponential:

$$\rho(t_w) = k_{AB}\ \exp(-k_{AB}\ t_w), \qquad (10.3)$$

where k_{AB} is the usual transition rate. This is the simplest ensemble description of a transition process, and many details are omitted (see Chapter 11).

The exponential distribution manifests itself in the accumulating population of state B, following Figures 10.4 and 10.5. Indeed the waiting-time distribution (10.3) directly predicts the early-time behavior of the state B probability: after all, at early times when there are only A → B transitions, $P_B(t)$ is the sum/integral of all transitions up to time t. Generally, then,

$$\text{Early times } \left(t < k_{AB}^{-1}\right): \quad P_B(t) \approx \int_0^t dt_w\ \rho(t_w), \qquad (10.4)$$

where the symbol "$\approx$" indicates this relation is exact in the $t \to 0$ limit (also assuming $t_b = 0$). Performing the integration for the exponential distribution (10.3) yields

$$P_B\left(t < k_{AB}^{-1}\right) \approx 1 - \exp(-k_{AB}t) = k_{AB}t + \cdots, \qquad (10.5)$$

so that we explicitly see the linear behavior with a prefactor given by the rate as depicted in Figure 10.5.

10.2.2 A THREE-STATE SYSTEM: ONE INTERMEDIATE

We turn our attention to the system schematized in Figure 10.7, which possesses a single intermediate state. Studying the apparently simple single-intermediate case will provide us with considerable insight into complex landscapes. By definition, an intermediate is separated by significant barriers from both states A and B. Thus, the timescale for relaxation within intermediate I is much less than the inverses of the rates $k_1' = k_{IA}$ or $k_2 = k_{IB}$. (The inverse rate is the average waiting time for a transition.)

Scheme AIB:

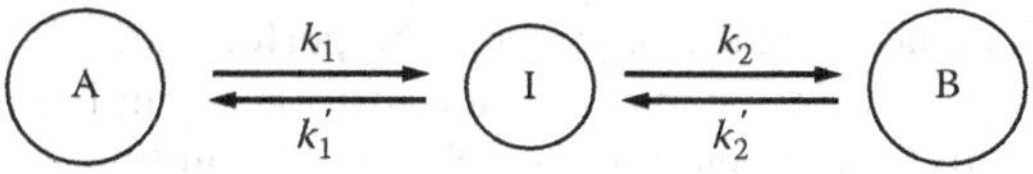

FIGURE 10.7 The "AIB" kinetic scheme, which has one intermediate. The presence of the intermediate drastically changes the observed waiting-time distribution for going from A to B. With all probability initialized in state A, there is no appreciable increase in state B's population until the intermediate has become significantly populated—which roughly requires an amount of time $\sim 1/k_1$.

The differential equations governing conformational change with an intermediate are quite similar to those used for the Michaelis–Menten analysis of catalysis (Section 9.8). For the AIB scheme of Figure 10.7, we have

$$\frac{dP_A}{dt} = -k_1 P_A(t) + k_1' P_I(t),$$

$$\frac{dP_I}{dt} = k_1 P_A(t) - k_1' P_I(t) - k_2 P_I(t) + k_2' P_B(t), \tag{10.6}$$

$$\frac{dP_B}{dt} = k_2 P_I(t) - k_2' P_B(t),$$

where we again use the conventions $k_1 = k_{AI}$, $k_1' = k_{IA}$, $k_2 = k_{IB}$, and $k_2' = k_{BI}$.

Such a system of equations can be used to fit the measured time evolution of a state's population—for example, $P_B(t)$—and hence determine individual rate constants. A poor fit can suggest that a different number of intermediates is present, or perhaps a different number of pathways, as discussed below.

We can analyze the system of kinetic equations (10.6) using approaches we employed earlier in Chapters 4 and 9. Equilibrium populations are easiest to determine: as usual, one uses forward and reverse rates to first determine the relative populations of the A-I and I-B state pairs; then, the constraint of all probabilities summing to one leads to the true fractional probabilities.

PROBLEM 10.4

Solve for the equilibrium populations of states A, I, and B.

We can also solve approximately for the evolving probabilities $P_A(t)$, $P_I(t)$, and $P_B(t)$, under the assumption that only state A is occupied initially—that is, $P_A(0) = 1$. Since state B does not directly receive probability from state A, at early times ($t < 1/k_1$) the intermediate becomes populated linearly with time: $P_I(t) = k_1 \int dt\, P_A(t) \approx k_1 t$, using $P_A \approx 1$ for short times. In turn, state B receives a tiny fraction of this probability so that we have

$$P_B(t) \approx k_2 \int dt\, P_I(t) \approx \left(\frac{1}{2}\right) k_2 k_1 t^2. \tag{10.7}$$

When one step of the conformational change is rate-limiting and has a rate much lower than the others, additional analyses can be performed. For instance, consider the case where $k_2 \ll k_1, k'_1$, which means that transitions from I to B are much slower than other steps. In this case, A and I will reach a quasi-equilibrium before significant transitions to B occur; after this, the system will behave as a single-barrier system (from I to B) with A and I maintaining their relative populations throughout.

PROBLEM 10.5

Behavior when the I $\rightarrow$ B transition in the AIB scheme of Figure 10.7 is rate-limiting. (a) Assuming no transitions to B can occur and $P_A(0) = 1$, what are the relative populations of A and I after a "long time"? (b) What is the timescale for reaching this quasi-equilibrium? (c) Derive and explain the approximately linear behavior in time expected for P_B after the quasi-equilibrium between A and I. (d) Sketch the behavior of $P_B(t)$ from $t = 0$ until equilibrium is reached. Note the timescale governing the apparent transient regime, and also indicate the effective rate for the linear regime.

Further discussion of this type of kinetic analysis can be found in many texts on physical chemistry and physical biochemistry. It is instructive to examine an intuitive method for solving the full system of equations (10.6), as in the following problem.

PROBLEM 10.6

Exact analytic solution of the kinetic equations (10.6). (a) Using the constraint $P_A(t) + P_I(t) + P_B(t) = 1$, show that the third equation for dP_B/dt can be removed from consideration. Write the two governing differential equations and be sure P_B is eliminated from them. (b) Note that since the equilibrium values P_j^{eq} are known for $j = A, I, B$ from Problem 10.4, we can conveniently define $\Delta P_j(t) = P_j(t) - P_j^{eq}$. Explain why it is reasonable to try exponential forms accounting for the two processes in the system—namely, $\Delta P_j = a_j e^{-\lambda_a t} + b_j e^{-\lambda_b t}$, with the same λ values for all species. (c) Noting that $dP_j/dt = d\Delta P_j/dt$, show that the number of equations and the number of unknowns is the same. Be sure to include equations for the initial conditions at $t = 0$.

Figure 10.8 illustrates the full probability evolution of state B when k_1 and k_2 are similar. Note that the presence of the intermediate means that there may not be a well-defined linear regime by which one can define a rate. However, an effective rate can be defined as the inverse of the time to reach half of the equilibrium population. This definition will account for the effects of the important timescales, regardless of whether one rate dominates. More carefully, one can define a rate based on analyzing a particular steady state or based on waiting ("first-passage") times as described below.

10.2.2.1   The Waiting-Time Distribution in the A-I-B System

To deepen our perspective, we can continue our earlier investigation of the distribution/ensemble of waiting times. Analyzing the waiting distribution at early times

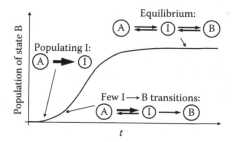

FIGURE 10.8 Probability evolution of state B, when there is an intermediate. The arrow widths indicate the total probability fluxes at various times—and not the rates, which are physical constants. The gradual increase in P_B contrasts with the linear behavior depicted in Figure 10.5 for the case of a direct transition with no intermediate. Complex biomolecular systems should be expected to have one or more intermediates.

for Scheme AIB in Figure 10.7 is straightforward. If t_1 is the waiting time for the A→I transition and t_2 the waiting time for I→B, then we want the distribution of $t_w = t_1 + t_2$. That is, t_w is the waiting time for the full A→B transition, and it is simply the convolution of two simple exponential distributions. (A convolution is the distribution of a sum of "increments," as we discussed way back in Chapter 2.) By taking a convolution, we are making the reasonable assumption that the t_1 value does not affect the t_2 value: while this could be false in principle, it is consistent with our definition of an intermediate being separated from other states by significant barriers.

We want to calculate the convolution of the waiting-time distributions of the two elementary steps: $\rho_1(t_1) = k_1 e^{-k_1 t_1}$ and $\rho_2(t_2) = k_2 e^{-k_2 t_2}$, with $0 \leq t_i < \infty$. By definition from Chapter 2, we have

$$\rho(t_w) = \int_{-\infty}^{\infty} dt_1\, \rho_1(t_1)\, \rho_2(t_w - t_1) \tag{10.8}$$

$$= \int_{0}^{t_w} dt_1\, k_1 e^{-k_1 t_1} k_2 e^{-k_2(t_w - t_1)}, \tag{10.9}$$

where the second line accounts explicitly for the fact that all times in the convoluted distributions cannot exceed t_w.

The integral in Equation 10.8 can be easily solved. The solution at first looks like a pure exponential, but when we perform a Taylor expansion for small t_w, we see clearly that it is not:

$$\rho(t_w) = \frac{k_1 k_2}{k_1 - k_2} \left[\exp(-k_2 t_w) - \exp(-k_1 t_w) \right] \approx k_1 k_2\, t_w. \tag{10.10}$$

This linear behavior for small t_w means that the distribution is exactly zero when $t_w = 0$, in sharp contrast to the exponential distribution (10.3) when there is no intermediate.

PROBLEM 10.7

Derive both parts of Equation 10.10.

Using Equations 10.4 and 10.10, we can derive the early-time probability for arrival in state B when there is an intermediate:

$$P_B(t) \approx \frac{1}{2}k_1 k_2 t^2.$$ (10.11)

This quadratic increase in P_B—which agrees with Equation 10.7—also contrasts sharply with the linear behavior of Equation 10.5. Figure 10.8 may be compared with Figure 10.5.

10.2.3 THE EFFECTIVE RATE IN THE PRESENCE OF AN INTERMEDIATE

In the presence of intermediates, it becomes possible to define different rate constants. The differences, unfortunately, are both subtle and important. The two principal rates can be termed "first-passage" rates, denoted k^{FP}, and steady-state rates, k^{SS}. Many cellular processes occur in quasi-steady states, whereas experiments often measure first-passage rates.

10.2.3.1 The First-Passage Rate

It is probably easiest to define the first-passage rate k^{FP}, which is based on waiting times. Indeed the "first-passage time" from A to B is simply a fancy way of saying the waiting time for such a transition. The rate k^{FP} is defined to be the inverse of the mean first-passage time, often abbreviated MFPT. One can imagine initiating a million systems in state A, and clocking the times at which they arrive to B: the average of these times is the MFPT, and the simple reciprocal is the rate. So far, so good.

Although it is fairly easy to understand the first-passage rate in the presence of intermediates, k^{FP} is not generally easy to calculate. At first glance, the problem doesn't seem so hard. After all, we just considered the convolution of A-to-I and I-to-B waiting times in Equation 10.8—and this calculation would seem to contain the information we need. Alas, no. Our derivation of Equation 10.8 applies solely to "early times" in the transition process—to times before significant reverse transitions from I back to A occur.

A more careful calculation of the MFPT $= 1/k^{FP}$ requires consideration of all possible pathways—that is, all possible sequences of states. In particular, a system can reach B by any of the following sequences: AIB, AIAIB, AIAIAIB,.... The reverses are essential to calculating the correct rate. In essence, as described in the article by David Wales, the MFPT is given by the weighted sum of first-passage times for all the possible sequences/pathways. The sum is weighted by the probability of each sequence of events. These probabilities are governed by the rates as you can show by completing the following problem.

PROBLEM 10.8

(a) Using the probabilistic definitions for the rates in the AIB scheme, calculate the conditional probability for returning to state A after arriving in I. (b) Write down a general expression for the probability of an arbitrary sequence of states in transitioning from A to B.

As we suggested above, an approximation to the MFPT can be obtained in a simpler way, as the time required for the state B probability to reach half its equilibrium value, if all probability has been started in A. Such a calculation would use the time-dependent solution for $P_B(t)$. Note that an exact MFPT calculation based on $P_B(t)$ would require a special solution of the differential equations with an absorbing boundary condition at state B: by definition, the FPT only includes first arrivals to state B, with no trajectories allowed to "bounce out."

10.2.3.2 The Steady-State Rate Is Easier to Calculate

Steady states are extremely important in the biological contexts of catalysis and locomotion by motor proteins.

In a steady-state scenario, we can perform a fairly simple analytical calculation of the rate k^{SS} for the AIB scheme. A steady-state rate can be defined and readily analyzed regardless of the landscape intervening between states A and B—that is, even if there are many intermediates. Here, we will only treat the single-intermediate case of Scheme AIB, but the interested reader can consult David Wales' article for more general calculations.

We imagine establishing a "feedback steady state" in which all probability arriving in state B is immediately transferred back to state A, as in Figure 10.9. No reverse transitions from B to I are allowed in this scenario. However, the feedback steady state does permit the precise monitoring of the probability/population arriving in B over time in terms of the population of state A. As we will see, this yields the same equation as the traditional definition of a rate. (Our assumptions are somewhat different from a Michaelis–Menten calculation where state B corresponds to an irreversibly formed product, which is not fed back to A.)

To proceed, we do a steady-state calculation on Equations 10.6 and set $dI/dt = 0$ while also forbidding transitions back from B by setting $k_2' = 0$ (refer back to Figure 10.7). We also add the probability increase which would have gone to B (i.e.,

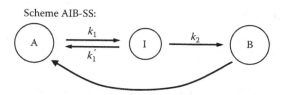

Scheme AIB-SS:

FIGURE 10.9 The steady-state scheme used to calculate an effective rate k_{AB}^{SS} for a system with an intermediate.

dP_B/dt) to state A. The modified kinetic equations are therefore

$$\frac{dP_A}{dt} = -k_1 P_A(t) + k_1' P_I(t) + \frac{dP_B}{dt} = 0,$$

$$\frac{dP_I}{dt} = k_1 P_A(t) - k_1' P_I(t) - k_2 P_I(t) = 0,$$

$$\frac{dP_B}{dt} = k_2 P_I(t) \rightarrow \text{ added to } \frac{dP_A}{dt}, \tag{10.12}$$

which can be compared to Equations 10.6. Because probability arriving to B is fed back to A, technically $P_B = 0$ for all t.

We can solve the middle equation (10.12) for P_I in terms of P_A and substitute the result into the (traditional) P_B equation. This yields $P_I/P_A = k_1/(k_2 + k_1')$ and thus

$$\frac{dP_B}{dt} = k_2 P_I(t) = \frac{k_1 k_2}{k_2 + k_1'} P_A(t) \equiv k_{AB}^{SS} P_A(t). \tag{10.13}$$

In other words, if our special steady state is established, the probability per unit time of an A→B transition is $k_{AB}^{SS} = k_1 k_2/(k_2 + k_1')$.

PROBLEM 10.9

(a) Use a similar analysis to derive the steady-state B→A rate. (b) Show that the ratio of the two rates yields the correct relative equilibrium probabilities—which result from the original Equations 10.6 (see Problem 10.4).

The steady-state rate expression can be probed to ensure it makes sense. First, note that k_{AB}^{SS} is always lower than k_1, as expected: the second transition from I to B can only slow down the process. We can also consider some extreme cases. If $k_2 \gg k_1'$, then $k_{AB}^{SS} \simeq k_1$, which correctly makes the A→I transition rate-limiting when the I→B transition is fast. On the other hand, if $k_2 \ll k_1'$, then the effective rate $k_{AB}^{SS} \simeq k_1 k_2/k_1'$ is determined by a balance among three processes: A→I, I→A, and I→B.

PROBLEM 10.10

Sketch one-dimensional energy landscapes corresponding to the cases analyzed above: (a) $k_2 \gg k_1'$, (b) $k_2 \ll k_1'$.

Our discussion of the single-intermediate case should make clear that the analysis of a system with multiple intermediates could be particularly complex if there is no dominant rate-limiting step.

10.2.3.3   Steady-State vs. First-Passage Rate: When It Matters

The presence of intermediates has begun to complicate the rate equations, as we have seen—and the expressions get a bit less intuitive, at least at first. The distinction between the first-passage (FP) and steady-state (SS) rates may seem unimportant, so we should understand when it matters.

Look carefully at the system sketched in Figure 10.10, and note particularly the off-pathway intermediate. The upper path is simplified in Figure 10.11. An off-pathway state is an energy basin or a set of basins which trajectories might "accidentally" explore on their way from A to B. After all, the trajectory does not "know" where it's going; it only follows dynamics. If an off-pathway state is accessible, it will be visited. In the complex landscapes of biomolecules, such off-pathway intermediates should be expected.

An off-pathway intermediate will have profoundly different effects on the SS and FP rates. In the FP case, all trajectories are initiated in A and we want to time their arrival at B. Some such trajectories, not knowing better, will waste time in the

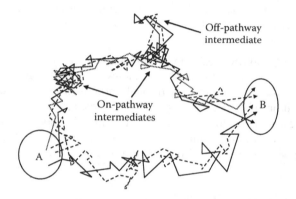

FIGURE 10.10   A schematic description of transition trajectories in a complex system. Two states, A and B, are connected by two channels, and each channel encompasses its own ensemble of trajectories. One channel has three fairly well-defined intermediates, one of which is "off pathway." This ensemble of different possible trajectories is simplified in a kinetic description.

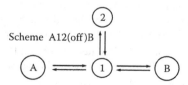

FIGURE 10.11   A kinetic scheme featuring an off-pathway intermediate. The first-passage rate from A to B could be significantly affected by the intermediate "2" because some trajectories will go there before reaching B. In a steady state, however, for every trajectory entering state 2, there will be one leaving.

off-pathway state, thus increasing the mean FP time. Steady-state trajectories will do the same—they must obey the same physics!—but the SS rate does not time arrival at B based on starting at A. Rather, the SS rate measures the probability of arriving at B regardless of when and where it came from: recalling Equation 10.13, the SS rate is defined by

$$k_{AB}^{SS} \equiv \left(\frac{dP_B}{dt}\right) \bigg/ P_A \quad (P_i \text{ in steady state}). \tag{10.14}$$

By the SS definition, furthermore, an equal number of trajectories will go into and out of every intermediate in every unit of time. Thus, it turns out the SS rate is unaffected by the presence of an off-pathway intermediate, as explicitly shown in the following problem.

PROBLEM 10.11

Show that the steady-state rate from A to B in the scheme of Figure 10.11 is unaffected by the presence of the off-pathway intermediate (state 2). Do this by explicitly calculating k_{AB}^{SS} in the presence and absence of state 2.

In fact, the FP and SS rates differ significantly whenever there are significantly populated intermediates, whether on- or off-pathway. On-pathway intermediates will slow the first-passage rate from A to B much more than the SS rate. Our elementary decomposition of the rate into an attempt frequency and success probability, Equation 4.3, corresponds to the SS picture. Equation 4.3 does not include "delay" times for slow-but-certain downhill-trending motions. In Chapter 11, we will revisit the distinction between SS and FP rates more rigorously from a trajectory-ensemble point of view.

10.2.4 The Rate When There Are Parallel Pathways

In some situations, parallel pathways may be present. Consider once again our ensemble of all possible transition trajectories from state A to B, as sketched in Figure 10.1 or Figure 10.10. Starting from state A, some fraction of the transitions will traverse the upper pathway, with the rest in the lower pathway. These two pathways differ in the configurations that occur during the transition. In general, there may be a number of pathways—though perhaps with one or a small number dominating.

Figure 10.12 sketches the kinetics broken down by pathway. In steady state, the rate k_i is the conditional probability per unit time to make a transition through pathway i, having started in state A. Because there is a transition probability k_1 via pathway 1 every second, as well as probability k_2 via pathway 2, the total steady-state rate is the sum of the individual pathway rates: $k_{AB} = k_1 + k_2$.

A slightly different description is obtained based on first-passage times. Assume that, in the FP description, $k_1^{FP} > k_2^{FP}$, so that the MFPT along pathway 1 is less than along pathway 2. This means that the presence of the second path will increase the overall average MFPT—and decrease the rate compared to k_1^{FP}. More precisely,

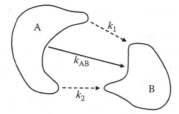

FIGURE 10.12 Parallel pathways and effective kinetics for a conformational change. In principle, a conformational change may occur through more than one pathway—that is, through a different set of intermediate configurations. Each pathway can be considered to have its own rate, and the net rate (in steady state) is the simple sum of the pathway rates: $k_{AB} = k_1 + k_2$.

noting that the probability of a given pathway is proportional to its individual rate, one finds that $k_{AB}^{FP} = \left(k_1^{FP} + k_2^{FP}\right)/2$. This is not really a different result from the steady-state case, merely a different way of describing it: The MFPT is simply the average waiting time for all transitions, whereas the steady-state rate for parallel paths describes whether the overall A-to-B flux will increase, without regard to how long it takes for the steady-state to be established.

10.2.5 IS THERE SUCH A THING AS NONEQUILIBRIUM KINETICS?

A nonequilibrium situation is one where the configuration-space (or velocity-space) distribution differs from that at equilibrium. That is, the populations of some regions of configuration space could differ from the equilibrium values; or even if the populations are not different, there could be a nonequilibrium steady state if there is a net flow of population from one region to another.

Yet, whether a system is in or out of equilibrium should not change the definitions of physical states or the rates among them. In other words, the kinetic description— embodied perhaps in a set of differential equations—should hold, so long as the physical states were appropriately defined to equilibrate rapidly, within each state. A nonequilibrium situation, then, should just be any "initial condition" that differs from equilibrium.

It only makes sense to speak of nonequilibrium kinetics if the initial discretization of configuration space was inadequate. This would mean there are relatively slow timescales for relaxation internal to one of the predefined "states," say A. Slow relaxation inside state A would mean that rates for transitions from A to other states would depend on the initial (nonequilibrium) preparation of state A.

It might seem that there is nothing to worry about here, as we are really saying that if the state definitions are adjusted properly, a kinetic description is always valid. Yet, in at least one important situation, protein folding, it is not so easy to define states. The denatured/unfolded state of a protein typically can only be defined by what it is not—that is, not folded. However, there are myriad ways for a protein not to be folded, and so there is no reason to expect that the denatured state is a physical state in the sense of exhibiting rapid internal equilibration.

10.2.6 Formalism for Systems Described by Many States

Our discussion of kinetics focuses on concrete examples of systems with one or two intermediates. Such cases help to build understanding of the most important concepts. However, it is important to appreciate that much more complex descriptions (recall Figure 10.2) can be treated using established matrix methods.

The official name for a kinetic description of a system with many states (or intermediates) is called a "master equation." For states $i = 1, \ldots, M$ and rate constants k_{ij}, the master equation employs the usual kinetic assumptions leading to

$$\frac{dP_i}{dt} = \sum_{j \neq i} P_j(t) \, k_{ji} - \sum_{j \neq i} P_i(t) \, k_{ij}. \qquad (10.15)$$

For two or three states, this precisely reduces to the kinetic equations we have already been considering. Note that some of the rates in the master equation may be zero, because transitions may not be possible between every pair of states.

The master equation is often used in discretized descriptions of biomolecular systems, but a classic discussion of the physics behind it can be found in van Kampen's book. The system of equations described by (10.15) is typically solved with matrix methods, which we shall not discuss. Nevertheless, the solutions are always linear combinations of decaying exponential terms, just as we found in the simple cases.

10.3 CONFORMATIONAL AND ALLOSTERIC CHANGES IN PROTEINS

We have already studied some thermodynamic aspects of allosteric conformational changes in Chapter 9. In particular, we examined some simple models describing the change in binding affinity at a second binding site following an initial binding event. Here, we will discuss the dynamics of the necessary conformational transitions and population shifts. Our discussion will be very general and fundamental to understanding allostery, but somewhat limited in its use of mathematics.

10.3.1 What Is the "Mechanism" of a Conformational Change?

How would we define the "mechanism" of a conformational transition? In other words, what would constitute a brief but informative description of a conformational transition in a complex system?

In a simple system of two states, A and B, it is fair to argue that the mechanism is embodied in the "transition state"—or the bottleneck—of the pathway, as described in Chapter 4. This would be the high-free-energy configurational state/region, which constitutes the main barrier of the transition.

Complex biomolecules, however, are likely to have many metastable intermediates and barriers separating two states of interest, and perhaps even multiple pathways. Figure 10.10 illustrates some of the ideas. No single state can describe the mechanism or pathway of such transitions. (Even though the chemical reaction in an enzymatic process may have a well-defined transition state, this will not be the case for the full process, which includes substrate binding and product release.) A good description of a full biomolecular transition process includes the set of intermediates and, ideally, also the transition states separating the intermediates. Yet such details really can only be learned from an almost impossibly detailed set of experimental data or, in principle, from the ensemble of transition trajectories discussed in Chapter 11.

10.3.2 INDUCED AND SPONTANEOUS TRANSITIONS

Sometimes a distinction is made between spontaneous conformational changes (e.g., folding) and those which are induced, perhaps by a binding event. Spontaneous transitions will occur when there is an equilibrium among two or more conformational states, or when a system is somehow prepared in only a subset of its states—and therefore out of equilibrium. "Induced" transitions, by definition, result from ligand binding or other "external" changes, which necessarily perturbs the equilibrium distribution of the receptor.

Do allosteric transitions "induced" by a binding event differ from spontaneous transitions? There are two levels to the answer: First, it would seem likely that many such transitions are not truly induced, but rather enhanced. As sketched in Figure 10.13, their frequency may be increased as the system adjusts to the new ligand-bound equilibrium. By the same token, however, the presence of a ligand will certainly perturb the underlying energy landscape, and generally the transition rate(s) as well.

On the second level, should we expect "binding-induced" transitions to exhibit qualitatively different physical properties? For instance, is an induced transition like a ball rolling downhill, as compared to a spontaneous activated transition over a barrier? This seems unlikely. For any large molecule, the roughness of the energy landscape

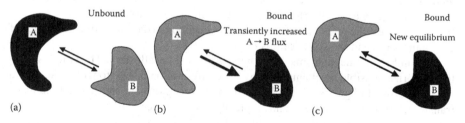

FIGURE 10.13 A quasi-kinetic view of allostery. Darker shading represents higher population, so population shifts from A to B upon binding. The arrow widths indicate the total flux in each direction (i.e., the population N_i multiplied by the rate k_{ij}) and the panels represent different time points. Binding changes the rate constants from k_{ij} to k'_{ij}, which leads to the observed shift in the populations. (a) Before binding, there is the usual equilibrium among states A and B, so the flux is equal in both directions: $N_A^{unbound} k_{AB} = N_B^{unbound} k_{BA}$. (b) Upon binding, there is a transient, nonequilibrium increase of flux from A to B: $N_A^{unbound} k'_{AB} > N_B^{unbound} k'_{BA}$. (c) After the transient relaxation period, the new equilibrium is established, with $N_B^{bound} > N_A^{bound}$ but $N_A^{bound} k'_{AB} = N_B^{bound} k'_{BA}$. It is important to note that this picture of allostery may be regarded as fundamental—that is, as defining allostery—and independent of the allosteric "mechanism," which concerns details of binding.

essentially guarantees that transition trajectories—"induced" or otherwise—will need to surmount significant and numerous energy barriers. In other words, the basic principles and dynamics governing both types of transitions are the same. Even though the A→B transition pathway and rate likely will be perturbed (as could the equilibrium distributions within the states), it is fair to say that "a transition is a transition is a transition"—it's the same underlying physics.

10.3.3 ALLOSTERIC MECHANISMS

To make our discussion of allosteric transitions somewhat more concrete, refer again to Figure 10.1, where we see a population shift from A to B accompanied by A→B transitions. While this picture makes sense, it simplifies the bidirectional dynamics that must always occur. That is, as shown in Figure 10.13, the initial unbound equilibrium between A and B is a consequence of transitions in both directions. The binding event perturbs the equilibrium and must transiently unbalance the fluxes. (Recall from Chapter 4 that we defined flux to be the total flow of population or probability, $N_i k_{ij}$, where N_i is the population of state i.) Once the fluxes rebalance, the new equilibrium is established.

Because of the flux–balance relation $N_A/N_B = k_{BA}/k_{AB}$, any binding event that shifts the conformational equilibrium also changes the rates. In physical terms, it is important to recognize that the rates change the populations, and not the other way around. That is, binding alters the system's dynamics, which in turn affects the equilibrium. This is shown in Figure 10.13.

The changes in the rates that cause the shifted equilibrium are not simple to analyze. In principle, it would be sufficient for just one of the rates to change the equilibrium ratio k_{BA}/k_{AB}. However, in reality this is not possible. In simple terms, recalling our double-well example of Figure 4.4, changing the barrier height will tend to affect both rates. More generally, a rate will change because the landscape has been altered (due to binding)—but the change in landscape necessarily affects both forward and reverse rates, if it changes either. (Although rigorously true, the proof of this point is beyond the scope of this book.)

A number of "mechanisms" have been proposed for allostery. The simplest way to categorize possible mechanisms is by the configurational state to which the ligand binds. That is, do ligands tend to bind when the receptor is in state A and "induce" transitions to state B? Or, do ligands typically bind to a preexisting state B and "trap" it there? There is the further possibility that ligands might bind to configurations neither in A or B.

There would not seem to be any statistical mechanics reason why one mechanism should be favored over the other—in general. On the other hand, it seems likely that the unbound equilibrium particular to a given system might have a strong influence. For instance, if $P_A(\text{unbound}) = 0.999$, it would seem reasonable to expect ligands to bind to configurations in state A—or, at least, this would be an evolutionarily sensible way to optimize the on-rate for binding. On the other hand, if state B is appreciably populated, it would seem that the ligand might bind to either state. If the unbound configuration space outside of states A and B has significant probability, then ligands surely might bind those configurations.

It is fair to say that nature does not worry about debates scientists may be having as to mechanistic theories of allostery. Rather, any mechanisms that is physically realizable and evolutionarily favorable can be expected to occur. If this kind of "promiscuity" prevents simple categorization of mechanism, so be it!

Finally, let us revisit Figure 10.13 and emphasize that it is mechanism independent. That is, regardless of the mechanism (how the ligand binds), allostery will necessarily shift the equilibrium distribution. And this population shift occurs, by definition, based on transiently nonequilibrium fluxes among states.

10.3.4 MULTIPLE PATHWAYS

On the whole, multiple pathways will not play a significant role in our discussion of conformational transitions, but a couple of questions are worth considering. First, are multiple pathways expected in biological systems? And second, though we cannot address it now, how can multiple pathways be defined from an ensemble of trajectories?

Usually, we think of enzymes, motor proteins, and other protein machines as performing finely tuned processes. We might translate this notion for the present discussion as the statement that a single pathway tends to dominate the transition(s). This seems quite plausible on the whole. A counterargument, however, arises from the expectation that proteins may also be evolved to be relatively insensitive to mutations. It is easy to imagine that multiple pathways, or at least significant pathway heterogeneity, would help dampen the effect of mutations. Nevertheless, such theories cannot be definitively addressed at present either experimentally or computationally.

Finally, we should touch on reverse transitions when there are multiple pathways. If one pathway is favored in the forward direction, is it necessarily favored in the reverse direction? The answer turns out to be yes—provided the conditions governing the system are the same in both directions. This can be seen by imagining two pathways are fully characterized by a dominant saddle point (energy barrier). In this case, the path with the lower barrier will exhibit the faster rate in both directions.

10.3.5 PROCESSIVITY VS. STOCHASTICITY

Some motor proteins are described as "processive," meaning they tend to be like man-made machines, with one step neatly following another. That such processivity should occur in a thermal environment is somewhat surprising given all we have learned. Indeed, essentially every process is reversible, even if the probability is small. However, for many cellular applications—locomotion, transport, and signaling—reversibility would not be desirable. Therefore, the cell has evolved many processive/mechanistic processes in which a system moves down in significant free energy increments, as sketched in Figure 10.14, making the process effectively irreversible. As is well known, such mechanistic processes need to have a source of free energy—which is often ATP, but also may be a trans-membrane proton gradient, GTP, or another molecule. One ATP molecule supplies about $20k_{B}T$ of free energy, which is certainly enough to make a process effectively

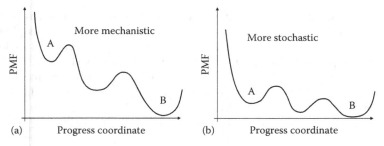

FIGURE 10.14 Mechanistic and stochastic landscapes. The categorization of a process as "mechanistic" or "stochastic" is not absolute, and is primarily governed by the free energy difference between the initial and final states. In the more mechanistic landscape (a), the large free energy difference between initial and final states leads to relatively large ratios of forward to reverse rates—that is, "mechanicity." By contrast, when there is a smaller net free energy difference as on (b), a much greater proportion of reverse steps can be expected—that is, "stochasticity." Note that in any "downhill" process where a decrease in free energy must recur repeatedly, it is necessary to add a source of free energy; in biological systems, this is often supplied by ATP.

irreversible! (Recall the truly free energetic nature of ATP's energy as discussed in Chapter 9.)

The fact that the net free energy difference must ultimately govern the degree of processivity means that some processes may be regarded as more stochastic—that is, with relatively frequent reverse steps. From Figure 10.14, it should be clear that reverse steps will be significantly likely when the A-to-B free energy difference is not much larger than $k_B T$. Similarly, even if the overall free energy difference is large (e.g., $\sim 20 k_B T$ for an ATP-based process), there could be many partial reversals when there are a sufficient number of intermediates between initial and final states. In the latter case, the overall ΔG must be shared additively among the intermediate sub-steps. This discussion has made use of the fact that, by definition, the system has enough time to cross the forward free energy barrier—which implies the reverse transition is possible if the equilibrium ΔG is not too large.

PROBLEM 10.12

Assume state A is $1 k_B T$ higher in free energy than state B, and only a single barrier separates them. What will be the ratio of forward to reverse rates? What if the ratio is $5 k_B T$?

More complete discussions of molecular machines, such as motors, and their use of energy can be found in the books by Howard, by Nelson, and by Phillips et al.

10.4 PROTEIN FOLDING

Protein folding is one of the most actively studied topics in the biophysics research community. Beyond its connection to a set of diseases, protein folding serves up a number of fascinating questions. It is critical to recognize first that there are two general protein-folding problems: (1) The first is the problem of predicting the three-dimensional folded structure of a specific protein, given its amino acid sequence. We shall not address this first problem, except glancingly in Chapter 12. (2) The second protein-folding problem inquires into the nature of the folding process—the rates, intermediates, and transition states—the "mechanism."

Our primary goal in this section is not to resolve long-standing controversies, but to explain briefly how the fundamental concepts of dynamics and kinetics apply in the context of protein folding. That is, the description of folding provided here is intended to enable you to see the problem in the larger context of statistical physics. We cannot hope to cover fully the thousands of papers that have been written on the subject.

10.4.1 PROTEIN FOLDING IN THE CELL

How do proteins really fold? This happens in more than one way, in fact. In the cell, proteins can begin to fold as soon as they are synthesized by the ribosome. That is, partially translated polypeptide chains will begin to explore configuration space as soon as they are able to. Of course, the chains don't really have a choice, since the translation process can take 10 s or more. Taking this a step further, Finkelstein and Ptitsyn suggest in their book that the sequences of some proteins have evolved to pause during translation, to allow proper folding of the initial secondary and/or tertiary structures.

On the other hand, it has also been established that many proteins will fail to fold spontaneously—at least on reasonable cellular timescales—without the aid of special helper proteins called "chaperones." Importantly, a protein that fails to fold properly is likely to adopt a compact structure that is not among the lowest in free energy: really, it is a misfolded protein and kinetically "trapped." A misfolded protein is delayed by barriers from accessing lower free energy regions of configuration space.

Chaperone proteins can be thought of as hydrophobic boxes, containers whose interior surfaces are coated with hydrophobic groups. Improperly folded proteins enter the chaperones. Inside, the misfolded protein can faster explore its configuration space because hydrophobic interactions with the chaperone container—which are not available in an aqueous environment—will reduce free energy barriers among the various folded and misfolded states. The faster exploration of configurations accelerates folding.

Already it is clear that "real" folding is complicated, and folding will happen quite differently for different proteins. In some cases, proteins may fold spontaneously, but aided somewhat by the order in which the chain is synthesized. Spontaneous unfolding and refolding can also occur. In other cases, spontaneous folding may be practically impossible.

Many experiments and theoretical/computational studies, however, have focused on spontaneous folding. In a test tube, many small proteins (re)fold spontaneously after they have been denatured by high temperature or extreme chemical conditions (perhaps low pH or high concentration of denaturing chemicals). While such folding experiments certainly do not model the full range of cellular folding processes, they have received the most theoretical attention because of their simplicity. Our goal is to understand the basics of this spontaneous *in vitro* protein folding in the context of the kinetic/dynamic framework we have already established.

10.4.2 THE LEVINTHAL PARADOX

The Levinthal "paradox" is not so much a paradox as a proof by contradiction. The essential point is that a protein would never be able to find the folded state if it performed a random search—that is, uniformly random—in the space of all possible configurations. Of course, we already know that nature uses dynamics, and does not bother with partition functions. Dynamics are governed by forces and the energy landscape, and are far from random.

Nevertheless, the back-of-the-envelope style calculation underlying Levinthal's paradox is worth going through. It starts by taking a very conservative view of the number of possible protein configurations—that is, underestimating the number—by assuming there are two possible configurations per amino acid. In a protein with 100 residues, there would then be $2^{100} \sim 10^{30}$ configurations available. If a protein could explore one configuration per picosecond (10^{-12} s), it would take 10^{18} s $> 10^{10}$ years to explore all configurations—assuming there were no repeats. Clearly, a protein does not have 10 billion years to fold!

Since measured protein-folding times tend to range from the μs to s range, the Levinthal paradox must be based on a completely wrong notion of how proteins explore configuration space. The "paradox" tells us how proteins do not fold. So how do proteins fold? This question has excited many decades of speculation and controversy. We shall try to explore some of the most basic principles governing folding, avoiding philosophical issues.

10.4.3 JUST ANOTHER TYPE OF CONFORMATIONAL CHANGE?

Our discussion of protein folding follows that of conformational change for a reason. At heart, protein folding is just another type of conformational change. Yet, while folding is necessarily subject to the same physical principles as other structural transitions, it differs in two important respects. First, experimental and computational studies are performed under a wide array of equilibrium and nonequilibrium conditions. As just one example sketched in Figure 10.15, a high-temperature unfolded state can be rapidly "quenched" to a folded temperature in a nonequilibrium process. Alternatively, the same system can be studied at a temperature where there is a dynamic equilibrium between folding and unfolding processes.

The second difference from a conventional conformational transition is that one of the principal states of the folding problem—the "denatured" or unfolded state—is unavoidably complex and far different from our usual conception of a physical state. The denatured state of a protein can be defined in different ways, and inevitably will exhibit relatively slow transitions internal to the "state." This contradicts the basic rules of state definition given in Section 10.1.1. Therefore, we will take a careful look at the unfolded state to begin to understand some of the complexities of the folding problem.

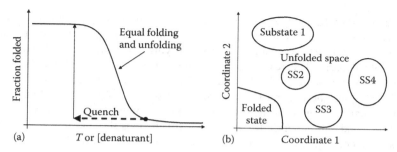

FIGURE 10.15 Different types of folding and unfolded states. Both experimentally and computationally, the folding process can be observed in different ways, based on different unfolded-state ensembles. For a given denaturant (e.g., temperature), each point along the horizontal axis of (a) corresponds to a different ensemble of unfolded configurations. Thus, folding observed during an equilibrium between unfolding and refolding should differ from that observed in a "quench"—that is, a sudden change of conditions. The dashed quench line in the figure sets up an (unstable) nonequilibrium situation where there is an anomalously high unfolded fraction under folding conditions; thus, rapid folding subsequent to the quench is expected, along the dotted line. (b) The configuration space of a protein is shown along with several unfolded sub-states (SS). Different unfolding conditions will exhibit different relative populations of the substates. Coordinates 1 and 2 could be any functions that increase with unfolding, such as the radius of gyration or the RMSD from a folded configuration.

10.4.4 What Is the Unfolded State?

The pathway(s) by which a protein folds generally should depend on where folding starts in configuration space. It is therefore very important to have a grasp of the unfolded "state"—or more properly, the unfolded ensemble.

Unfortunately, there is no standard definition of the unfolded state or ensemble. While one can argue that there is a (somewhat) ideal choice from a theoretical point of view, the fact is that folding experiments are performed from a variety of (equilibrium and nonequilibrium) preparations. Proteins can be denatured via temperature, pH, and chemical denaturants. They can be renatured via a sudden nonequilibrium "quench" to folding conditions, or folding can be observed in an equilibrium between folded and unfolded states (see Figure 10.15). There is no reason to expect that the ensemble of unfolded configurations would be the same in any two distinct experimental conditions.

Why will different conditions lead to different unfolded ensembles? This is not hard to see. Assume first that we have agreed upon classification of configuration space into folded and unfolded parts: that is, given any configuration, we can say whether it is unfolded or folded. Now consider the distribution of configurations that have been classified as unfolded, as sketched in Figure 10.15b. The weight (Boltzmann factor) associated with each configuration clearly will change depending on the temperature. It will change, similarly, with the concentration of denaturant—since these molecules certainly have nontrivial interactions with the protein.

From a purely theoretical point of view, there is a fairly natural definition of the unfolded state—albeit one that is difficult to probe experimentally. Assume that there is an agreed upon folded temperature of interest, T—for example, room temperature or body temperature. For the equilibrium ensemble at T, governed by the usual Boltzmann factor, we can again imagine classifying configuration space into folded and unfolded parts. The resulting equilibrium distribution of unfolded structures seems to be the most theoretically natural definition of the unfolded ensemble.

The key point here is not to memorize the list of possible unfolded ensembles, but to maintain awareness of the need to specify the conditions defining the unfolded ensemble—and subsequently to "filter" any conclusions through this specification. As a simple consequence, two folding experiments on an identical protein may not be studying the same process, due to differences in the unfolded ensembles.

10.4.4.1 Structurally Speaking...

What types of structures can be expected in the unfolded state? As you might guess, the answer depends on the degree and type of denaturation. However, mild denaturation to a point where most proteins in a sample/ensemble are no longer folded has some typical characteristics. The configurations tend to be compact rather than extended, and tend to maintain a high degree of secondary structure—even some tertiary structure. In short, some of the folded structure remains. To a first approximation, it is reasonable to imagine that secondary structure elements are simply a bit scrambled, with a bit of unraveling here and there.

10.4.5 MULTIPLE PATHWAYS, MULTIPLE INTERMEDIATES

Figure 10.16 depicts the basic "picture" one can expect for protein folding, generally. Because the space of unfolded configurations is so large and diverse—as schematized

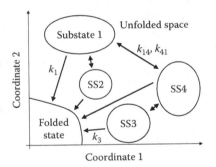

FIGURE 10.16 Multiple-pathway kinetic scheme for protein folding. Because unfolded space is so structurally diverse—that is, there are so many ways for a protein to be unfolded—a multiplicity of folding pathways should be expected. A number of on- and off-pathway intermediates can also be expected, again because of the complexity of the unfolded state. However, the observed rate will depend on the experimental conditions—both the "nature" of the denaturing and whether an equilibrium or nonequilibrium study is performed. Despite the complicated details, a general expectation is that the largest rate will dominate for a set of parallel pathways connecting the same states. In the figure, a one-sided arrow is meant to suggest the reverse rate is negligible. Only some of the rates are labeled to avoid complicating the figure.

by the variety of substates—one can expect a variety of folding pathways. Transitions among the various unfolded substates should also be possible in general. The relative populations among the substates will be determined by the external conditions (temperature, pH, denaturant concentration) as well as the "history" of the system (whether in equilibrium or recently quenched). These are the gory details.

Given all the rate constants present in the system, what determines the overall folding rate? In the simplest picture, we can assume all the rates among substates k_{ij} are much faster than the folding rates from individual substates k_i. This rapid equilibration among substates means that the fastest substate folding rate max$\{k_i\}$ will govern the overall folding rate, since the population that folds will quickly be replenished from other unfolded substates. On the other hand, if most of the unfolded population must first pass through an "intermediate" substate before folding (e.g., from SS1 to SS2 in Figure 10.16 if the direct rate k_1 is small), then we must apply the same considerations as in our discussion of conformational changes with intermediates.

Given the preponderance of nonequilibrium folding studies, it is worthwhile to emphasize that the observed overall folding rate depends, in principle, on the precise type of unfolding employed. Figure 10.17 graphically demonstrates the difference between the equilibrium case and a nonequilibrium example. In simple terms, depending on the particular nonequilibrium and equilibrium unfolded ensembles, a system put out of equilibrium may first have to relax into equilibrium and then fold. That is, it may be necessary to proceed from panel (b) of Figure 10.17 to (a) before significant folding occurs. The observed rate, which will depend on the various fluxes from individual substates, may change during a transient equilibration phase. Such considerations would be unimportant if the unfolded state could rapidly equilibrate—but

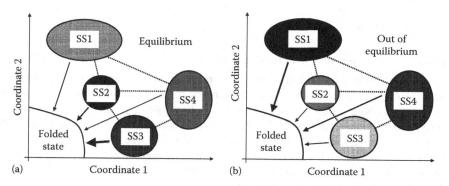

FIGURE 10.17 The difference between folding from nonequilibrium and equilibrium conditions. Panel (a) shows the relative equilibrium populations of the various unfolded substates (SS1, SS2,...), with darker shading denoting a larger population. The equilibrium populations are those observed under some fixed set of conditions—which could be mostly or partly folded. The arrow widths in both panels represent the net flux proceeding along the indicated pathway, while the dotted lines remind us of the dynamic equilibrium among substates. Panel (b) shows that unfolded substates will be differently populated after a quench to a nonequilibrium state. In turn, the fluxes in individual pathways and the observed overall folding rate may be different.

this may be unlikely given that large barriers are expected between compact, unfolded structures.

The example of an off-pathway intermediate can help illustrate the difference between equilibrium and nonequilibrium folding. In equilibrium unfolding, which is a steady state, an off-pathway intermediate has no effect on the observed (steady-state) rate: it will "absorb" as many transitions as it "emits" (see again Section 10.2.3). Out of equilibrium, however, the situation could be quite different. If the off-pathway substate needs to increase in population to reach equilibrium, it will tend to absorb population that could have folded.

10.4.6 TWO-STATE SYSTEMS, Φ VALUES, AND CHEVRON PLOTS

While the preceding discussion has hopefully communicated the basic statistical physics of protein folding, several other topics warrant special mention.

10.4.6.1 Two-State Folding

Two-state folding refers to the case, frequently discussed in the literature, when a single timescale appears to control a highly "cooperative" transition from the denatured to the folded state. By definition, such proteins lack significant intermediates, and hence are either folded or unfolded. This is sometimes described as "all or none" folding. Such simple behavior should only be expected from single-domain proteins, because additional quaternary structure in a protein is highly likely to complicate the kinetics with significant intermediates and additional timescales.

Based on our preceding discussion, it should be clear that the observation of two-state behavior implies that proteins have been prepared in a denatured state characterized by a single dominant basin or sub-state. This may not be possible, however, even for some single-domain proteins. Indeed, two-state behavior is not always observed.

10.4.6.2 Kinetics and Structure: Φ Values

If the three-dimensional folded structure of a protein is known, and if folding rates can be measured for different amino-acid mutations, then we can infer quite interesting information about structural aspects of the folding pathway. The Φ-value analysis developed by Alan Fersht, and well described in his book, uses a fairly straightforward logic for two-state folders assumed to possess a single significant transition state (i.e., free energy barrier). That is, an asymmetric double-well free energy landscape is assumed (our old friend, Figure 4.4), with unfolded/denatured (D) and folded/native (N) basins separated by a transition state ($\ddagger$).

The logic is that if the relative free energy for two of the three states—that is, D and $\ddagger$ or N and $\ddagger$—remains unchanged based on one or several localized mutations, then it is likely the structure in those two states is the same near the mutation site(s).

In practice, the Φ analysis is based on standard folding measurements. In many cases, two of the following quantities can be measured—k_{ND}, k_{DN}, and ΔG_{ND}—from which the third can be determined using the standard relation among rates and equilibrium free energy differences. If these measurements are performed both on wild-type (wt) and mutant (mut) proteins, then we can define

$$\Delta\Delta G_{ij} = \Delta G_{ij}(\text{mut}) - \Delta G_{ij}(\text{wt}),\qquad(10.16)$$

where (i,j) are $(D, \ddagger)$ or $(N, \ddagger)$. The barrier heights ΔG_{ij} values are estimated from the rate measurement using Arrhenius expressions based on Equation 4.2, namely, $k_{DN} \propto \exp(-\Delta G_{D\ddagger}/k_B T)$ and $k_{ND} \propto \exp(-\Delta G_{N\ddagger}/k_B T)$.

Fersht defined a convenient indicator-style function, called the "Φ value," as

$$\Phi_F = \frac{\Delta\Delta G_{D\ddagger}}{\Delta\Delta G_{DN}} = \frac{\Delta\Delta G_{D\ddagger}}{\Delta\Delta G_{D\ddagger} - \Delta\Delta G_{N\ddagger}}.\qquad(10.17)$$

When $\Phi_F = 0$, then the free energy difference from the denatured to the transition state remained unchanged due to the mutation under consideration: this suggests the particular site is denatured in both the transition and denatured state. If $\Phi_F = 1$, then both the native and transition states shifted equally due to the mutation and the site likely is folded in the transition state. Other values of Φ_F are more difficult to interpret.

PROBLEM 10.13

Sketch a free energy diagram to illustrate the two extreme cases just described.

The key point about Φ-value analysis is that three ingredients enable structural inferences about the transition state: (1) knowledge of a predetermined protein structure, (2) the ability to generate site-specific mutations, and (3) the ability to measure rates. Perhaps we should add to this list the fourth, theoretical ingredient—namely, the Arrhenius picture for the connection of the rate to the transition state.

10.4.6.3 Chevron Plots

The "chevron" plot is a well-known depiction of the way unfolding processes gradually overwhelm folding as increasing denaturant is added to a protein solution. A chevron is a "V" shape, and the chevron plot shows the behavior of $\ln(k_{obs})$ as a function of denaturant concentration. In the case of two-state folding, following Equation 10.2, the observed relaxation rate is the sum of the forward and reverse rates: $k_{obs} = k_{DN} + k_{ND}$, again with D = denatured/unfolded and N = native/folded.

The chevron-shaped behavior reflects a series of fairly simple facts: (1) at low denaturant, the folding rate is dominant, $k_{DN} \gg k_{ND}$, so that $k_{obs} \simeq k_{DN}$; (2) as denaturant is gradually increased, folding remains dominant but $k_{obs} \simeq k_{DN}$ decreases; (3) at some intermediate level of denaturant, k_{obs} reaches a minimum, and (4) subsequently climbs with further increase of denaturant (unfolding conditions) when $k_{obs} \simeq k_{ND}$, the unfolding rate. In some special cases, the "arms" of the chevron are spectacularly linear, suggesting a simple underlying model of the effects of denaturant (see Fersht's book for details and additional references).

The "surprising" minimum in a chevron plot reflects nothing more than the fact that the observed inverse relaxation time k_{obs} changes from the folding to the unfolding rate, as emphasized in the discussion of Equation 10.2.

10.5 SUMMARY

A kinetic description divides configuration space into states among which rate constants fully characterize transition processes. Such a description is most useful if a the system of interest naturally possesses a small number of states, each of which can equilibrate rapidly internally. For the folding problem, the denatured state may often not fulfill this criterion. For both allosteric conformational transitions and protein folding, nonequilibrium concepts become particularly important. Some initiating event—ligand binding or a sudden quench to folding conditions—may place the system out of equilibrium, and careful application of kinetic principles is necessary to understand the resulting relaxation to the appropriate equilibrium populations. Steady-state descriptions tend to be simpler.

FURTHER READING

Atkins, P. and de Paula, J., *Physical Chemistry*, 7th edition, W.H. Freeman, New York, 2002.

Cantor, C.R. and Schimmel, P.R., *Biophysical Chemistry, Part III*, W.H. Freeman, New York, 1980.

Chodera, J.D., Singhal, N., Swope, W.C., Pitera, J.W., Pande, V.S., and Dill, K.A., *Journal of Chemical Physics*, 126:155101, 2007.

Cui, Q. and Karplus, M., *Protein Science*, 17:1295–1307, 2009.

Fersht, A., *Structure and Mechanism in Protein Science*, W.H. Freeman, New York, 1999.

Finkelstein, A.V. and Ptitsyn, O.B., *Protein Physics*, Academic Press, Amsterdam, the Netherlands, 2002.

Frauenfelder, H., Sligar, S.G., and Wolynes, P.G., *Science*, 254:1598–1603, 1991.

Hill, T.L., *Free Energy Transduction and Biochemical Cycle Kinetics*, Dover, New York, 2004.

Howard, J., *Mechanics of Motor Proteins and the Cytoskeleton*, Sinauer Associates, Sunderland, MA, 2001.

McQuarrie, D.A. and Simon, J.D., *Physical Chemistry*, University Science Books, Sausalito, CA, 1997.

Nelson, P., *Biological Physics*, W.H. Freeman, New York, 2008.

Phillips, R., Kondev, J., and Theriot, J., *Physical Biology of the Cell*, Garland Science, New York, 2009.

van Kampen, N.G., *Stochastic Processes in Physics and Chemistry*, 3rd edition, Elsevier, Amsterdam, the Netherlands, 2007.

Wales, D.J., *Energy Landscapes*, Cambridge University Press, Cambridge, U.K., 2003.

Wales, D.J., *International Reviews in Physical Chemistry*, 25:237–282, 2006.

11 Ensemble Dynamics: From Trajectories to Diffusion and Kinetics

11.1 INTRODUCTION: BACK TO TRAJECTORIES AND ENSEMBLES

In this chapter, we shall take the final fundamental theoretical step in the statistical physics of biomolecules. We have already discussed how equilibrium phenomena are caused by dynamics, and we have carefully constructed an ensemble view of equilibrium behavior. We will now extend that ensemble view to dynamics. Typical biomolecular systems of interest, after all, do not perform dynamical processes just once but millions, billions, or trillions of times—or perhaps more. In a single living cell, further, there generally will be many copies of each molecule, although the number will fluctuate as the cell responds to external conditions or proceeds through the cell cycle. An ensemble picture, therefore, is perfectly natural.

As we stressed in Chapter 4, dynamics describes both "external" (real-space) and "internal" (configuration-space) motions. Examples include spatial diffusion and conformational changes. The ideas and formalism of this chapter apply equally to both.

The essential idea of this chapter is that, in the same way that equilibrium behavior requires description by an ensemble of configurations, dynamical behavior can only be specified fully by an ensemble of trajectories. A trajectory $\mathbf{r}^{\text{traj}}$ is the sequence of configurations $\mathbf{r}^N$ through which a system passes as it undergoes ordinary dynamics. We will see that in analogy to the equilibrium case, the probability of a trajectory can be written as a Boltzmann factor. Roughly speaking, a (scalar) function E^{traj} describes the effective "energy" of a trajectory, so that "low energy" trajectories are most probable and higher temperature permits more fluctuations:

$$\text{prob}\left(\mathbf{r}^{\text{traj}}\right) \propto \exp\left[-E^{\text{traj}}\left(\mathbf{r}^{\text{traj}}\right)/k_{\text{B}}T\right]. \qquad (11.1)$$

This chapter will be devoted to clarifying these ideas for trajectories and applying them to both new and familiar situations.

11.1.1 WHY WE SHOULD CARE ABOUT TRAJECTORY ENSEMBLES

The reason we care about trajectory ensembles is really quite simple: the complete trajectory ensemble contains the answer to every possible question about a system—mechanistic, structural, kinetic, you name it. This ensemble, after all, should contain every possible event and behavior at every timescale. To put it another way,

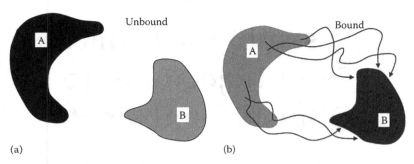

FIGURE 11.1 Trajectories of conformational change resulting from allosteric binding. Before binding, state A is predominantly occupied, while binding induces a population shift to state B. Thus, the ensemble must shift along trajectories (arrows) to state B. A variety of different trajectory paths can be expected due to (1) diversity in the initial ensemble, (2) diversity in the final ensemble, and (3) general thermal noise/stochasticity.

all the behavior we can observe results from trajectory ensembles. Even repeated experiments on single molecules reflect such an ensemble.

Trajectory ensembles—also called path ensembles—are extremely difficult to probe, but initial studies have been undertaken. It is fair to say that single-molecule experiments represent the frontier in the laboratory exploration of trajectory ensembles. Computationally, such ensembles have been obtained in detail for some small molecules and small proteins.

Yet even if there were no hope of ever "seeing" trajectory ensembles, they are so fundamental to nonequilibrium statistical mechanics, that we would be remiss not to study them anyway. In essence, all other dynamic and equilibrium descriptions constitute different averages/simplifications/projections of the trajectory ensemble. We will explore the connection between the path ensemble and more conventional "observables": (1) discrete states, (2) reaction rates, (3) transition states and reaction coordinates, (4) multiple pathways, as well as (5) diffusion and related phenomena. We have already seen in Chapter 10 that simpler kinetic descriptions are very useful in understanding allostery and protein folding; here we will understand the underlying trajectory-ensemble picture. As one example, consider again the process of allostery shown schematically in the now familiar Figure 11.1: the nonequilibrium transfer of population from one state to another is accomplished by dynamical trajectories, which typically will traverse important intermediates.

11.1.2 ANATOMY OF A TRANSITION TRAJECTORY

Before discussing the ensemble itself, we'll focus on a single transition trajectory in a toy model, as shown in Figure 11.2. The trajectory shows typical behavior of transitions over a single high barrier: the waiting time between events, t_w, is much longer than the time used to cross the barrier, t_b. That is, the trajectory tends to diffuse around in the initial state until a rare sequence of thermal fluctuations sends it over the barrier. This separation of timescales—which we saw long ago for butane in Figure 1.3—will prove very useful for analyzing transition dynamics.

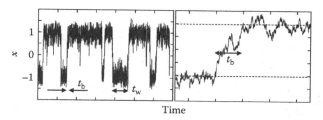

Time

FIGURE 11.2 Trajectory anatomy: the two principal timescales of a single-barrier transition. A trajectory for a one-dimensional double-well system is shown on the left, with one example transition depicted on the right. The first timescale is the waiting time t_w in the initial state before a transition, which is essentially the inverse of the transition rate, k. The second, much shorter timescale, is the barrier-crossing time $t_b \ll t_w \sim k^{-1}$, which is fairly independent of the rate.

What is the consequence of the long waiting time in the initial state prior to the transition? In many cases of interest, this means that the trajectory "forgets" its past. Any correlation with the previous history of the trajectory (e.g., which state it came from in a multistate system) is lost. Such history-independence suggests the single-barrier transitions will be a Poisson process, occurring with a constant probability per unit time. Once a trajectory enters a state, its behavior is governed by that state.

11.1.3 THREE GENERAL WAYS TO DESCRIBE DYNAMICS

Now that we have gotten our feet wet a bit, it's time to develop the general ideas that will inform the rest of the chapter. One essential idea is that dynamics can be described in a variety of (correct) ways that differ in their degree of averaging. Figure 11.1, for instance, shows a shift in the state populations, which can be more fully described in terms of a shift in full configuration-space distribution $\rho(\mathbf{r}^N)$. (Recall that $\mathbf{r}^N$ is the configuration of an N-atom molecule or system.) Figure 11.1 also shows that the change is brought about by the detailed dynamics—in fact by an ensemble of trajectories.

More generally, Figures 11.3 and 11.4 show that there is a hierarchy of descriptions of dynamics. Each trajectory is a list of configurations at a series of time points. With an ensemble of trajectories, we therefore can histogram configurations at any specific time point to get the configurational distribution particular to that time. Because we can do this at any time point, we can also track the evolution of the distribution with time. We can simplify further by summing over the probability density found in each of a set of discrete states to get a discretized description as a function of time. Each of these transformations can be considered as a type of averaging. More averaging leads to simpler descriptions.

11.1.3.1 Most Detailed: Ensemble of Trajectories

We have already sketched the trajectory ensemble picture, for which the fundamental description is the Boltzmann factor of Equation 11.1. We can be a bit more precise in our trajectory description, however. In particular, even though a physical trajectory

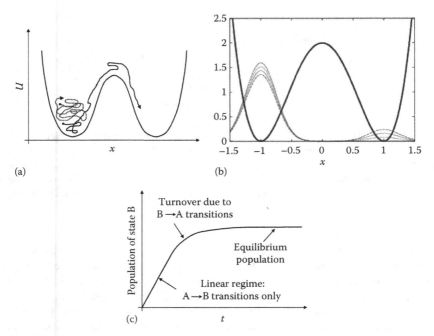

FIGURE 11.3 Three views of double-well dynamics. Panel (a) shows different dynamical trajectories that might occur in this system. Most trajectories that start in the left well (state A) will remain there, as does the dashed path. Occasionally, one will cross the barrier into the right well (state B), as shown by the solid path—and the fraction that cross per unit time is essentially the rate. Panel (b) illustrates the "distribution evolution" picture: the initial distribution/ensemble solely occupies the left well and shifts to the right over time. The shift in probability results from dynamical trajectories such as those shown at left. The rate constant for this transition can be computed from either the trajectory or distribution picture. Panel (c) shows the total probability accumulating in state B over time, in the discrete-state picture. The A-to-B rate is given by the initial linear slope in this two-state system, which reflects the transition probability per unit time. The probability "turns over" at later times as reverse (B to A) trajectories start to effect the final relaxation to equilibrium (long-time plateau value).

is continuous, it is simplest to assume a trajectory consists of a sequence of configurations at discrete time steps, $t = 0, \Delta t, 2\Delta t, \ldots$. (We can make Δt as small as necessary.) A trajectory is therefore the list of configurations

$$\mathbf{r}^{\text{traj}} = \left\{ \mathbf{r}^N(t=0),\ \mathbf{r}^N(\Delta t),\ \mathbf{r}^N(2\Delta t), \ldots \right\}$$

$$\equiv \left\{ \mathbf{r}^N_0,\ \mathbf{r}^N_1,\ \mathbf{r}^N_2, \ldots \right\}. \tag{11.2}$$

How can we begin to think about ensembles of trajectories? One simple example of the validity of the Boltzmann-like factor (11.1) comes from deterministic (e.g., Newtonian) dynamics. If the dynamics are deterministic, then the initial configuration (and set of atomic velocities) of every trajectory fully determines the remainder of

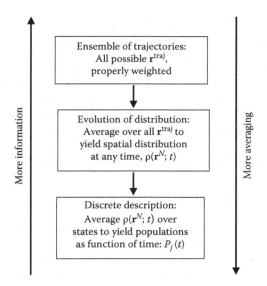

FIGURE 11.4 A hierarchy of descriptions of dynamics. The most complete description is the full statistical mechanical ensemble of trajectories. Each trajectory $\mathbf{r}^{traj}$ traverses a different path—that is, a set of configurations $\mathbf{r}^N$ at a series of time points. In the next simplest description, all individual trajectories can be averaged over to yield the probability distribution $\rho(\mathbf{r}^N)$ at every time t—denoted $\rho(\mathbf{r}^N; t)$. This distribution evolves with time based on the landscape, the dynamics, and the initial distribution. Finally, if physical states are known, all the probability in each state can be summed at each time t. After suitable analysis, this discrete/state picture can yield the transition rates among states.

the trajectory. Therefore, the distribution of initial configurations/velocities fully determines the ensemble of subsequent trajectories. Although arbitrary distributions of initial configurations are possible, in practice initial distributions commonly will be proportional to an ordinary Boltzmann factor.

11.1.3.2   Medium Resolution: Evolution of the Probability Distribution

The "medium resolution" description averages over all trajectories to yield the probability distribution at every point in time. Imagine starting a million trajectories at $t = 0$. This corresponds to some initial distribution $\rho(\mathbf{r}^N, t = 0) \equiv \rho_0(\mathbf{r}^N)$, which perhaps is confined to a single state. If each trajectory evolves dynamically, then at $t = \Delta t$, the distribution also evolves to $\rho(\mathbf{r}^N, \Delta t)$. The process continues, leading to $\rho(\mathbf{r}^N, 2\Delta t)$ and so on. Each ρ is a regular distribution—that is, probability density function—normalized over the space of configurations $\mathbf{r}^N$.

Why is the evolving ρ description less detailed than the trajectory ensemble? The essence is that we can no longer follow the progress of any individual trajectory over time. Compared to a discretized kinetic description employing states, however, the distributions have more information—namely, the distribution of probability within each state. As discussed in Chapter 10, not every system can be described well in terms of a simple kinetic model.

We can formalize the process of obtaining configurational distributions $\rho(\mathbf{r}^N, t)$ from the trajectory ensemble. We want to average over all trajectories weighted by the Boltzmann factor (11.1) in analogy to a potential-of-mean-force (PMF) calculation. After all, according to Equation 11.2, the configuration at a particular time t can be seen as one of many coordinates in a trajectory. In further analogy to a PMF calculation, many different trajectories may lead to the same configuration at time t. Mathematically, then, the distribution of configurations $\mathbf{r}^N(t)$ at time t is the weighted average over all trajectories sharing the same $\mathbf{r}^N$ value:

$$\rho(\mathbf{r}^N; t) \equiv \rho_t(\mathbf{r}^N) \propto \int d\mathbf{r}^{\text{traj}}\, \rho_0\left(\mathbf{r}_0^N\right) \exp\left[-E^{\text{traj}}\left(\mathbf{r}^{\text{traj}}\right)/k_B T\right] \delta\left(\mathbf{r}^N - \mathbf{r}^N(t)\right) \quad (11.3)$$

$$= \text{Sum over probabilities of traj. ending at } \mathbf{r}^N \text{ at time } t,$$

where $\mathbf{r}^N(t)$ is the configuration reached by the trajectory $\mathbf{r}^{\text{traj}}$ at time t. The initial distribution ρ_0 is necessary because E^{traj} does not normally include information about the initial conditions. The details of calculating a "path integral" like Equation 11.3 are not important now. Rather, what is important is the fact that the configurational probability distribution at any time point can indeed be obtained from the ensemble of trajectories. Our present equation can be compared to that for a general PMF given in Equation 6.8.

11.1.3.3 Discrete Description: Kinetics among States

We just saw that the configurational probability distribution at any time can be obtained from the trajectory ensemble by integrating over all paths. Now, to obtain a kinetic description based on discrete states, we can perform further averaging over the configurational distributions. Again, more averaging means lower resolution. The discrete picture has the least information, but is often easiest to understand.

The discrete-state description, as we saw in Chapter 10, requires state probabilities as functions of time: $P_j(t)$. To obtain the probability of state j at any time, we need to sum/integrate over the configuration volume V_j defining state j. This can be formalized as

$$\text{prob}(\text{state } j; t) = P_j(t) = \int_{V_j} d\mathbf{r}^N\, \rho(\mathbf{r}^N; t) = \int d\mathbf{r}^N\, h_j(\mathbf{r}^N)\, \rho(\mathbf{r}^N; t). \quad (11.4)$$

The second (equivalent) integral over all space selects out state j using an indicator function h_j

$$h_j(\mathbf{r}^N) = \begin{cases} 1 & \text{if } \mathbf{r}^N \text{ is in state } j, \\ 0 & \text{otherwise.} \end{cases} \quad (11.5)$$

The particular configuration-space volume of a state can be defined in many ways, perhaps based on one or more energy basins, or perhaps based on root-mean-squared distance (RMSD) from a reference configuration. (RMSD is defined in Equation 12.11.)

Once the state populations (i.e., probabilities) as functions of time are known, the rates among the states can be calculated. Following Chapter 10, this can be done starting with an initial distribution confined solely to one state, say i, and calculating the probability in the target state of interest, say j. In other words, we calculate the conditional probability to be in state j as a function of time, having started in state i. For a two-state system, as depicted in the last two panels of Figure 11.3, the slope of the linear regime of this growing conditional probability yields the rate. This can be written as

$$k_{ij} \simeq \frac{\mathrm{prob}(j; t \mid i; t=0)}{t} \qquad \text{(Linear regime; two states).} \qquad (11.6)$$

That is, the slope is the conditional probability per unit time—that is, the rate k_{ij}. When there are more than two states, the rate calculation is much more difficult, as discussed in Chapter 10.

The kinetic picture can also be derived directly from the trajectory-ensemble picture. This should not be surprising, since the ensemble is the most complete description of a dynamical process. Jumping directly from the ensemble to kinetics is particularly important in this chapter, because the more conceptually advanced mathematics required for the distribution-evolution picture will be delayed until chapter's end.

We can apply our usual (fundamental) probabilistic outlook to calculate the conditional probability of a state as a function of time. Roughly, from an ensemble of trajectories initiated in state i, we need only count the fraction of trajectories arriving in state j as a function of time. We can do this using an indicator function, $h_j(\mathbf{r}^N)$, as in Equation 11.4. The conditional probability to be in state j at time t having started in i at $t = 0$ is thus

$$\mathrm{prob}(j; t \mid i; t=0) \propto \int d\mathbf{r}^{\mathrm{traj}}\, h_i\!\left(\mathbf{r}_0^N\right) \rho_0\!\left(\mathbf{r}_0^N\right) e^{-E^{\mathrm{traj}}\left(\mathbf{r}^{\mathrm{traj}}\right)/k_{\mathrm{B}}T}\, h_j\!\left(\mathbf{r}^N(t)\right) \qquad (11.7)$$

$$= \text{Sum probabilities of trajectories ending in } j \text{ at time } t.$$

This equation is equivalent to Equation 11.4 aside from the condition of starting in state i, as can be seen by examining Equation 11.3.

11.2 ONE-DIMENSIONAL ENSEMBLE DYNAMICS

To help us understand trajectory ensembles, we will study a simple example in detail. In particular, we focus on one-dimensional dynamics, so instead of $\mathbf{r}^N$ we have only x.

How do transition paths vary in one dimension? There would seem to be only one way to travel from A to B in one dimension. However, there is still an incredible range of dynamical variation. Because forward and reverse steps of different lengths are allowed, trajectories will vary dramatically in how long they take to make a transition—as well as by the "local" velocity at every point in time during a transition. This is suggested by Figure 11.2.

Below, we derive an explicit probabilistic description of one-dimensional trajectories. From a mathematical standpoint, the derivation protocol here will carry over directly to complex molecular systems. The details of the equations will be different, but the main ideas are the same.

11.2.1 DERIVATION OF THE ONE-DIMENSIONAL TRAJECTORY ENERGY: THE "ACTION"

The idea of "trajectory probabilities" sounds complicated, but here you can see—based on a familiar example from Chapter 4—that the necessary math and physics are not beyond us. We will derive the explicit "trajectory energy" E^{traj} of Equation 11.1 for the important case of Brownian dynamics. (Note that E^{traj} is proportional to the more technical quantity, the "action.")

When we studied basic dynamics in Chapter 4, we considered the Langevin equation, which embodies a common form of stochastic dynamics. For the case when the Langevin dynamics are overdamped ("Brownian"), we also considered a discretization procedure $(dx/dt \rightarrow \Delta x/\Delta t)$, which led to a simple recipe for performing simulations. That same basic recipe will be used here to help derive E^{traj}.

The basic equation (4.14) describing how to go from one position $x_j = x(j\Delta t)$ to the next x_{j+1} is fairly simple, and we will write it in two equivalent ways to make our subsequent work easier:

$$x_{j+1} = x_j + \frac{\Delta t}{m\gamma}\left[-\frac{dU}{dx}\right]_{x_j} + \Delta x^{\text{rand}}, \qquad (11.8)$$

$$x_{j+1} - x_j \equiv \Delta x_{j+1} = \Delta x_{j+1}^{\text{det}} + \Delta x^{\text{rand}}, \qquad (11.9)$$

where we have denoted the deterministic part of the step, due to the force, as

$$\Delta x_{j+1}^{\text{det}} = -\left(\frac{\Delta t}{m\gamma}\right)\frac{dU}{dx}\bigg|_{x_j}. \qquad (11.10)$$

Recall that $-dU/dx$ is simply the force, and Δx^{rand} is a random number with zero mean and variance $2k_B T \Delta t/(m\gamma)$. The most natural choice for the form of the Δx^{rand} distribution is a Gaussian, and we will adopt that convention here.

Based on Equation 11.8, we can in fact derive E^{traj} and the action specific to this type of simulation. That is, we can explicitly write an equation for the probability of any trajectory to occur in a simulation that is based on Equation 11.8. Since each step has a deterministic part (proportional to the force) and a random part chosen from a Gaussian, we can expect that the probability distribution for the whole trajectory will contain a product of Gaussians.

To proceed, we will consider a one-dimensional trajectory denoted by

$$x^{\text{traj}} = \{x(t=0), \ x(\Delta t), \ x(2\Delta t), \ \ldots, x(n\Delta t)\}$$

$$= \{x_0, \ x_1, \ x_2, \ \ldots, x_n\}. \qquad (11.11)$$

The probability of such a trajectory is the probability of going from x_0 to x_1, multiplied by the probability of going from x_1 to x_2, and so on. Put into equation form, we have

$$\text{prob}\left(x^{\text{traj}}\right) = \rho_{\Delta t}(x_1|x_0) \cdot \rho_{\Delta t}(x_2|x_1) \cdots \rho_{\Delta t}(x_n|x_{n-1})$$

$$= \rho_{\Delta t}(\Delta x_1) \cdot \rho_{\Delta t}(\Delta x_2) \cdots \rho_{\Delta t}(\Delta x_n), \tag{11.12}$$

where $\rho_{\Delta t}(x_j|x_{j-1}) = \rho_{\Delta t}(\Delta x_j)$ is the conditional probability of getting to x_j from x_{j-1} in time Δt. Note that any trajectory can be specified equivalently by the points themselves $\{x_0, x_1, \ldots, x_n\}$ or by the initial point and the deviations between positions $\{x_0, \Delta x_1, \Delta x_2, \ldots, \Delta x_n\}$.

In the second line of Equation 11.12, it must be remembered that $\rho_{\Delta t}(\Delta x_j)$ depends not only on the increment Δx_j but also on the starting point of the increment, x_{j-1}. After all, the probability of making an "uphill" step resulting in an energy increase is very different from that of a downhill step. The value x_{j-1} implicitly contains the information on the potential surface in that region—that is, whether a given Δx_j is uphill or downhill.

Equation 11.12 is very general and abstract, but it can be evaluated explicitly for overdamped dynamics since the random aspect is fully described by Gaussians. In particular, we can write the probability distribution for an arbitrary step Δx_j based on Equation 11.9, which is a Gaussian with a mean displaced according to the forced step Δx^{det}:

$$\rho_{\Delta t}(\Delta x_j) = \frac{1}{\sqrt{2\pi\sigma^2}} \exp\left[\frac{-\left(\Delta x_j - \Delta x_j^{\text{det}}\right)^2}{2\sigma^2}\right], \tag{11.13}$$

where the variance is

$$\sigma^2 = \frac{2k_B T \Delta t}{(m\gamma)}, \tag{11.14}$$

as discussed earlier. If we now substitute the single-step distribution (11.13) into the distribution for whole trajectory (11.12), we obtain

$$\text{prob}\left(x^{\text{traj}}\right) = \left(\frac{1}{\sqrt{2\pi\sigma^2}}\right)^n \exp\left[\frac{-\sum_{j=1}^{n}\left(\Delta x_j - \Delta x_j^{\text{det}}\right)^2}{2\sigma^2}\right], \tag{11.15}$$

where we have omitted the distribution of the initial point to simplify our discussion.

Now we know the "energy" of a trajectory E^{traj} or, more properly, "the action." After all, Equation 11.15 contains all the information we need to specify the Boltzmann factor of Equation 11.1 for discretized overdamped dynamics. Substituting σ^2 from Equation 11.14 to obtain the hoped-for $k_B T$ dependence, we find

$$\text{Overdamped dynamics:} \quad \text{Action} \propto E^{\text{traj}} = \frac{m\gamma}{4\Delta t}\sum_{j=1}^{n}\left(\Delta x_j - \Delta x_j^{\text{det}}\right)^2. \tag{11.16}$$

The conventional action differs from our E^{traj} by a constant, unimportant in this context, which adjusts the units. Furthermore, be warned that taking the continuum limit ($\Delta t \to 0$) in the context of a discrete stochastic model such as underlies Equation 11.15 involves very nontrivial stochastic calculus. (Stochastic trajectories are intrinsically "jumpy" and not differentiable in the usual way. You can read more, and find references, about these issues in the papers by Adib and by Zuckerman and Woolf.)

Note that the normalization factors for the Gaussians, $(2\pi\sigma^2)^{-n/2}$, do not appear in Equation 11.16 because—unlike Δx_j and Δx_j^{det}—the normalization is independent of the particular positions x_j under consideration. That is, E^{traj} only includes the dependence on trajectory coordinates.

The bottom line is that we have indeed written down the probability of a trajectory in Equation 11.15, and furthermore it takes a Boltzmann-factor-like form with the action playing the role of an effective trajectory energy as in Equation 11.1.

11.2.2 Physical Interpretation of the Action

The physical meaning of the action given in Equation 11.16 is straightforward. In the simplest sense, just as for an equilibrium Boltzmann-factor distribution, the lowest "energy" E^{traj} corresponds to the most probable trajectory. More generally, any average over trajectories will be an ensemble average. If you think about it, moreover, you will see that we could define the path entropy that would play an analogous role to the equilibrium entropy. Similarly, a free energy barrier can be more or less entropic—but these interesting topics are beyond the scope of this book.

The action of Equation 11.16 also provides more concrete physical understanding of one-dimensional dynamics. Recall from Equation 11.10 that Δx^{det} is proportional to—and, critically, has the same sign as—the force. Thus the action for a given step varies with the square of $(\Delta x - \Delta x^{\text{det}})$, which is the difference between the actual step taken and one proportional to the force. When a trajectory is climbing uphill, the force always points downhill, making the uphill steps necessary for a transition the least likely of all. A transition event must possess many unlikely uphill steps almost in succession, since too many downhill steps will erase any upward progress, making it extremely unlikely (see again Figure 11.2).

The barrier height, which tends to be correlated with the magnitude of the force (and hence Δx^{det}), will have a strong influence on the likelihood of a transition. Although we will not derive the Arrhenius factor here, the exponential form of trajectory probabilities makes the exponential sensitivity of kinetics on barriers seem very plausible.

11.2.2.1 Including the Initial Distribution ρ_0

If we account for the fact that we may only know the initial point x_0 of a trajectory statistically, via its distribution ρ_0, then the full probability of a trajectory should be written as

$$\text{prob}(x^{\text{traj}}) \propto \rho_0(x_0) \exp\left[-E^{\text{traj}}\left(x^{\text{traj}}\right)/k_B T\right]. \tag{11.17}$$

In the most interesting cases, the distribution of the initial configuration $\rho_0(x_0)$ will not be an equilibrium distribution. In equilibrium, as we know from Chapter 4, trajectories simply will go back and forth—with equal flux—between any pair of states. More on this below.

11.3 FOUR KEY TRAJECTORY ENSEMBLES

Although there is only one way to be in equilibrium (recall Chapter 4), there are many ways to be out of equilibrium. Here we will consider different trajectory ensembles that reflect this fact. In broad terms, there are two main ways to be out of equilibrium—based on a nonequilibrium initial distribution or based on a steady state—and there is a corresponding trajectory ensemble for each. One can also define an equilibrium trajectory ensemble. The fourth type of ensemble actually is a subset of any of the other three, namely, the transition path ensemble.

We can think about how these different ensembles relate to the single basic equation (11.1) for trajectory probability. To specify the path ensemble fully, we not only need to know the "action" or effective energy $E^{\text{traj}}(\mathbf{r}^{\text{traj}})$, which gives the probability of a trajectory $\mathbf{r}^{\text{traj}}$, but we also need to specify the governing conditions. In the same way that differential equations admit of multiple solutions until initial or boundary conditions are specified, the same is true for trajectory ensembles.

11.3.1 INITIALIZED NONEQUILIBRIUM TRAJECTORY ENSEMBLES

Most simply, we can specify a trajectory ensemble based on an initial condition. That is, following Equation 11.17, we can specify the configuration-space distribution at $t = 0$, which is $\rho_0\left(\mathbf{r}_0^N\right)$. In other words, the initialized trajectory ensemble consists merely of the set of trajectories resulting from configurations distributed according to ρ_0. Such an initial condition—if it is interesting—typically will differ from the equilibrium distribution. Starting from equilibrium, after all, simply will lead to trajectories "buzzing" back and forth and balancing the effects of one another. By contrast, the most interesting dynamical situations occur when one asks how a system will relax to equilibrium after it is "perturbed"—or slammed or blasted—to a nonequilibrium state. The most typical nonequilibrium initial condition will be when a system is prepared in a single dominant configurational state. Examples include the unbound or unfolded states of a protein described in Chapter 10.

Unlike most of the other ensembles to be considered, the initialized trajectory ensemble typically will consist of a set of trajectories of equal length (i.e., equal duration or number of time steps). In other words, starting from the initial distribution ρ_0, trajectories are included up to some time t. This readily enables an "apples-to-apples" comparison of trajectory probabilities according to the distribution $\rho_0\left(\mathbf{r}_0^N\right) \exp[-E^{\text{traj}}(\mathbf{r}^{\text{traj}})/k_B T]$, where $\mathbf{r}^{\text{traj}}$ implicitly has $t/\Delta t$ time steps.

11.3.2 STEADY-STATE NONEQUILIBRIUM TRAJECTORY ENSEMBLES

The steady-state trajectory ensemble is of fundamental importance. We have already discussed steady states using a kinetic description in Chapter 10. Approximate steady-state conditions are important in the laboratory and in the cell. Recall, for instance, from our discussion of the Michaelis–Menten description of catalysis occurring with an excess of substrate. Motor proteins in the presence of an excess of "fuel" will also be in a steady state.

It is worth emphasizing that the steady-state picture is necessarily statistical: it applies to the average behavior of an ensemble of molecules. The steady-state trajectory ensemble consists of dynamical trajectories occurring when a system is held in a steady state. Unlike the initialized ensemble, the distribution of configurations will never vary in time.

Echoing our concern with rate constants, one particular steady state is of primary interest: the case when systems that successfully transition to product state B are immediately converted back to reactant state A. In Chapter 10, we phrased this in terms of probability from B being fed back to A. Now, we can say that trajectories that reach B are restarted in A. In biomolecular systems, this usually occurs based on the unbinding and rebinding of a ligand.

Steady-state trajectories may have a variety of durations. To see this, imagine a steady state has been established, and that we observe the trajectories entering state B. Although there will be a constant number entering B per unit time, if we trace back the history of the trajectories, each will have left state A at a different time. Thus, there is no simple mathematical formula for weighting steady-state trajectories. Instead, somewhat complex boundary conditions must be invoked, which are beyond the scope of our discussion.

11.3.3 THE EQUILIBRIUM TRAJECTORY ENSEMBLE

Since the beginning of the book, we have emphasized that equilibrium may be viewed from an ensemble-average point of view, or based on time averages. We invoked the image of making a movie (or molecular dynamics trajectory) of a molecular system. Configurations visited in a long movie, we argued, would necessarily have the same distribution seen in a "relaxed" (equilibrium) set of many configurations at any single time.

We can accordingly define the equilibrium trajectory ensemble in two ways: (1) a single very long trajectory that has visited all states many times or (2) a set of equal-duration shorter trajectories initiated from an equilibrium ensemble of configurations. These two ensembles will exhibit the same observable behavior for all phenomena that occur on timescales less than the length of the trajectories. Description (2) can be seen as a special case of the initialized ensemble started from an equilibrium distribution— that is, started from $\rho_0\left(\mathbf{r}_0^N\right) \propto \exp\left[-U\left(\mathbf{r}_0^N\right)/k_B T\right]$. A more complete discussion of the equilibrium trajectory ensemble can be found in the paper by Hummer.

11.3.4 TRANSITION PATH ENSEMBLES

A transition-path ensemble might be termed a "sub-ensemble," which can be derived from any of the trajectory ensembles described above. It consists of transition events only. That is, given any other trajectory ensemble, the transition path ensemble consists of continuous trajectory segments that start in state A (however defined) and end in state B (also arbitrary). There will be no A or B configurations except at the very beginning and end of each trajectory.

As in the steady-state ensemble, each trajectory will have a different duration. The mathematical description also requires complex boundary conditions, which we shall not describe here. The interested reader can consult the review article by Bolhuis et al.

11.3.4.1 Examples of Transition Path Ensembles

It is always useful to examine concrete examples, and here we will examine three transition-path ensembles calculated from simulations. We will look at a toy model and two molecular models.

The very simplest example is a one-dimensional double well, such as we encountered in the "mini-project" of Chapter 4 (see also Figure 11.3). This is a smooth and bistable potential, described by $U(x) = E_b[(x/d)^2 - 1]^2$. The data shown in Figure 11.5 were generated from overdamped simulation using a barrier height of $E_b = 7k_BT$ and $d = 1$. Even in this apparently trivial one-dimensional system, the transition path ensemble is evidently quite complex, with significant variation among transition events. More information about the details of these events can be found in the paper by Zhang, Jasnow, and Zuckerman.

Transition events in the butane molecule exhibit similar behavior, as seen in Figure 11.6. Of course, there are no intermediates for the chosen transition, so the similarity is not surprising.

We finally consider protein transition trajectories, based on a highly simplified alpha-carbon model of one domain of calmodulin. (Details of the model can be found in the 2004 paper by Zuckerman.) Calmodulin opens and closes based on calcium binding and thus transmits the "signal" of calcium's presence (open/holo state) or absence (closed/apo state) to other proteins via subsequent binding or unbinding events. Two transition events are depicted in Figure 11.7. Importantly, calmodulin's 72 alpha carbons indicate that the configuration space is 216 dimensional! The

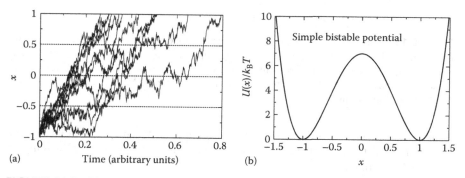

(a) Time (arbitrary units)

(b) x

FIGURE 11.5 The simplest transition path ensemble, from a toy bistable potential. (a) Ten representative transition trajectories are shown from overdamped simulation on the one-dimensional symmetric double-well potential in (b). The transition trajectories are limited to the region $-1 < x < 1$, which is an arbitrary choice. There is substantial heterogeneity among the transitions, both based on the total duration of the events and the distribution of local velocities along the paths.

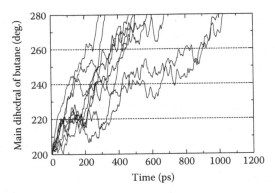

FIGURE 11.6 A transition path ensemble for butane. As in the toy model of Figure 11.5, there is substantial variation in the durations of the events.

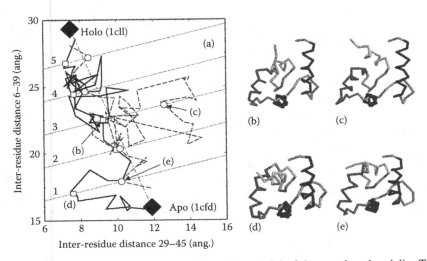

FIGURE 11.7 Transition paths in an alpha-carbon model of the protein calmodulin. Two representative transition events are shown in (a), projected onto the space of two inter-residue distances. Specific intermediate structures are shown in panels (b)–(e). Experimental structures are also labeled in (a) with large diamonds. (From Zuckerman, D.M., *J. Phys. Chem. B*, 108, 5127, 2004. With permission.)

projections down to two dimensions thus only hint at the true complexity of the transitions, but the heterogeneity of the ensemble is clear, as is the possibility of metastable intermediates.

11.4 FROM TRAJECTORY ENSEMBLES TO OBSERVABLES

The trajectory ensemble contains so much information that it is not surprising that we can perform many interesting averages—over time and space, and in some creative ways. A very rich understanding of dynamics can result from exploring such averages.

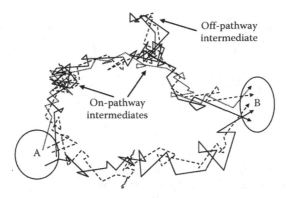

FIGURE 11.8 Schematic description of a trajectory ensemble, again. Two states, A and B, are connected by two channels, and each channel encompasses its own trajectory ensemble. One channel has three fairly well-defined intermediates, one of which is "off pathway." The full trajectory ensemble encompasses all channels and all dwell periods in the intermediates.

As a schematic example, consider a trajectory ensemble connecting two stable states, as sketched in Figure 11.8. First, in principle there may be more than one "reaction channel"—that is, more than one pathway by which a process may occur. Second, there may be metastable intermediate states, at which trajectories tend to pause. These intermediates can be directly on a pathway for conformational change or "off pathway." Metastable on-pathway intermediates are of particular interest for drug-targeting. If an intermediate is sufficiently stable, then the protein will spend a significant fraction of its time in that state before completing the transition process—perhaps enabling inhibition via small-molecule binding to the configurations in the state. All this information and more is contained in the trajectory ensemble.

11.4.1 CONFIGURATION-SPACE DISTRIBUTIONS FROM TRAJECTORY ENSEMBLES

Perhaps the most basic type of information that can be derived from the trajectory ensemble is the configuration-space probability distribution $\rho_t(\mathbf{r}^N)$, which may change with time t. Equation 11.3 has already described how to obtain this most fundamental "observable" in the context of an initialized trajectory ensemble. In general, ρ will change over time unless it started out in equilibrium or in another steady state. Figure 11.9 gives a schematic description of the configuration-space distribution derived from a steady-state system.

A different projection onto configuration space is the distribution of configurations in the transition-trajectory ensemble. In this distribution, configurations that are visited more frequently, or for longer periods of time, during transitions will have larger weights. This distribution would clearly highlight metastable intermediates, even if their net probability was small compared with that of the initial and final states. See the paper by Hummer for further details.

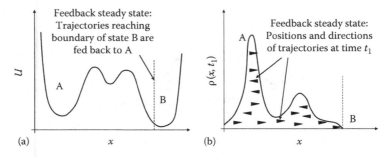

FIGURE 11.9   The configurational distribution derived from a steady-state path ensemble. The panel (a) depicts a one-dimensional potential with an intermediate, and the dashed vertical line indicates the boundary defined for state B. Trajectories reaching the boundary are fed back into state A. Therefore, as shown in (b), there are no trajectories (and no probability) to the right of the boundary. The panel (b) also shows a snapshot in time of the steady-state path ensemble. The arrow heads schematically indicate the location and direction of each trajectory at time t_1. Arrow heads are separated vertically for clarity—the system is only one dimensional.

11.4.2   FINDING INTERMEDIATES IN THE PATH ENSEMBLE

Metastable intermediate states are a key part of describing the "mechanism" of a biomolecular process. If indeed it is true that the trajectory ensemble includes all possible information about a system, then it must include "knowledge" of the intermediates. However, a careful description of the definition of any state, let alone one that is metastable, is not trivial (see for example the article by Chodera et al.). Approximate approaches, on the other hand, are not hard to conceive. For instance, projections onto appropriate coordinates—as schematized in Figure 11.9—will yield peaks at potential intermediates.

It is also possible to search for intermediates directly from a trajectory ensemble, using dynamical information. After all, states are fundamentally defined by their kinetic properties: roughly, a state is a region of configuration space that does not permit rapid escape. The precise definition of "rapid" will depend on the system, but key timescales can be inferred from the ensemble of transition trajectories.

Operationally, one might proceed as follows, based on a transition path ensemble. First, one can define a basic timescale from the average "length" (duration) of the trajectories. Then, trajectory segments can be sought that tend to dwell in a small region of configuration space for a time of similar order (see Figure 11.7). Such dwelling can be assessed based on any similarity measure, such RMSD. What constitute a "small" region of configuration space can be gauged by variation within the trajectory ensemble.

11.4.3   THE COMMITMENT PROBABILITY AND A TRANSITION-STATE DEFINITION

Besides the configurational distributions described above, there are other trajectory averages that can usefully be projected onto configuration space. A good example is the "commitment probability" $\Pi_B(\mathbf{r}^N)$, the fraction of trajectories initiated from configuration $\mathbf{r}^N$ that will reach a predefined target state (B) before returning to an initial state (A). Near state B, the commitment probability will be close to one, while near A, it will be close to zero.

Note that Π_B is a dimensionless probability, and not a normalized density. Also note that Π_B cannot be directly associated with any timescale for reaching state B, since the complexities of the landscape have been averaged out.

PROBLEM 11.1

Sketch a one-dimensional energy landscape containing points with $\Pi_B \simeq 1$, but which will reach state B fairly "slowly."

The commitment probability Π_B leads to a simple definition of the "transition state." In Chapter 4, we schematically suggested the transition state was the peak of the (free) energy landscape—that is, the critical barrier of a transition. That simple picture breaks down in a complex landscape, where there may be many barriers. However, one can define the transition state as the set of configurations $\mathbf{r}^N$ for which $\Pi_B(\mathbf{r}^N) = 0.5$. In this conception, the transition state is the set of configurations from which trajectories have an equal chance of "reacting" to B or returning to A.

The $\Pi_B = 0.5$ definition has the advantage of simplicity and generality, but surely it fails to convey the potential richness of a landscape with several important intermediates and saddle points. At the same time, it must be emphasized that every averaging or projection process removes information. The key is to be aware of which information is most important in a given situation. For instance, in a complex protein with multiple intermediates, it probably makes more sense to attempt to determine the key intermediates, rather than the $\Pi_B = 0.5$ hypersurface.

11.4.4 PROBABILITY FLOW, OR CURRENT

We can also consider probability flows. That is, from the trajectory ensemble, one can calculate how much probability tends to flow into (and out of) a given volume of configuration space in a steady state. Further, one can calculate where that probability tends to flow, and whence it originated. These flows are called the current.

Such probability flows can be defined based on either an initialized or a steady-state trajectory ensemble. Even though the probability distribution of configurations ρ_t remains unchanged in a steady state, there is generally a flow of probability from the source state (A) to the sink state (B). There is no flow in equilibrium, by definition: indeed, the lack of flow distinguishes equilibrium from other steady states, which may have the same configurational distribution.

The mathematical formulation of the current will be described briefly when we discuss diffusion and the Smoluchowski equation below. Further information on this topic can be found in books by Gardiner, by Risken, and by van Kampen.

11.4.5 WHAT IS THE REACTION COORDINATE?

We have discussed reaction coordinates before, but we should briefly revisit this important topic. Recall that, loosely speaking, a reaction coordinate is a single variable or a set of variables along which the PMF can be calculated to give a reasonable representation of the dynamics. The "commitment probability" (described above) should increase monotonically along the reaction coordinate.

In some cases, simple choices for the reaction coordinates may be sufficient for an intuitive grasp of the system. For instance, the distance between a pair of critical atoms or residues is often used to describe a conformational change in a biomolecule. Yet from a theoretical standpoint, we would like to stand on firmer ground.

A key issue is whether there is a unique reaction coordinate for any given system. If not, then following our discussion of PMFs, different coordinate choices can lead to different apparent barrier heights—and hence different implied rates. There certainly are choices that are guaranteed to be unique and fulfill the desired characteristics. For instance, the commitment probability Π_B described above could serve as a reaction coordinate: those configurations with 1% chance of transitioning, followed by those with 2%, and so on. While formally appropriate, note that Π_B typically is extremely difficult to calculate and obscures certain important physical information, such as the locations of intermediate states. Additional information, such as the configuration-space distribution in the transition-path ensemble, could be used to augment the Π_B coordinate.

A more informative description would come from the analysis of probability flows (directions and magnitudes) described above. Such a description would seem to be more fundamental, because it is less averaged. However, this is an advanced research-level topic, and the interested reader should search the current literature for more information.

11.4.6 FROM TRAJECTORY ENSEMBLES TO KINETIC RATES

We have already discussed somewhat the process of "translating" the trajectory picture into states and kinetics. Now, we want to step beyond the kinetic view in two ways, however. First, the trajectory picture will reveal a fundamental relationship between first-passage and steady-state kinetics—one which is almost invisible from the states-and-rates picture. Second, we'll use this relation to see how rates can be calculated directly from the steady-state trajectory ensemble.

Although in Chapter 10 we argued that the steady-state and first-passage rates are different (indeed they are not the same), the first-passage rate is actually given by the flux in the steady state. To the author's knowledge, this was first pointed out in Hill's book.

The connection between the feedback-steady-state flux (probability per unit time) entering state B and the first passage rate, k_{AB}^{FP}, can be derived via a thought experiment. Imagine we have set up a large number of systems in a steady state, so that a constant fraction of the systems are entering state B per unit time. This fraction is the steady state current or flux, Flux(A → B, SS)—refer to Chapter 4. Each individual system, however, does not "know" it is in a steady state. Rather each system simply executes unbiased dynamics and arrives at state B via an "ordinary" trajectory, after which it is immediately relaunched in A.

Based on this steady state consisting of ordinary trajectories, we can define the first-passage time (FPT) in a backward way. For every trajectory that arrives to B, we can probe the trajectory ensemble to see how much total time has passed since it was fed back to A. (Importantly, after being launched in A, a trajectory may leave and reenter A many times before reaching B. We want the total time after the initial "launch.") This total time is nothing other than the FPT, and we can average over all trajectories to get k_{AB}^{FP} as the inverse of the mean FPT (MFPT). But now we know the mean time the trajectories spend making the transition and we also have a constant rate of trajectories reaching state B. We can therefore calculate the fraction of trajectories that reach B in an arbitrary time interval Δt as $\Delta t/\text{MFPT} = \Delta t\, k_{AB}^{FP}$. Well, this defines the flux into B as

$$\text{Flux}(A \to B, \text{SS}) = k_{AB}^{FP}. \tag{11.18}$$

The difference between the flux and the steady-state rate, k_{AB}^{SS} is a factor of the state A probability in the steady state (see Equation 10.13). The "proportionality constant" P_A is in turn influenced by the whole system—including off-pathway intermediates, because the overall probability has to be normalized.

Based on the equivalence (11.18), we have two ways of calculating the first-passage rate from the steady-state trajectory ensemble. First, we can calculate the average trajectory length as the MFPT. Second, we can compare trajectories at subsequent time points to calculate the flux entering state B in that time interval.

11.4.7 MORE GENERAL DYNAMICAL OBSERVABLES FROM TRAJECTORIES

There is much more dynamic/mechanistic information in the path ensemble than simply rates. Consider, for example, an allosteric system. We might like to know how a binding event in one part of a protein affects other parts of the system. The trajectory ensemble provides the perfect tool for examining such issues because it permits the computation of arbitrary correlations and averages. Of particular interest is the form $\langle f\left(\mathbf{r}^N(t),\, \mathbf{r}^N(t+\tau)\right)\rangle$; here f could be a function quantifying fluctuations and τ is any time separation of interest. The ability to probe time differences permits the examination of causal relationships, which depend on time-ordering by definition.

11.5   DIFFUSION AND BEYOND: EVOLVING PROBABILITY DISTRIBUTIONS

We will take our last look at diffusion, and recast it into the language of an evolving probability distribution—which is just a spreading Gaussian. Despite the simplicity of Gaussians, in the bigger picture, we are employing a new type of statistical/probabilistic description. We shall see how to apply the evolving-distribution description more generally, to spatially varying landscapes—that is, where U is not a constant.

Bear in mind that the diffusional picture of evolving probability distributions applies both to real and configuration space. Our representative coordinate x could be an ordinary Cartesian coordinate or an internal coordinate. To put it another way, we could be discussing the spreading of ink in a cup of water or protein folding.

11.5.1 Diffusion Derived from Trajectory Probabilities

We have discussed diffusion more than once already in Chapters 2 and 4. What was simple then remains simple now: diffusion is described by a Gaussian distribution with a width that grows as the square root of time. It is instructive to confirm that the same result emerges from the fundamental trajectory-ensemble picture. We will perform our calculations in one dimension for simplicity; the generalization to two or three dimensions is not difficult.

Simple diffusion is just a special case of the overdamped dynamics that are simulated using Equation 11.8 or Equation 11.9. Specifically, simple diffusion is defined to be overdamped motion on a spatially constant energy surface, meaning the force is always zero. Therefore, the deterministic step Δx^{det} in Equation 11.9 is always zero—see Equation 11.10. This means that we can immediately write down the probability of a diffusive trajectory by setting $\Delta x^{\text{det}} = 0$ for all j in Equation 11.15. If we further assume that all trajectories are initiated from a single point $x_0 = 0$, which is equivalent to setting $\rho_0(x_0) = \delta(x_0)$, we find

$$\text{Diffusion:} \quad \text{prob}[x^{\text{traj}}] = \left(\frac{1}{\sqrt{2\pi\sigma^2}} \right)^n \exp\left[-\frac{\sum_{j=1}^{n} \left(\Delta x_j\right)^2}{2\sigma^2} \right], \tag{11.19}$$

which is just a product of Gaussians with zero mean and width σ.

The difference between Equation 11.19 and our previous description of diffusion is that we have now written down the probability for an arbitrary individual diffusive trajectory. This is more detailed than our previous description via the configurational probability distribution—that is, the distribution of x values—which evolves over time by spreading. Although we can simply set time to be $t = t_n = n\Delta t$ in Equation 11.19, that in itself does not yield the probability distribution at time t. To see this, note that Equation 11.19 depends on the n step sizes $\Delta x_j = x_j - x_{j-1}$.

To obtain the ordinary spatial probability density for a position x at a given time t_n, we need to integrate over all possible trajectories reaching the point $x_n = x$ at time $t = t_n$. In other words, we need a "path integral," such as Equation 11.3. Fortunately, the path integral for diffusion is one of the very simplest you will encounter. Operationally, we want to integrate over every possible first position x_1 (which occurs at time $t_1 = \Delta t$ by definition), and over every possible second position x_2 at $t_2 = 2\Delta t$, and so on. The appropriate integral for the probability density at time t_n is therefore

$$\rho(x; t = n\Delta t) = \rho_n(x) = \int dx_1 \, dx_2 \ldots dx_{n-1} \, \text{prob}[x^{\text{traj}}]. \tag{11.20}$$

Although Equation 11.20 in fact applies to any type of one-dimensional stochastic dynamics, we will specialize to the case of diffusion by employing the diffusive trajectory probability (Equation 11.19) as the integrand, obtaining

$$\text{Diffusion:} \quad \rho_n(x_n) = \int dx_1 \, dx_2 \ldots dx_{n-1} \left(\frac{1}{\sqrt{2\pi\sigma^2}} \right)^n \exp\left[-\frac{\sum_{j=1}^{n} \left(\Delta x_j\right)^2}{2\sigma^2} \right], \tag{11.21}$$

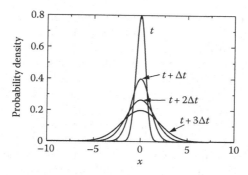

FIGURE 11.10   The diffusive evolution of a probability distribution. The initial distribution is narrow, but diffusion always leads to a spreading and flattening over time. The evolution of a purely Gaussian distribution is shown over three equal time intervals. Starting with $\rho(x; t)$, the distribution evolves to $\rho(x; t + n\Delta t)$, with $n = 1, 2, 3$.

which you should recognize as a simple convolution of Gaussians, noting that $\Delta x_j = x_j - x_{j-1}$.

PROBLEM 11.2

(a) Fully evaluate the diffusive path integral (Equation 11.21) for the simple case $n = 2$, which is a single convolution. (b) Explain why your result is expected based on our basic discussion of convolutions in Chapter 2.

Performing the integration of Equation 11.21 yields the familiar result

$$\rho_n(x_n) = \frac{1}{\sqrt{2\pi n \sigma^2}} \, e^{-(x_n - x_0)^2/2n\sigma^2}, \qquad (11.22)$$

which is just a Gaussian that spreads with time, where time is represented by $n = t/\Delta t$. See Figure 11.10.

11.5.2   DIFFUSION ON A LINEAR LANDSCAPE

Our preceding derivation, which hopefully was useful "mental exercise," did not lead to any new result. We want to proceed to somewhat more difficult systems and also to a different view of dynamics based on evolving probability distributions. A step in this direction is to consider stochastic dynamics on an inclined energy landscape, specified in one dimension by $U = -\hat{f} x$, where $\hat{f}$ is a constant force. The trajectory probability for this potential can be determined easily by noting from Equation 11.10 that the deterministic step in overdamped/Brownian dynamics is simply a constant,

$$\Delta x^{\text{det}} = \frac{\hat{f} \Delta t}{m\gamma}. \qquad (11.23)$$

We can obtain the corresponding trajectory probability density by substituting Equation 11.23 in Equation 11.15. In turn, this probability is used as the integrand of the path integral for $\rho_n(x_n)$, namely, Equation 11.20. It will be left as an exercise to write down and perform the necessary integration. Conceptually, however, the situation is simple. The distribution of x_n values is a convolution of Gaussians, each of which has the same variance as before, σ^2, but now with a mean shifted by the constant amount $\hat{f}\Delta t/m\gamma$. We know the shifted means won't affect the final variance, which will therefore be the simple sum $n\sigma^2$. The overall mean of the x_n distribution will, however, be shifted at every time step by the constant $\hat{f}\Delta t/m\gamma$.

Based on the arguments just given, we can write down the result of the path integral as a shifted Gaussian with the usual diffusive variance:

$$\rho_n(x_n) = \frac{1}{\sqrt{2\pi n\sigma^2}} \exp\left\{-\left[(x_n - x_0) - \hat{f}n\Delta t/m\gamma\right]^2 / 2n\sigma^2\right\}. \qquad (11.24)$$

The new feature of this distribution, by comparison with Equation 11.22 is the "drifting" of the Gaussian, which is characterized by the drift velocity $\hat{f}/m\gamma$. Figure 11.11 illustrates the diffusion-with-drift behavior.

This simple case of a diffusive system with constant drift can help form our intuition for what will happen when there is a spatially varying landscape. The probability always spreads and drifts but, in general, the drift will depend on the local slope/force. We will briefly treat this general case in our discussion of the Fokker–Planck picture.

PROBLEM 11.3

(a) Write down the path integral defining $\rho_n(x_n)$ for the landscape defined by $U = -\hat{f}x$ and the initial condition $\rho_0(x_0) = \delta(x - x_0)$. (b) Perform the integration for the case $n = 2$ by hand and show you obtain the correct result.

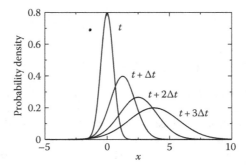

FIGURE 11.11   Diffusive evolution of a probability distribution subject to a constant force. The initial distribution both spreads and drifts. The rate of spreading depends on the diffusion constant, but the drift rate depends only on the force. Starting with $\rho(x; t)$, the distribution evolves to $\rho(x; t + n\Delta t)$, with $n = 1, 2, 3$. As in the case of "pure diffusion," a distribution that is initially Gaussian remains Gaussian.

11.5.3 The Diffusion (Differential) Equation

We are now ready to become more intimately acquainted with the probability-evolution point of view. In particular, we will see that fairly simple differential equations can describe the way the probability distribution at a certain time, $\rho_n(x)$, evolves to a future time, $\rho_{n+k}(x)$. This provides a mathematical means for equilibrium to be "self-healing" (Chapter 4) in the probability-distribution picture: a nonequilibrium distribution relaxes over time to equilibrium. Similar behavior occurs for other steady states.

Because time will again be treated as a continuous variable, t, we will want to rewrite some of our results, starting with Equation 11.22, for the diffusive probability distribution at time $t = n\Delta t$. If we recall from Equation 11.14 that the variance σ^2 is proportional to Δt, we see that the combination $n\Delta t$ occurs naturally in the expression for ρ_n. Instead of using σ^2, we will employ the diffusion constant D, which was defined in Chapter 4 to be

$$D = \frac{\sigma^2}{2\Delta t}. \tag{11.25}$$

Equation 11.22 can then be rewritten as

$$\rho_n(x) = \rho(x;t) = \frac{1}{\sqrt{4\pi Dt}}\, e^{-[x(t)-x_0]^2/4Dt}, \tag{11.26}$$

which is simply a continuous-time version of our earlier characterization. We have continued to assume an initial delta-function distribution at x_0.

At first, we will simply write down the differential equation governing the time evolution of $\rho(x;t)$. Afterward, we will study the equation to gain more insight. Fick's law, which is the governing differential equation, is given by

$$\frac{\partial \rho}{\partial t} = D\frac{\partial^2 \rho}{\partial x^2}, \tag{11.27}$$

in one dimension. Fick's law indicates that the probability of an x value will increase if the second derivative (a.k.a. curvature) of the distribution is positive, but decrease if the curvature is negative. Of course, the overall distribution must remain normalized at all times, so probability increases must be compensated by decreases in other regions. It is remarkable that such a simple equation can account for that.

It is important to verify that Fick's law is indeed correct, as is called for in the next problem.

PROBLEM 11.4

Substitute $\rho(x;t)$ from Equation 11.26 into Equation 11.27. Perform the indicated differentiations and check that equality is obtained.

We can understand the functional form of Fick's law a bit more deeply. The key question is this: Why does the time-derivative of the density depend on the second spatial derivative and not the first? We can answer the question by considering a

Gaussian-like uni-modal distribution. We know that diffusion should cause such a distribution to spread and flatten with time. Furthermore, we know that the process of diffusion is "dumb": it has no knowledge of where probability "should" go. (Recall candy mixing in Chapter 1.) Rather, by definition, diffusion is a process with an equal chance of moving in any direction.

To understand the origin of the second derivative in Equation 11.27, imagine focusing on a small region somewhere in the right tail of $\rho(x;t)$—which, again, is nothing more than $\rho(x)$ at time t. Assume first that this region is exactly linear, with no curvative (i.e., $\partial^2\rho/\partial x^2 = 0$). Let's divide the region into three equi-sized subregions and see what happens to the probability at the center. One acceptable model for diffusion in one dimension is that, for any subregion, half the probability moves left and half moves right (in an appropriate time interval). Under this model, the linearity of ρ guarantees that the probability of the middle region remains unchanged: the small incoming probability on one side is exactly compensated by a larger amount from the other side. Only a positive curvature ($\partial^2\rho/\partial x^2 > 0$) can make ρ increase at a certain point, because the incoming probability will be greater than the outgoing. A negative curvature will lead to the opposite effect.

PROBLEM 11.5

This problem provides more details on the argument just given. (a) Assuming $\rho(x;t)$ is linear in x over a certain interval, show that $\rho(x;t+\Delta t) = \rho(x;t)$ in that region. Use a sketch and equations to justify your conclusion. Assume, as above, that diffusion works by dispersing all probability in an x interval in time Δt, with half the probability moving left one interval and half right. (b) The same result can be obtained using a different model for diffusion. Assume that in time Δt, one-third of probability remains where it started, so that one-third moves left and one-third moves right. Show again that ρ does not change with time if it is linear in x. (c) Show that with a nonzero curvature, $\rho(x;t)$ changes under both models of diffusion just considered.

The diffusion equation can also be understood in terms of probability current or flow. The current $J(x;t)$ is defined to be the net probability flowing (to the right) past the position x at time t per unit time. In one-dimensional diffusion ($U = $ const.), the current is given simply by the negative of the gradient, $J(x;t) = -D\partial\rho/\partial x$. Fick's law can then be recast to say that the change in probability with time at a given x is the derivative of the current there: $\partial\rho/\partial t = -\partial J/\partial x$. Introductory discussions emphasizing the current can be found in the books by Nelson and by Phillips et al.

Finally, it is worth emphasizing that after infinite time—that is, after a long time compared to all timescales of the system—diffusion will always yield the equilibrium distribution. In a flat landscape, $U(x) = $ const., this will be a purely uniform distribution. Thus, a drop of ink in a glass of water will ultimately color all the water.

PROBLEM 11.6

Consider a finite system in one dimension with $0 < x < L$. (a) Based on Fick's law, show that $\rho(x;t) = $ const. is a stationary (unchanging) distribution.

(b) Using a sketch, explain why any nonuniform distribution must evolve to become uniform. (c) Even if the initial distribution is exactly linear (i.e., $\rho(x; 0) \propto x$) in the full space, $0 < x < L$, explain why the system will evolve to uniformity. Hint: Consider the edges.

On the biological side, the book by Phillips et al. provides a good discussion of how the three-dimensional diffusion equation can be used to determine whether a particular biological reaction (e.g., catalytic process) is rate-limited by diffusion or by the rate of the reaction.

11.5.4 FOKKER–PLANCK/SMOLUCHOWSKI PICTURE FOR ARBITRARY LANDSCAPES

Our preceding discussion of pure diffusion considered only the very simplest (and most boring) cases. Of course, every interesting system has a spatially varying potential—particularly when we consider its configuration space.

The evolution of the probability distribution can still be described by a differential equation, even for a nontrivial potential, but that description is more complicated. For full Langevin dynamics, the applicable description is the Fokker–Planck equation, which we will not discuss; the interested reader can refer to the books by Gardiner and by Risken. In the case of overdamped dynamics, the simpler Smoluchowski equation governs the time evolution of the distribution:

$$\frac{\partial \rho}{\partial t} = D \frac{\partial^2 \rho}{\partial x^2} - \frac{D}{k_B T} \frac{\partial}{\partial x} \left[\left(-\frac{dU}{dx} \right) \rho \right], \qquad (11.28)$$

where $U(x)$ is the potential energy as usual.

The difference from the pure diffusion equation (11.27) is clear: there is an additional term that depends on both the potential and the density. This new term is often called the "drift" term because it accounts for motion of the overall probability distribution—e.g., of the peak in a uni-modal ρ. Using the product rule of differentiation to generate two drift terms, one of them leads to accumulation of probability in energy basins (where $d^2U/dx^2 > 0$), and the other will tend to move the "leading" edge/tail of the distribution in the direction of the force. The drift and diffusion terms act simultaneously on ρ, and their effects may partially cancel one another. In equilibrium or a steady state, of course, all terms will exactly cancel as ρ remains constant.

PROBLEM 11.7

Show that overdamped dynamics preserve equilibrium. That is, show that if $\rho(x, t) \propto \exp[-U(x)/k_B T]$ at some point in time, that $d\rho/dt = 0$ in the Smoluchowski equation (11.28).

The constant-force case of Section 11.5.2 is also readily confirmed using Equation 11.28, as shown in the next problem.

PROBLEM 11.8

Show that Equation 11.24 satisfies the Smoluchowski equation (11.28) with $t = n\Delta t$ and $x = x_n$.

To sum up, there are three key ingredients to describing the time evolution of a probability distribution. First, the probability tends to spread, diffusively. Second, the presence of forces means that probability will tend to move downhill in energy. Third, the normalization of the distribution needs to be preserved at all times. (Probability can't disappear or appear over time, unless special boundary conditions are being analyzed.) The Smoluchowski equation provides all three characteristics.

For completeness, we note that the Smoluchowski equation can also be written in terms of the current J. As in pure diffusion, we have $\partial \rho / \partial t = -\partial J / \partial x$, except now the current depends not only on the distribution ρ but also on the potential. Specifically, $J(x;t) = -D\partial \rho / \partial x - D(\partial U / \partial x)\rho(x;t)/k_B T$. The current must depend on U: if it did not, then any nonuniform ρ would cause current to flow, even in equilibrium!

11.5.4.1 Examples of Evolving Probability Distributions

The example of the one-dimensional double-well potential will prove instructive for the general case. If all the probability starts out in one well, perhaps the left, how will it evolve? As depicted in Figure 11.12, the probability gradually "leaks" over. If the barrier is high, the probability will "equilibrate" locally in each well long before global equilibrium occurs. The probabilities of the two states must be equal in (global) equilibrium, of course. The rate of "leaking" is nothing other than the standard rate k_{AB}, and is governed primarily by the barrier height—recall Equation 4.2—rather than the diffusion constant. The diffusion constant, which incorporates the friction constant γ of overdamped dynamics, enters the prefactor in a rate factorized according to Equation 4.3.

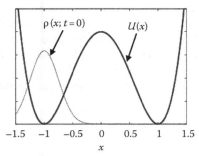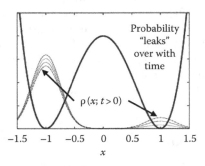

FIGURE 11.12 The "Smoluchowski evolution" of a distribution initialized in a single state (left panel). As time evolves, a small fraction of the probability "leaks" to the right. This leakage is governed by the usual rate, k_{AB}. The figure assumes the barrier is high compared to $k_B T$, so that the distribution within each well is distributed according to the Boltzmann factor, locally. Because the total probability is conserved, the left-well population decreases as the right increases.

The rate k_{AB} can be approximated more precisely in terms of the diffusion constant and other parameters, as originally described by Kramers. (See, for instance, the books by Gardiner and by Risken.) We will not derive Kramers' theory, which is based on steady-state boundary conditions in the Smoluchowski equation. However, the following problem will provide some physical insight.

PROBLEM 11.9

In Chapter 4, we saw that a rate could generally be described as the product of an attempt frequency and transition probability. For the double-well potential, the transition probability can be taken as the usual Arrhenius factor. If $U(x) = E_b[(x/a)^2 - 1]^2$, where a is a lengthscale, try to construct an approximation for the attempt frequency. You should describe your logic clearly, and your answer should have the correct units of inverse time.

Although the "leaking" probability of Figure 11.12 sounds rather unphysical, we know that underlying this picture are dynamical trajectories. A small amount of probability leaking over a barrier is nothing more than a small fraction of trajectories—increasing with time—which successfully makes the transition (look again at Figure 11.3). Indeed, as shown in the book by Chaikin and Lubensky, the Fokker–Planck/Smoluchowski type of equations can be derived directly from the governing dynamical equations of motion—for example, from the Langevin equation. The descriptions are equivalent, except that the Fokker–Planck/Smoluchowski differential equations have averaged out the details of specific trajectories.

The same leaking-probability picture applies to biomolecules as well. Figure 11.13 shows how a distribution initially confined to one state of calmodulin (as modeled in the 2004 paper by Zuckerman) relaxes and leaks over to the other available state. The ideas we are discussing are indeed quite general.

11.5.5 THE ISSUE OF HISTORY DEPENDENCE

In this book, we have exclusively discussed "Markovian" dynamics where the future depends only on the present configuration of the system. For instance, in overdamped Brownian dynamics, the configuration at time step j depends only on the situation at time $j - 1$. Even in full Langevin dynamics or Newtonian molecular dynamics, the future depends only on the current position in phase space—that is, on the current configuration and momenta of all atoms.

Somewhat confusingly, there appear to be important situations when the past— that is, the "history"—also matters. Consider Langevin or Newtonian dynamics, described only in configuration space. If the momenta (velocities) are ignored in this way, future configurations of the system appear to depend on the previous two configurations, with the earlier time essentially supplying the velocity information. More generally, whenever degrees of freedom are omitted from the equations describing a system, you can expect that they will affect the dynamics (for the remaining coordinates) in a complicated way. In this sense only, "history dependence" is an important practical part of theoretical descriptions of motion.

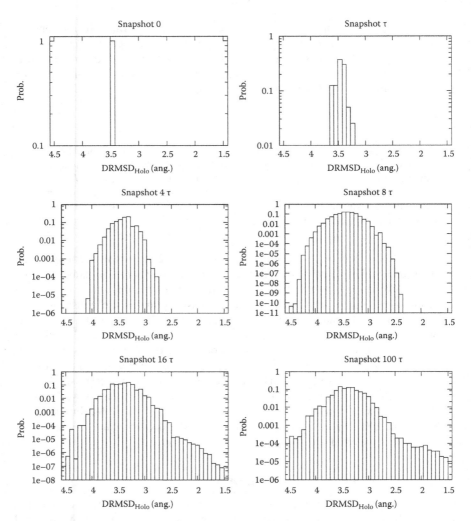

FIGURE 11.13 The "Smoluchowski evolution" for the protein calmodulin. The initial histogram is confined to a single bin. As in the toy model, an increasing fraction of probability leaks to the right over time—and also somewhat to the left. The figure describes the same model used in Figure 11.7. Roughly, state "B" (holo) is reached when DRMSD<1.5 Å. (From Zhang, B.W. et al., *Proc. Natl. Acad. Sci. U.S.A.*, 104, 18043, 2007. With permission.)

For instance, the phenomenon of "hysteresis" (e.g., in magnetic systems) can be understood in the same terms. The key is that typical descriptions of hysteresis do not reflect the precise configuration of a system, but rather projections/averages—for example, the average magnetization.

Fundamentally, there is no history dependence in nature. No fundamental law of motion—classical or quantum mechanical—can depend on the past, unlike approximate descriptions. In other words, molecular systems cannot store information beyond their present "configuration." The configuration of a system, however, generally will be more complicated than you realize. Beyond including momenta, true molecular configurations would also encompass all nuclear and electronic degrees of freedom—quarks, perhaps! Such a full phase-space description (of an isolated system, or of the whole universe) must completely describe the future—even if it is of little practical use.

11.6 THE JARZYNSKI RELATION AND SINGLE-MOLECULE PHENOMENA

At the time of this writing, single-molecule experiments in biophysics are burgeoning in prominence. Fortunately for us, the underlying physics is based on trajectories. In fact, we have already covered some of the topics fundamental to understanding single-molecule phenomena.

A common type of single-molecule experiment stretches a biomolecule (e.g., protein or DNA) by attachments at two parts of a molecule. The stretching force—applied by laser tweezers or atomic-force microscopes—can be described as resulting from an "external potential," U^{ext}. This external potential describes all forces in addition to the usual potential U describing the molecule in isolation. To keep our discussion general and avoid specifying the attachment points, we will simply say that the potential depends on the entire molecular configuration (although it might be just the distance between two atoms). Further, by construction, the external potential changes with time as the molecule is pulled, and so we have $U^{\text{ext}} = U^{\text{ext}}(\mathbf{r}^N; t)$.

Because the pulling is fairly rapid and the molecule typically does not have time to adapt fully to the changing force, single-molecule systems are usually never in equilibrium. However, the work, $\hat{W}$, done in each stretch of the molecule can be measured. In our language, the work is a function of the trajectory, $\hat{W} = \hat{W}(\mathbf{r}^{\text{traj}})$, because the force and extension at every time reflect the detailed molecular configuration. We now refer to work done on the system, so $\hat{W} = -W$ compared to Chapter 7. Details of calculating work from a trajectory can be found in the papers by Jarzynski and by Hummer and Szabo.

A powerful trajectory-averaging analysis was developed by Chris Jarzynski, which yields equilibrium information from nonequilibrium pulling trajectories. (We already know there are intimate connections between dynamics and equilibrium.) The initial extension of the molecule is described as the "a" condition, corresponding to a particular static choice $U_a^{\text{ext}}(\mathbf{r}^N)$ and distribution $\rho_a \propto \exp\{-[U(\mathbf{r}^N)+U_a^{\text{ext}}(\mathbf{r}^N)]/k_B T\}$. Similarly, the final condition is described as "b," with $U_b^{\text{ext}}(\mathbf{r}^N)$. The partition functions for the a and b states are defined in the usual way, for instance, by $Z_a = e^{-F_a/k_B T} \propto \int d\mathbf{r}^N \exp\{-[U(\mathbf{r}^N) + U_a^{\text{ext}}(\mathbf{r}^N)]/k_B T\}$.

The Jarzynski relation tells us that the (usual) equilibrium free energy difference between these conditions is given according to

$$e^{-(F_b - F_a)/k_B T} = \left\langle e^{-\hat{W}/k_B T} \right\rangle$$

$$\propto \int d\mathbf{r}^{\text{traj}} \, \rho_a\left(\mathbf{r}_0^N\right) \exp\left[-E^{\text{traj}}(\mathbf{r}^{\text{traj}})/k_B T\right] \exp\left[-\hat{W}(\mathbf{r}^{\text{traj}})/k_B T\right]. \quad (11.29)$$

Here E^{traj} is the effective path energy for the "prevailing conditions"—which include the changing external potential that forces trajectories to end in the b condition, having started from a. The normalization constant for the average $\langle e^{-\hat{W}/k_{\text{B}}T} \rangle$ is the effective partition function for the process, namely, the same integral without the Boltzmann factor of $\hat{W}$.

Although a derivation of the Jarzynski relation is beyond the scope of this book (see the papers by Jarzynksi and by Hummer and Szabo), the main point is that trajectory-ensemble concepts are at the very heart of single-molecule studies. Especially pertinent are trajectory probabilities, dwells in intermediate states, and projections onto configurational coordinates, all of which have been discussed above.

11.6.1 Revisiting the Second Law of Thermodynamics

The Jarzynski relation also provides a "microscopic" or full statistical mechanics description of the second law of thermodynamics. In Chapter 7, we learned that $\hat{W} \geq \Delta F = F_{\text{b}} - F_{\text{a}}$. The equality (11.29) tells us this and more. First using only concavity/convexity arguments (e.g., Problem 7.14) for the exponential function, one can show that $\langle e^{-\hat{W}/k_{\text{B}}T} \rangle \geq e^{-\langle \hat{W} \rangle/k_{\text{B}}T}$. This leads to the usual statement of the second law, but now with the work explicitly averaged: $\langle \hat{W} \rangle \geq F_{\text{b}} - F_{\text{a}}$. More interestingly, we see that for some individual trajectories, the work can actually be less than the free energy change—so long as the equality (11.29) is maintained. It is only on average that the work is limited by the second law, which is a remarkable "microscopic" insight into a macroscopic law. This echoes a fact we have stressed throughout this book: free energies always describe average behavior.

11.7 SUMMARY

Dynamical trajectories can be treated in much the same way as configurations are treated in equilibrium statistical mechanics. A probability can be assigned to any trajectory, implying the existence of an ensemble of trajectories. Observed phenomena thus correspond to a suitably weighted average over all possible trajectories. The trajectory probability is quantified by the "action," which we derived explicitly for one-dimensional overdamped dynamics. In contrast to equilibrium, different trajectory ensembles can be defined based on initial or boundary conditions. Precisely as in our earlier kinetic analyses, one can study a steady state, for instance, or the evolution of a system initially confined to a single state.

The trajectory ensemble is rich with information and can be analyzed (i.e., averaged) in many ways to yield important physical quantities. Transition rates can be calculated, based on the fraction of transitioning trajectories in different intervals of time. Candidates for reaction coordinates can be studied, including the "commitment probability" for trajectories originating from a particular configuration to complete a forward transition. The configuration-space probability distribution at different time points can be calculated from the trajectory ensemble. Furthermore, differential equations—the diffusion equation and the related Fokker–Planck/Smoluchowski equations—describe this evolution of the distribution.

FURTHER READING

Adib, A.B., *Journal of Physical Chemistry B*, 112:5910–5916, 2008.

Bolhuis, P.G., Chandler, D., Dellago, C., and Geissler, P., *Annual Review in Physical Chemistry*, 59:291–318, 2002.

Chaikin, P.M. and Lubensky, T.C., *Principles of Condensed Matter Physics*, Cambridge University Press, Cambridge, U.K.

Chodera, J.D., Singhal, N., Swope, W.C., Pitera, J.W., Pande, V.S., and Dill, K.A., *Journal of Chemical Physics*, 126:155101, 2007.

Gardiner, G.W., *Handbook of Stochastic Methods*, 2nd edition, Springer, Berlin, Germany, 1985.

Hill, T.L., *Free Energy Transduction and Biochemical Cycle Kinetics*, Dover, New York, 2004.

Hummer, G., *Journal of Chemical Physics*, 120:516, 2004.

Hummer, G. and Szabo, A., *Proceedings of the National Academy of Sciences, USA*, 98:3658–3661, 2001.

Jarzynski, C., *Physical Review Letters*, 78:2690–2693, 1997.

Nelson, P., *Biological Physics*, W.H. Freeman, New York, 2008.

Phillips, R., Kondev, J., and Theriot, J., *Physical Biology of the Cell*, Garland Science, New York, 2009.

Risken, H., *The Fokker-Planck Equation*, 2nd edition, Springer, Berlin, Germany, 1996.

van Kampen, N.G., *Stochastic Processes in Physics and Chemistry*, 3rd edition, Elsevier, Amsterdam, the Netherlands, 2007.

Zhang, B.W., Jasnow, D., and Zuckerman, D.M., *Journal of Chemical Physics*, 116:2586–2591, 2002.

Zhang, B.W., Jasnow, D., and Zuckerman, D.M., *Proceedings of the National Academy of Science, USA*, 104:18043–18048, 2007.

Zuckerman, D.M., *Journal of Physical Chemistry B*, 108:5127–5137, 2004.

Zuckerman, D.M. and Woolf, T.B., *Physical Review E*, 63:016702, 2001.

12 A Statistical Perspective on Biomolecular Simulation

12.1 INTRODUCTION: IDEAS, NOT RECIPES

Many books have been written about simulation methods (see 'Further Reading'), and it would hardly be reasonable to cover the methodology comprehensively in a single chapter. Nevertheless, there are a handful of key ideas underlying all simulation methods. These are ideas that will be critical for the researcher trying to choose among a large array of techniques—as well as to anyone wishing to understand what simulation results really mean. Naturally, we will stick to the statistical point of view, which is essential to biophysics.

In fact, our statistical considerations will directly suggest ways to answer some interesting questions regarding biomolecular simulation. What has simulation done for us lately? How long is long enough? Does the type of dynamics matter? Which results can be trusted? Are those fancy methods worth it? And on the subject of fancy methods, we will explain the two basic statistical mechanics techniques that form the core of numerous modern algorithms.

12.1.1 Do Simulations Matter in Biophysics?

Although the importance of computer simulations is widely accepted in the fields of chemistry, physics, and engineering, the value of simulation in molecular biophysics has been debated. In fact, the contributions have been significant. The author is biased (making his living performing simulations), but the evidence seems fairly clear.

Perhaps most basic is the often overlooked fact that we would have no protein structures without computers. This is true for both x-ray crystallography and NMR (nuclear magnetic resonance) techniques. X-ray studies employ extensive numerical calculations, including the direct use of molecular simulation in structure refinement. Molecular simulation is even more fundamental to NMR structure determination, where experimental data are directly incorporated as energy terms in simulations that yield the final structures.

Molecular dynamics (MD) simulations are now employed extremely broadly in the study of biophysics and biochemistry. Although the results are typically incomplete (due to sampling problems, as discussed below), even a partial picture of bimolecular motions can be highly informative. On a more biophysical level, simulations have clearly demonstrated the importance of significant structure fluctuations extending well beyond the quasi-static picture implied in experimental structures.

That is, if simulations have taught us one thing, it's that there's no such thing as "the structure" of a protein. Rather, an ensemble description is generally necessary.

12.2 FIRST, CHOOSE YOUR MODEL: DETAILED OR SIMPLIFIED

Model selection is a critical part of simulation. The choice is almost always dictated by practical issues. The basic conundrum, to which we shall refer many times, is that the more accurate the model, the less "affordable" it is to simulate. Figure 12.1 illustrates the point schematically (see also the review by van Gunsteren et al.).

While we may be able to imagine many types of models, typically one does not have the time or resources to write software from scratch, which embodies our "perfect" ideas. It is therefore more pertinent to understand what is available. In classical "molecular mechanics" (MM), all-atom models are by far the most common choice for biomolecular simulation, but the range of choices is rapidly growing. In particular, many "coarse-grained" (CG) (simplified) models are available, as are quantum-mechanical (QM) software packages. The reader is advised to consult the Internet and current literature for the most current information.

From a statistical mechanics point of view, the fairly clear choice would be to use a model that is as detailed as possible. In statistical terms, "as detailed as possible" means the most accurate model for which 10–100 independent samples can be obtained (see Section 12.7). After all, statistical mechanics is absolutely predicated on an ensemble picture, which translates to the need for statistically reliable sampling. For most biomolecular systems, unfortunately, this rules out atomistic models and suggests simplification should be pursued (see again Figure 12.1).

Of course, biochemistry evidently runs on atomistically specific interactions and chemistry. Indeed, the accurate description of a chemical reaction where covalent bonds are formed or broken intrinsically requires quantum mechanics. With these

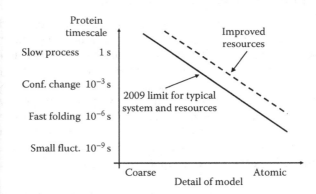

FIGURE 12.1 Timescale limitations of conventional dynamics simulation. Most interesting biology happens at least on the μs timescale if not well beyond (ms – s) but conventional atomistic simulations currently require substantial resources to reach even the μs scale. The situation is worse for large systems, such as membrane proteins. However, the capacity gradually improves over time with hardware advances: that is, the limiting line will shift to the right.

facts in mind, it is natural to perform simulations based on models incorporating whatever chemical details are necessary. The user must carefully consider the priorities of the given project.

We now sketch some of the common models used in biomolecular simulation.

12.2.1 ATOMISTIC AND "DETAILED" MODELS

Standard MM potential energy functions, which commonly are called "forcefields" (Section 5.3.3), describe molecules based on interactions among classical—that is, non-quantum—atoms. These forcefields employ many types of interaction terms: van der Waals repulsions and attractions typically modeled by the Lennard-Jones potential, electrostatic interactions, plus a host of terms to describe forces acting on specific types of angles. Each atom is characterized by many different parameters, typically reflecting the atoms with which it interacts and the type of bonding. The common forcefields have been parameterized over decades by careful comparison with experiments and QM calculations, but they must still be recognized as imperfect approximations to the full QM reality of biochemistry. Despite many well-appreciated weaknesses, it is not advisable for the novice to adjust parameters due to the complex interdependencies in the parameterization process.

PROBLEM 12.1

One of the most widely studied flaws of molecular (classical) mechanics forcefields is the lack of polarization. Explain what polarization means and why the phenomenon is dependent on configuration.

It is possible to perform a limited amount of QM computation on biomolecules, and there are two strategies. The first is to perform careful QM calculations on entire molecules based on a small number of configurations of the atomic nuclei. The second is to treat only a limited region of a system—for example, a binding site—with QM methods and the rest with classical forcefields. The latter approach is called QM/MM, and needless to say, requires great care in modeling the boundary between the QM and MM regions. See, for instance, the article by Senn and Thiel.

12.2.2 COARSE GRAINING AND RELATED IDEAS

The process of simplifying or coarse graining takes many forms. In general, a CG model is one where fewer degrees of freedom are used than actually are present. As you may recall from Chapter 4, in fact, coarse graining is intrinsic to statistical mechanics: in fact, our basic theory uses a single parameter, temperature, to model everything in the universe except what we think is important. Thus, the real issue is the degree of coarse graining in any given model.

Coarse graining in biomolecular simulation typically takes one (or more) of a few forms: (i) lest we forget, standard atomistic forcefields have ignored degrees of freedom associated with electron density; (ii) solvent molecules are sometimes treated by "implicit" or continuum models that assign a solvation free energy to configurations of the designated solute molecule(s); and (iii) groups of atoms, even entire amino acids or groups of amino acids, are modeled as single often-spherical interaction centers. As in the case of QM/MM calculations, coarse graining can be performed in different ways on different parts of a system—as exemplified in the use of implicit solvent with an atomistic solute.

Although "traditional" coarse graining—items (2) and (3) above—has been employed since the earliest days of molecular simulation, there is less standardization than for atomistic MM models. Perhaps because of their relative simplicity compared to atomistic models, many CG models are employed in software custombuilt by individual investigators. Review articles by Tozzini and by Ayton et al., as well as the book by Voth, describe some of the approaches that have been used.

In general, you should expect that even the most carefully parameterized, CG model will have important limitations. To appreciate the problem in a quick-and-dirty way, observe that biology employs 20 amino acids along with myriad posttranslational modifications and QM chemistry to do its work. How much simplification do you then expect is possible, which will still allow quantitative predictions? It is probably more reasonable to expect to draw rougher, more qualitative conclusions from the study of CG models. For instance, one can expect that larger-scale motions are better described than small-scale behavior in CG models—with the caveat that atomic details can have large effects in some cases.

PROBLEM 12.2

Assuming a CG model is designed to mimic the behavior of an atomistic model, explain how the potential of mean force (PMF) strategy might be useful. Which specific PMF would be used?

12.3 "BASIC" SIMULATIONS EMULATE DYNAMICS

The most basic of type of simulation directly simulates dynamics of the molecules of interest. If the configuration of a system is denoted by $\mathbf{r}^N$, then a dynamics simulation makes a "movie"—a series of "snapshots" of the system at succeeding times—$\mathbf{r}^N(t_0)$, $\mathbf{r}^N(t_1)$, $\mathbf{r}^N(t_2)$, ... MD simulation, which we discussed in Chapter 1, uses Newton's laws to follow a system's time evolution and to generate these snapshots. Alternatively, the explicitly stochastic Langevin dynamics (Chapter 4)—ordinary or overdamped—can be used. Any of these methods, if run long enough, will also generate the correct distribution of configurations according to the Boltzmann factor, $\exp[-U(\mathbf{r}^N)/k_B T]$.

Dynamics simulations are "basic" in the sense of attempting literally to reproduce molecular behavior. If the dynamics are correct, the correct distribution must be obtained with sufficient computation. (After all, nature only knows dynamics, not distributions.) As we discussed in Chapter 4, MD simulations cannot be considered exact by any means, but they are indeed a useful conceptual reference point. If we consider the stochastic Langevin dynamics (LD) algorithms to be approximations to MD, then we can see that even approximate algorithms can produce the correct equilibrium behavior (distribution). Furthermore, as we will see below, "Monte Carlo (MC)" simulations also produce the correct distribution, and in many cases can be considered to model physical dynamics.

PROBLEM 12.3

Explain one or more criteria you could use to determine the time step $\delta t = t_j - t_{j-1}$ in a dynamics simulation. Your criteria could be based on physical principles or on the analysis of simulation data.

12.3.1 TIMESCALE PROBLEMS, SAMPLING PROBLEMS

Current atomistic biomolecular simulations, for the most part, are not even close to being fully sampled in a statistical sense. The basic issue is timescales, as sketched in Figures 12.1 and 12.2. The slow timescales could correspond to typical waiting times for transitions over the largest barriers, as sketched Figure 12.2. After all, the basic goal of equilibrium simulation is to sample the important states in correct proportion. A simulation determines those proportions only by repeated crossing of the barriers. Table 12.1 provides some specific values of timescales.

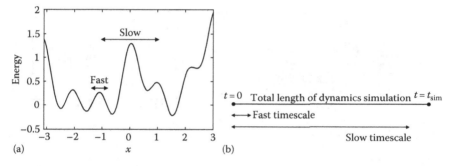

FIGURE 12.2 A rough energy landscape and a schematic depiction of some of the associated timescales. (a) Low barriers can be crossed rapidly while larger barriers require exponentially longer timescales. (b) The total length of a simulation, ideally, should be many times longer than the longest timescale. However, this is often not the case, as suggested in the hypothetical comparison of the simulation length against molecular timescales. Indeed, many atomistic simulations of large biomolecular systems are significantly shorter than the longest timescales.

TABLE 12.1
Timescales Believed to Characterize
Typical Motions in Proteins

Protein Timescales

Covalent bond vibrations	<1 ps
Unhindered side-chain rotations	ps – ns
Allosteric transitions	µs – s
Protein folding	µs – s

Note: The last two entries cover a range of 6 orders of magnitude, and the overall range is 12 orders.

Some systems may not be characterized by a single governing barrier, or even a small number of them. However, every biomolecular system can be expected to have a "complex" energy landscape with many minima. A system with a very large number of low barriers would also require substantial time for a trajectory to "diffuse" throughout the configuration space.

Sample trajectories from molecular simulations are shown in Figures 12.3 (butane) and 12.4 (the large membrane protein rhodopsin). There are a number of critical lessons contained in this data: (1) Even a simple molecule like butane exhibits a range of timescales spanning orders of magnitude: compare the bond angle with the dihedrals. Fortunately, butane is small, permitting good sampling. (2) A well-sampled trajectory must exhibit many transitions or fluctuations among the significantly populated states/regions of configuration space.

The rhodopsin data, from a simulation containing 40,000 atoms, suggests other important points. (3) Apparently fast motions are generally "coupled to"—correlated with—slower motions. This is not surprising given all the interactions in potential

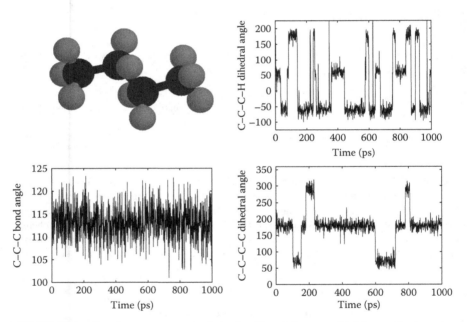

FIGURE 12.3   Simple system, complex behavior. Even in butane, which is usually characterized solely based on its central (C–C–C–C) dihedral, a multiplicity of timescales are evident when different coordinates are examined. The C–C–C–C dihedral indeed exhibits the slowest behavior, while a C–C–C–H dihedral characterizing the spinning of one of the end methyl groups undergoes more rapid transitions. The motions of a C–C–C bond angle are on an altogether faster scale. Because the three states of butane are known and fully characterized by the central dihedral, the trajectory shown can be considered "long": each state is visited at least twice. The precision of population estimates increases—that is, the uncertainty decreases—with an increasing number of visits/transitions. (From Grossfield, A. and Zuckerman, D.M., *Annu. Rep. Comput. Chem.*, 5, 23, 2009. With permission.)

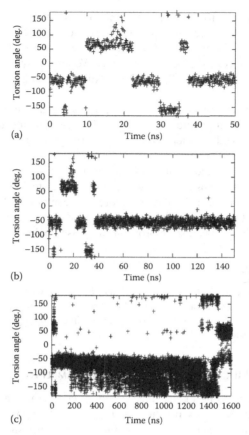

(a)

(b)

(c)

FIGURE 12.4 Big system, big problems. The plots show trajectories of a single dihedral angle from a 40,000-atom supercomputer simulation of an atomistic representation of the membrane protein rhodopsin. Even the several transitions observed in (a) the initial 50 ns are no guarantee of good sampling, as clearly demonstrated by the longer trajectories in (b) and (c). The data exemplify a very general phenomena in which apparently fast motions are, in fact, coupled to very slow motions occurring among other coordinates. The apparently changing distribution of the torsion angle indicates that the overall configuration-space distribution has not been well sampled in this microsecond-scale simulation. (From Grossfield, A. and Zuckerman, D.M., *Annu. Rep. Comput. Chem.*, 5, 23, 2009. With permission.)

energy functions. (4) Simulations of large biomolecular systems are rarely well sampled. Worse yet, (5) a trajectory that seems sufficient after some finite length may well be proven inadequate based on additional simulation.

To think about sampling in statistical terms, recall from Chapter 2 that, generally speaking, error decays as $1/\sqrt{N}$, where N is the number of independent samples. Any decent statistical sample should have at least $N \sim 10$ and preferably $N \sim 100$ to get on the order of $1/\sqrt{100} = 10\%$ precision. In a simulation context, N roughly corresponds to the number of times relevant conformational motions have been seen.

What are typical N values for biomolecular simulations? Well, biological motions are known to take from microseconds up to seconds and beyond, yet typical atomistic simulations (in the year 2009) span <1 μs—and often are only ~ 10 ns. In other words, typical simulations never even achieve $N = 1$. We are many orders of magnitude from obtaining full sampling (see Table 12.1 for timescale ranges for selected protein motions).

The same dismal conclusion must be reached whether one discusses MD, LD, or MC when applied to atomistic models. For this reason, many statistically minded investigators have turned to alternative models and methods. Some of these new approaches are described below.

You may want to conclude that simulation is a hopeless endeavor. However, we have already discussed the absolutely key role of simulation techniques in determining protein structures. Even short simulations, moreover, can offer valuable insights into fast local motions of proteins and nucleic acids. And one can hope that simulation algorithms and methods may also improve, as discussed below.

PROBLEM 12.4

Look up Moore's law and estimate the number of years it will take for computer speeds to increase by six orders of magnitude. Note that six orders of magnitude is a conservative estimate of the current shortfall in computing speed for many biomolecular systems.

12.3.2 Energy Minimization vs. Dynamics/Sampling

Energy minimization is one of the most common computational procedures, so it is important to understand its practical and theoretical importance. For practical purposes, minimization is simply indispensable for molecular simulations. Most significantly, energy minimization is necessary before running a dynamics simulation for a large molecule. Otherwise, even starting from an experimental structure, the forces will be extremely large and lead to numerical instabilities. (Recall that dynamics algorithms typically assume the force is constant over a simulation time step.) Even though an MM forcefield may have been used in preparing the experimental coordinates, the chances are about zero that the parameters used for that purpose will be identical to those of a later simulation!

From a theoretical—that is, fundamental—point of view, the value of a particular energy-minimized configuration for a large molecule is fairly unclear. Firstly, a configuration so generated will almost certainly depend on the configuration from which the minimization process was started. Imagine the landscape of Figure 12.2, except with the exponentially worse complexity of a large biomolecule. It is therefore highly unlikely (almost statistically impossible) that the minimized configuration will be the global energy minimum. Yet, more importantly, even if it were the global energy minimum, the importance of such a configuration to the overall ensemble—that is, to the observed populations—is wholly unknown without good sampling or reliable estimates of the energies and entropies of all other minima! In sum, the most that can be said about an energy minimum is that... it's a minimum of the energy.

12.4 METROPOLIS MONTE CARLO: A BASIC METHOD AND VARIATIONS

MC methods are practical in many situations when dynamics are not—for instance, in models where forces and dynamics are ill defined. But perhaps more importantly, the MC technique can be used to devise numerous advanced algorithms.

12.4.1 SIMPLE MONTE CARLO CAN BE QUASI-DYNAMIC

The Metropolis MC technique must be considered "basic" both because of its prevalence in so many types of specialized biomolecular simulation and because simple MC simulations have a dynamical character. That is, like "true" dynamics simulations, simple MC also generates a sequence of configurations, each evolved from the previous one. In many cases, moreover, this sequence of configurations reflects basic aspects of dynamics: configurations are more likely to move in a direction that lowers the energy, but energy increases are possible with Boltzmann-factor probability.

The MC procedure generates its sequence of configurations by probabilistically accepting or rejecting a random trial move from the previous configuration. A simple trial move might consist of displacing a single atom by a small distance. If the move is rejected, the system remains in its old configuration for another step of the "trajectory." Much more complicated trial moves are possible, but even the apparently simple procedure of changing a bond or dihedral angle requires accounting for Jacobian factors (Chapter 5).

We can formulate the acceptance criterion more precisely by first adopting some notation. Given the current or "old" configuration $\mathbf{r}_{\text{old}}^N$, a trial move is attempted to a configuration $\mathbf{r}_{\text{try}}^N$. The trial move is accepted based on the Boltzmann factor, so that in typical simple cases, the probability for acceptance is

$$P(\text{accept}) = \begin{cases} 1 & \text{if } \Delta U < 0, \\ e^{-\Delta U/k_B T} & \text{otherwise}, \end{cases} \qquad (12.1)$$

where the change in potential energy has been denoted $\Delta U = U\left(\mathbf{r}_{\text{try}}^N\right) - U\left(\mathbf{r}_{\text{old}}^N\right)$. Thus, trial moves that lead to decreases in energy are always accepted, while energy increases may be accepted.

MC simulations can use standard forcefields normally employed for biomolecular dynamics. However, because Equation 12.1 does not require forces, MC can also be applied to discontinuous models—for example, with "square" energy wells.

In practical terms, to accept a trial move with probability $P < 1$, one can use a computer program or function to generate (approximately) random numbers distributed uniformly between 0 and 1. A random number is generated and the move is accepted if the number is less than P.

The physical reason why MC simulation works—that is, produces the equilibrium distribution—is sketched in Figure 12.5. Properly balanced transitions in every local area must eventually lead to overall balanced transitions in the full configuration space, and hence to equilibrium. In further detail, the two following problems

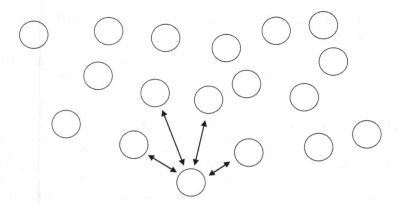

FIGURE 12.5 MC algorithms balance local transitions. Each circle in this schematic represents an energy basin, and the arrows depict transitions allowed from one particular basin based on an MC trial move. If the correct relative probabilities are ensured locally via a Metropolis criterion, and if every basin is accessible somehow, then an MC simulation will eventually produce the correct equilibrium distribution. This basic explanation holds even when MC is used in extended spaces based on multiple copies of a given system (i.e., in "exchange simulation").

explain rejection and the meaning of the MC criterion. They will greatly deepen your understanding of simple MC simulation.

PROBLEM 12.5

By considering a system with only two configurations, which differ in energy, explain why the "old" configuration must be counted again whenever a trial move is rejected. You must show this is necessary for obtaining the correct equilibrium distribution.

PROBLEM 12.6

By considering both the forward $\left(\mathbf{r}_{\text{old}}^N \to \mathbf{r}_{\text{try}}^N\right)$ and reverse $\left(\mathbf{r}_{\text{try}}^N \to \mathbf{r}_{\text{old}}^N\right)$ move, show that the criterion (Equation 12.1) indeed leads to the correct Boltzmann-factor ratio of probabilities. Until you do this exercise, it is hard to see why energy decreases can *always* be accepted and still preserve the Boltzmann factor.

12.4.2 THE GENERAL METROPOLIS–HASTINGS ALGORITHM

Although it is far from obvious from our preceding discussion of basic MC, the approach is incredibly general and flexible. Perhaps most interestingly, the MC method can be adapted to simulations involving multiple temperatures, models, and/or resolutions. "Metropolis–Hastings" MC, which we will now explore, applies to any distribution on any variable of any dimensionality.

We'll want somewhat more general notation for the new scheme. Let's call our (possibly vectorial) variable $\mathbf{x}$, and its un-normalized distribution (weight function) $w(\mathbf{x})$. That is, $w(\mathbf{x}) \propto \rho(\mathbf{x})$, and remember that a Boltzmann factor is a classic example of an un-normalized distribution. Our "old" configuration will be $\mathbf{x}_o$ and the "new"/trial configuration will be $\mathbf{x}_n$.

The MC method generates "configurations" $\mathbf{x}$ according to w based on a flux–balance principle (Chapter 4). We can consider flux-balanced transitions between configurations $\mathbf{x}_o$ and $\mathbf{x}_n$, which are governed by the rates k_{on} (not to be confused with an "on rate") and k_{no}. Following Chapter 4, we must have the following balance:

$$w(\mathbf{x}_o)\, k_{on} = w(\mathbf{x}_n)\, k_{no}, \tag{12.2}$$

where any overall normalization of w would cancel out, and is therefore unnecessary. As always, the rates k are conditional probabilities per unit time for the indicated transition.

The "trick" of MC simulation is to separate each transition rate into two parts: the conditional probability for generating a (trial) move k^{gen} and the conditional probability to accept the move k^{acc}. Explicitly, we write

$$k_{on} = k^{gen}(\mathbf{x}_o \to \mathbf{x}_n)\, k^{acc}(\mathbf{x}_o \to \mathbf{x}_n) \qquad k_{no} = k^{gen}(\mathbf{x}_n \to \mathbf{x}_o)\, k^{acc}(\mathbf{x}_n \to \mathbf{x}_o). \tag{12.3}$$

If the factorizations (12.3) are substituted into the balance condition (12.2) and rearranged, we can obtain the critical ratio of acceptance rates:

$$R \equiv \frac{k^{acc}(\mathbf{x}_o \to \mathbf{x}_n)}{k^{acc}(\mathbf{x}_n \to \mathbf{x}_o)} = \frac{w(\mathbf{x}_n)\, k^{gen}(\mathbf{x}_n \to \mathbf{x}_o)}{w(\mathbf{x}_o)\, k^{gen}(\mathbf{x}_o \to \mathbf{x}_n)}. \tag{12.4}$$

The fully general MC approach is implemented by proposing trial moves according to k^{gen} and accepting them with probability $\min[1, R]$. Our earlier simple criterion (Equation 12.1) is a special case of Equation 12.4.

PROBLEM 12.7

Show that the simple acceptance criterion (Equation 12.1) is a special case of using $\min[1, R]$, with a symmetric k^{gen}—that is, where $k^{gen}(\mathbf{x} \to \mathbf{y}) = k^{gen}(\mathbf{y} \to \mathbf{x})$.

12.4.3 MC VARIATIONS: REPLICA EXCHANGE AND BEYOND

Now the magic starts. Using only the simple formulation of Equation 12.4 and the $\min[1, R]$ acceptance probability, an amazing range of algorithms can be derived. The basic idea was introduced by Swendsen and Wang; see also the article by Frantz et al.

To define an algorithm, we need to specify three things: the variable $\mathbf{x}$, the distribution embodied in w, and the trial moves. For an exchange algorithm, we will consider two separate systems a and b combined (conceptually) in a single larger system. That is, we choose $\mathbf{x} = (\mathbf{r}_a^N, \mathbf{r}_b^N)$ and

$$w\left(\mathbf{r}_a^N, \mathbf{r}_b^N\right) = \exp\left[-U\left(\mathbf{r}_a^N\right)/k_B T_a\right] \exp\left[-U\left(\mathbf{r}_b^N\right)/k_B T_b\right], \tag{12.5}$$

where U is the usual potential energy of a molecular system but $T_a < T_b$ are two different temperatures. Trial moves will be of the exchange form, with $\mathbf{x}_o = (\mathbf{r}_a^N, \mathbf{r}_b^N)$ and $\mathbf{x}_n = (\mathbf{r}_b^N, \mathbf{r}_a^N)$, so that the two configurations are swapped in the $\mathbf{x}$ vector.

In fact, the above describes the famous "replica exchange" algorithm for two temperatures. Two independent simulations are run at the temperatures T_a and T_b according to any algorithm consistent with a Boltzmann factor distribution (e.g., MD). The combined system of the two simulations will indeed exhibit the distribution (12.5). To aid sampling, however, occasional swaps between the configurations of the two systems can be attempted, with the acceptance criterion as described in the problem below. The acceptance criterion, moreover, can be used in a replica-exchange simulation of any number of temperatures run in parallel, so long as configuration-swaps are always performed in a pairwise manner.

In practical terms, an exchange simulation will only be useful if the two levels (i.e., temperatures) are close enough so that exchange trial moves are indeed accepted. Typical replica exchange simulations will consist of many replicas at a series of increasing temperatures, as shown in Figure 12.6. Even when many levels are included, only pairwise exchanges need to be attempted, which will ensure overall correct sampling.

PROBLEM 12.8

Derive the acceptance criterion appropriate to the distribution (12.5), when configuration-swap trial moves are symmetrically generated. Consult the literature or a simulation textbook to confirm your result.

12.4.3.1 The Beyond: Hamiltonian Exchange and Resolution Exchange

Once the basic idea of replica exchange is well understood, it's not hard to see how to generalize other situations. For instance, in replica exchange, the same potential function is typically used for all temperatures, but this is not strictly necessary. Thus,

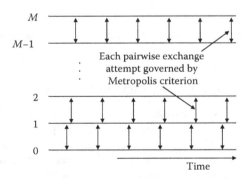

FIGURE 12.6 A ladder of "replicas" in exchange simulation. In the most common type of replica-exchange simulation, each level represents a different temperature—$T_0 < T_1 < \cdots < T_M$—with the hope that higher temperature permits faster barrier crossing. The exchange criterion described in the text governs pairwise exchange attempts. As in ordinary MC, if these "local" exchanges are handled correctly, the global distribution is correct. The weakness of exchange simulation is that, by definition, every level of the ladder has a different distribution that may not contribute significantly to the distribution of interest—for example, to the target temperature T_0 in temperature-based replica exchange.

we can generalize the distribution (12.5) to depend on arbitrary "Hamiltonians" or potential energy functions:

$$w\left(\mathbf{r}_a^N, \mathbf{r}_b^N\right) = \exp\left[-U_a\left(\mathbf{r}_a^N\right)/k_B T_a\right] \exp\left[-U_b\left(\mathbf{r}_b^N\right)/k_B T_b\right], \quad (12.6)$$

where U_a and U_b are now different energy functions. The temperatures T_a and T_b could be the same or different. For instance, if U_a is the desired potential, one could try to design U_b so that it is smoother or less expensive to calculate. These goals can be achieved in many ways—for instance, by adjusting the van der Waals interaction terms or by eliminating long-range electrostatics.

As in temperature-based replica exchange, the fraction of accepted moves will be very sensitive to the choices of the potential functions and temperatures. Configurations in one ensemble must have at least modest "importance" (i.e., reasonably significant Boltzmann factors) in the opposite ensemble. This is the requirement for "overlap," which is common to many advanced algorithms.

We can also derive "resolution exchange" simulation, in which systems with different numbers of degrees of freedom are run in parallel. When fewer degrees of freedom are used to model a given system, this implies lower resolution: for example, one could simulate both all-atom butane (C_4H_{10}) and a simplified low-resolution version with only four "carbon" atoms. This approach is implemented by splitting the full set of coordinates for the detailed system "a" into a coarse subset Φ_a and the remainder ξ_a, so that $\mathbf{r}_a^N = (\Phi_a, \xi_a)$. The coarse system "b" is defined to only depend on the coordinates Φ_b, implying the full set of coordinates for the combined a-b system is $\mathbf{x} = (\Phi_a, \xi_a, \Phi_b)$. The corresponding distribution is then

$$w(\mathbf{x}) = w(\Phi_a, \xi_a, \Phi_b) = \exp\left[-U_a(\Phi_a, \xi_a)/k_B T_a\right] \exp\left[-U_b(\Phi_b)/k_B T_b\right], \quad (12.7)$$

where the temperatures of the two systems, T_a and T_b, are arbitrary. Trial moves consist of swapping only the common Φ variables between the systems: $\Phi_a \leftrightarrow \Phi_b$.

PROBLEM 12.9

Derive the "resolution exchange" acceptance criterion and check the literature to see that you have obtained the correct result.

In the literature, you can find further examples of creative uses of the Monte Carlo idea.

12.5 ANOTHER BASIC METHOD: REWEIGHTING AND ITS VARIATIONS

The reweighting idea is even simpler than Metropolis MC, and it also has many applications. Assume, as usual, that we wish to generate configurations $\mathbf{r}^N$ distributed according to the Boltzmann factor—that is, $w(\mathbf{r}^N) = e^{-U(\mathbf{r}^N)/k_B T}$. Suppose further, however, that we are only able to generate configurations according to some other distribution w'. Thus, in a "volume" of configuration space $d\mathbf{r}^N$, we will generate a number of configurations proportional to $w'(\mathbf{r}^N)d\mathbf{r}^N$, although we want $w(\mathbf{r}^N)d\mathbf{r}^N$. Therefore, to correct the ensemble, we can assign a weight "wt" to each configuration according to

$$\text{wt}(\mathbf{r}^N) = \frac{w(\mathbf{r}^N)}{w'(\mathbf{r}^N)}. \qquad (12.8)$$

Note that when this weight is multiplied by the actual distribution sampled, w', we obtain the desired distribution w.

PROBLEM 12.10

Consider an ordinary six-sided cubic die painted with the numbers 1, 2, 3, 4, 4, 4—that is, where "4" occurs three times. (a) Write down the discrete distribution $w'(j)$ for $j = 1, 2, 3, 4$. (b) Suppose that you actually want to sample a uniform distribution of the $j \leq 4$ values, with $w(j) = $ constant. What weight should be assigned to each value to recover a uniform distribution?

A simple example of reweighting for physical systems is to use a sampling or generating distribution at some different temperature, T', so that $w'(\mathbf{r}^N) = \exp\left[-U(\mathbf{r}^N)/k_BT'\right]$. Another alternative is to use a different potential energy function, possibly at a different temperature. It may be desirable to use a less expensive energy function and/or a higher temperature to facilitate sampling, in some cases. In other words, just with the simple formulation (Equation 12.8), there are already a host of possibilities.

From a practical standpoint, however, the method will yield very noisy data—if not useless data—unless the two distributions w' and w are very similar. This is another way of saying that the configurations generated in the w' ensemble must be important in the w ensemble. Once again, we have the requirement for "overlap" (see Figure 12.7).

12.5.1 REWEIGHTING AND ANNEALING

One interesting way to use the reweighting idea is in the context of "annealing," which means slow cooling. Thus, you can imagine first generating an ensemble at a high temperature T_M and then reweighting it into successively lower temperatures $T_{M-1}, T_{M-2}, \ldots$. This is closely related to the replica exchange scheme of Figure 12.6. This simple scheme, as it turns out, is equivalent to reweighting directly from the maximum to the minimum temperature—and hence one expects a severe overlap problem if the extreme temperatures are very different.

There is a simple but practical solution to the problem in this case. In an algorithm proposed by Huber and McCammon and later by Neal, one alternates reweighting and "relaxation" stages in which each configuration in the ensemble is simulated at the (constant) current temperature. That is, after the temperature is changed to, say, T_j and the weights calculated according to Equation 12.8, then some kind of dynamics

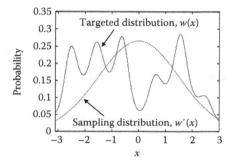

FIGURE 12.7 Reweighting from one distribution to another. The targeted distribution $\rho(x) \propto w(x)$ may differ significantly from the distribution used to sample it, $w'(x)$. If few configurations from w' are important in w, based on a necessarily finite sample, the reweighted configurations may not be valid—that is, there may not be sufficient "overlap" between the distributions. Ideally, one designs w' to be as similar to w as possible.

or MC simulation at temperature T_j is launched from each configuration and run for a short time. The modified ensemble consists of the final relaxed configurations—one from each short simulation—along with the weights. In this way, the ensemble becomes more adapted to each successively lower temperature, ameliorating the overlap problem. Lyman and Zuckerman showed this procedure can be applied to biomolecular systems.

12.5.2 POLYMER-GROWTH IDEAS

The reweighting idea can be used to generate ensembles of molecular configurations without any kind of dynamics at all—not even MC. The ideas come from algorithms for growing polymers. See, for instance, the paper by Garel and Orland.

In our case, imagine that the full set of molecular coordinates $\mathbf{r}^N$ is divided into a set of n molecular fragments, so that $\mathbf{r}^N = (\mathbf{s}_1, \mathbf{s}_2, \ldots, \mathbf{s}_n)$, with $\mathbf{s}_i$ being the coordinates for fragment i. Also assume an ensemble of each fragment has been prepared in advance, with configurations distributed according to the Boltzmann factor of U_i for each fragment i. The potential U_i will be assumed to correspond to all interactions among atoms internal to fragment i. If we consider the configuration space of our full system and imagine selecting one configuration at random from each fragment ensemble, we will have configurations for the whole molecule distributed according to

$$w'(\mathbf{r}^N) = w(\mathbf{s}_1)\, w(\mathbf{s}_2) \ldots w(\mathbf{s}_n), \qquad (12.9)$$

where $w(\mathbf{s}_i) = \exp[-U_i(\mathbf{s}_i)/k_B T]$ is the Boltzmann factor for fragment i.

On the other hand, we really want configurations distributed according to the Boltzmann factor of the full potential $U(\mathbf{r}^N)$, which is given by the fragment terms, plus all interactions terms U_{ij} among fragments i and j:

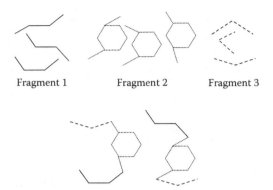

FIGURE 12.8 Polymers constructed by fragments. A library of each molecular fragment is generated in advance. For example, fragments can correspond to amino acids. The fragment configurations can be assembled into full molecular configurations, which must be reweighted to account for interactions among fragments.

$$U(\mathbf{r}^N) = U(\mathbf{s}_1, \ldots, \mathbf{s}_n) = \sum_i U_i(\mathbf{s}_i) + \sum_{i<j} U_{ij}(\mathbf{s}_i, \mathbf{s}_j). \quad (12.10)$$

As usual, the corresponding desired distribution is specified by $w(\mathbf{r}^N) = \exp[-U(\mathbf{r}^N)/k_B T]$.

Now that the two distributions w and w' have been specified, one can perform reweighting using one of the schemes we have discussed. Importantly, one does not have to add all the U_{ij} interactions at once. Instead, any set of terms can be added at a time, perhaps followed by annealing (Figure 12.8).

12.5.3 REMOVING WEIGHTS BY "RESAMPLING" METHODS

In some situations, you may want to remove the weights attached to an ensemble and create an "ordinary" ensemble where each configuration is weighted equally. Indeed, in communicating with nonquantitative scientists or nonscientists, it could be very confusing to talk about configurations with different degrees of importance. There may also be instances where you find yourself with a weighted ensemble consisting of a very large number of configurations, which you simply want to reduce in size.

"Resampling" provides a way to remove weights from an ensemble—without altering the statistical distribution. (Although many sophisticated types of resampling are possible, particularly for multi-stage reweighting procedures, we shall only discuss the simplest approach.) Because the weight of a configuration is proportional to its probabilistic contribution to the ensemble, the idea is simply to select configurations with probabilities proportional to their weights. However, weights derived from Equation 12.8 can be bigger or smaller than one, so we first need to normalize the weight values. One therefore includes a configuration $\mathbf{r}_j^N$ in the unweighted ("resampled") ensemble if a new random number is less than $\mathrm{wt}\left(\mathbf{r}_j^N\right)/\max(\mathrm{wt})$, where $\max(\mathrm{wt}) = \max_j\left\{\mathrm{wt}\left(\mathbf{r}_j^N\right)\right\}$ is the largest weight among the whole weighted ensemble.

The book by Liu is a good starting point to learn resampling methods.

12.5.4   CORRELATIONS CAN ARISE EVEN WITHOUT DYNAMICS

We started this chapter with a discussion of dynamical simulations, noting that correlations persist along the sequence of the trajectory. In biomolecular systems, these correlations can be very long lived, possibly persisting throughout the whole trajectory. The situation is the same in MC simulations, where correlations also arise along the sequence of configurations.

Unfortunately, non-dynamical simulation methods also tend to induce correlations. Let's consider an example of "growing" a molecule from two fragments, following the ideas of Section 12.5.2. One could do this starting from 100 configurations of each fragment. However, if we make an ensemble of the full molecule based on all 10,000 possible combinations of the fragments, these configurations are correlated: after all, there are sets of 100 configurations that share an identical configuration for the first fragment. Such correlations will persist even after reweighting and/or resampling.

12.6   DISCRETE-STATE SIMULATIONS

In Chapter 10, we described molecular systems in terms of a small number of states and discussed the kinetic behavior among those states. It is also possible to describe systems using a large number of more structurally specific states—specifically, each state can correspond to a single energy basin. Of course, both energy and entropy are necessary to characterize the equilibrium probability of each basin. As studied in detail by David Wales, among others, if every minimum of a system can be identified, one can determine whether direct transitions are possible among every pair of states. The saddle points (i.e., barriers in a high-dimensional space) will determine which transitions are possible and provide the basis for estimating forward and reverse rates for each pair.

A "kinetic MC" simulation can be run based on the states and rates. In essence, one can assume a Poisson process (recall Chapter 2) for such single-basin transitions, with the average waiting time for a basin-to-basin transition given by the reciprocal of the corresponding rate constant. By calculating a series of waiting times and basins, a discrete-state trajectory is generated.

As has been noted by Wales, however, the number of basins is expected to increase exponentially with the number of atoms in a system. Therefore, such a description of a large protein would have to be employed with great care. On the other hand, if properly handled, the limitations of a discrete description might be considerably less than a more detailed continuum description.

12.7   HOW TO JUDGE EQUILIBRIUM SIMULATION QUALITY

Whenever an experiment is performed—whether numerically or in the laboratory—it is fundamental to the good practice of science to make an attempt to understand the shortcomings of the resulting data. It is notoriously difficult to obtain statistically reliable data from biomolecular simulations, as we have discussed already.

Here, we would like to sketch some of the basic issues involved in assessing sampling. For further details, you can consult the review by Grossfield and Zuckerman.

12.7.1 VISITING ALL IMPORTANT STATES

In brief, the basic goal of equilibrium simulation could be described as visiting all important states with the correct probabilities/populations. However, in general, it is impossible to know whether a simulation has visited all the important states. There may be states that have never been visited: imagine that the central barrier in Figure 12.2 may never be crossed. Sometimes one may have prior information (e.g., from experimental structures) about important states, but typically one doesn't know much more than what the trajectory "sees." Therefore, there is an intrinsic uncertainty in all molecular simulation data as to whether one or more states have been completely missed. The rhodopsin data in Figure 12.4 show why this is such a worry in atomistic simulations.

The analysis suggestions below thus all carry a caveat about self-consistency: one can analyze statistical quality of a simulation only by assuming all important states have been visited at least once.

12.7.2 IDEAL SAMPLING AS A KEY CONCEPTUAL REFERENCE

If you really want to understand simulation, don't skip this section. Although we will discuss a purely hypothetical calculation, the "ideal" case is critical for understanding the limits—and the possibilities—of molecular simulation.

Consider the case of idealized (perfect) equilibrium sampling, where the only goal is to generate an equilibrium ensemble distributed according to the Boltzmann factor. What would a perfect equilibrium simulation be like? We can imagine an ideal computer program that generates a sequence of configurations $\mathbf{r}^N$ completely independent of one another but distributed according to $\exp[-U(\mathbf{r}^N)/k_B T]$. Thus, these configurations do not embody the dynamics of the system, but only its equilibrium distribution. The configurations are completely uncorrelated. This is "ideal sampling," and it is a key reference point.

12.7.3 UNCERTAINTY IN OBSERVABLES AND AVERAGES

Consider the task of analyzing a dynamical trajectory or essentially dynamical data, such as from simple MC. Let's assume we want to calculate the average of some "observable" f—that is, we want to calculate $\langle f(\mathbf{r}^N) \rangle$ and also estimate the associated statistical error. The function f could be almost anything: the energy, the distance between two atoms, or the RMSD (root-mean-squared deviation) to a fixed reference configuration.

The first thing to do is to plot f as a function of time and inspect this "time trace" visually. You would like to see that f fluctuates many times about some mean value. If you can see that f is drifting, surely your estimate of $\langle f \rangle$ is not reliable. Recall the contrasting examples of butane and rhodopsin in Figures 12.3 and 12.4.

If indeed f exhibits the requisite (probably irregular) oscillations about some mean value, then you can examine its behavior in more detail. First, you can calculate the autocorrelation time as described in Chapter 4. This correlation time should be much shorter than your total trajectory length. Next, to get a precise estimate of the statistical uncertainty in $\langle f \rangle$, you probably want to use the statistical block-averaging technique as described in the review by Grossfield and Zuckerman. Block averaging uses estimates of $\langle f \rangle$ from different parts of the simulation and checks that they are statistically independent.

What if your data result from a non-dynamical method—perhaps from an exchange simulation or a polymer-growth technique? You must be extremely careful, as subtle correlations can arise within a given simulation. The safest course is to perform multiple independent simulations using initial conditions that are as different as possible. (If the results depend on the initial conditions, then you have not performed good equilibrium sampling.) Based on multiple independent estimates of $\langle f \rangle$, you can use the standard error of the mean described in Chapter 2.

12.7.4 OVERALL SAMPLING QUALITY

12.7.4.1 The RMSD Trajectory Can Give a Global View

One of the most commonly measured properties in a biomolecular simulation is the RMSD of the simulated configurations as a function of time. We will define RMSD mathematically below, but the first point about it is to take the name seriously: the RMSD is a deviation or, in simpler language, a distance. That is, given any configuration in a trajectory, you cannot measure the absolute RMSD, but rather the RMSD from some reference structure. The RMSD is typically measured from the initial structure of a simulation, which itself likely was based upon an experimental structure.

Mathematically, the RMSD has a bit of an unusual definition, because it involves minimization of itself. Carefully examine the definition of RMSD for a configuration $\mathbf{r}^N = (\mathbf{r}_1, \mathbf{r}_2, \ldots, \mathbf{r}_N)$ of an N-atom system, as compared to the reference configuration $\mathbf{r}^N_{ref} = (\mathbf{r}^{ref}_1, \mathbf{r}^{ref}_2, \ldots)$:

$$\text{RMSD}\left(\mathbf{r}^N, \mathbf{r}^N_{ref}\right) = \min \left\{ \left[\frac{1}{N} \sum_{i=1}^{N} |\mathbf{r}_i - \mathbf{r}^{ref}_i|^2 \right]^{1/2} \right\}, \quad (12.11)$$

where the minimization is performed over all possible "alignments" of the two configurations $\mathbf{r}^N$ and $\mathbf{r}^N_{ref}$—that is, over all possible relative translations and rotations. In other words, the RMSD is defined to be the deviation based on the best possible alignment.

The very first step in your overall analysis of a dynamical or simple MC simulation should be to plot the RMSD as a function of time. As with any other observable, you would like to see that the RMSD clearly fluctuates repeatedly about a mean value. Of course, some transient "relaxation" period is inevitable, but the RMSD should not show any overall drift as a function of time after the initial transient. By now you should recognize that any drift rules out reliable equilibrium sampling.

12.7.4.2 Quantitative Analysis of Dynamical Trajectories

As just described, the examination of the RMSD trajectory is a qualitative analysis, which is good at ruling out good sampling. What if your trajectory passes the RMSD visual test and you want more quantitative information? Specifically, although your trajectory likely contains millions or billions of time steps, how many of the "frames" of your movie are statistically independent?

A correlation time (Section 4.7) can reveal how long it takes the trajectory to forget its history and generate a new statistically independent configuration. However, the problem with conventional autocorrelation times for specific observables—for example, for the energy or a certain angle—is that they might not be sensitive to important slow timescales. Look again at Figure 12.3. In assessing the overall sampling quality, one would like to assay an observable that reports on the global quality of a simulation. Lyman and Zuckerman proposed a method for doing just this, essentially by using an observable that reflects the overall configuration-space distribution, as encoded in the populations of small regions of configuration space. The resulting "structural decorrelation time," when compared to the total simulation time, can give a global characterization of sampling quality. The analysis applies to dynamics or simple MC simulations.

12.7.4.3 Quantitative Assessment of Arbitrary Simulation Methods

Some simulation methods are entirely non-dynamical and others, such as exchange methods, will exhibit highly nontrivial time correlations. Therefore, an analysis method is required that does not probe time-dependent behavior.

To see what must be done in such cases, we should think back to the basic goal of equilibrium sampling—namely, to generate the populations/probabilities of all important physical states. This statement is not quite as casual as it sounds: because the state populations reflect partition functions internal to the states (Chapter 3), one cannot obtain the correct state populations without good sampling within each state. To put that another way, if good state populations have been obtained, then the more local sampling within states must have been adequate. In terms of dynamical timescales, it is generally much easier to sample within a state, which is a fast process, as compared to the much slower process of repeatedly visiting different states to probe their relative populations.

While we cannot go into great detail here, we shall point out how an analysis can be performed if (approximate) physical states have been established for the system being simulated. As discussed earlier, for non-dynamical methods, multiple independent simulations should be run from widely different starting configurations. This is the only way to remove correlations as much as possible. From each independent simulation, the set of state populations should be estimated. The variance in these different estimates for each state reflects the sampling quality.

12.8 FREE ENERGY AND PMF CALCULATIONS

There are several different types of free energy calculations, and we will touch briefly on the most important ones. As in the case of equilibrium sampling methods, there are simply too many specific approaches to detail. At least one entire book, in fact, has been devoted to free energy calculations—edited by Chipot and Pohorille. Therefore, we will try to convey only the main qualitative aspects of free energy calculations.

Here, we will refer to the free energy by F, assuming a constant-volume calculation, but the discussion carries over directly for constant-pressure calculations of G. The differences between these free energies were described in Chapter 7. We will also focus solely on free energy differences and the related PMF. Although it is possible to estimate absolute free energies, that is, partition functions, this is a more advanced topic for which the reader should consult the literature.

You should recall from Chapter 6 that the PMF is properly understood as a conditional free energy or a free energy function of specified coordinates. The centrality of the PMF in describing biomolecular behavior makes it important to consider computational means for estimating a PMF.

12.8.1 PMF AND CONFIGURATIONAL FREE ENERGY CALCULATIONS

To understand the basics of estimating free energies for different states—or the PMF for any arbitrary coordinate—think back to the reason we have been studying statistical mechanics in the first place. If our model (potential energy function) is good, statistical mechanics should be able to predict the observed populations of various configurational states. As we know from Chapter 3, a free energy difference is simply related to the logarithm of the observed population ratio. For instance, for two states A and B with population ratio p_A/p_B, we have

$$F_A - F_B = -k_B T \ln\left(\frac{p_A}{p_B}\right). \tag{12.12}$$

Put another way, if your simulation can sample relative populations, then by definition, it can also determine free-energy differences.

A similar characterization applies to the PMF (Chapter 6). Assume a histogram from a simulation determines the distribution along some coordinate y to be proportional to $w(y)$. We can again use the logarithmic relation to determine the PMF according to

$$\mathrm{PMF}(y) = -k_B T \ln w(y) + \mathrm{const.,} \tag{12.13}$$

where the additive constant is irrelevant, as usual, and may be set to 0. Recall that a PMF can describe any set of "ordinary" coordinates or an arbitrary function of the coordinates, such as the RMSD from some reference configuration. Also, for some more complex cases such as the calculation of a radial distribution function, the histogram w may require special normalization, as discussed in Chapter 6.

If our simulation exhibits the correct populations of the various states, then we need only use the relation (12.12) to determine all the pertinent free energy differences—or Equation 12.13 for the PMF.

12.8.1.1 Advanced Tricks for Configurational Free Energy Calculations

There are many specialized techniques for calculating configurational free energies and PMFs. The main reason such "tricks" are necessary is the usual sampling problem—which is somewhat accentuated from the free energy perspective. Consider a complex landscape such as in Figure 12.2, which will contain regions of both low and high probability, with low-probability barriers between basins of special interest. In an ordinary sampling scheme, with configurations generated in direct proportion to the Boltzmann factor, the barriers will be visited infrequently. As a consequence, the free energy (or PMF) associated with these barriers will be imprecisely determined relative to the basins, regardless of the sampling quality. Thus, the primary goal of specialized configurational free energy techniques is to add precision to the probability estimates of less-populated regions, although the methods may also attempt to accelerate overall sampling.

The essence of the means by which rarer regions can be sampled with better precision lies in the reweighting idea (Section 12.5) (see also Figure 12.7). Reweighting uses one distribution to sample another, so for free energy/PMF calculations, the goal would be to employ alternative distributions that sample rarer regions more frequently. This can be done in a "global" way by increasing the temperature, for instance. If there is a particular coordinate of interest, special restraining potentials along that coordinate can be used—see the work by Kumar et al. For example, to ensure sampling near a barrier at the value x_b of the coordinate x, an extra term of the form $(k/2)(x - x_b)^2$ can be added to the physical potential energy $U(x, y, \ldots)$. To compute true free energies, reweighting must be used to correct for such biasing.

12.8.2 THERMODYNAMIC FREE ENERGY DIFFERENCES INCLUDE ALL SPACE

By definition, configurational free energy differences describe population differences in different conformational states or regions of configuration space. However, it is often of interest to calculate more "traditional" free energy differences, defined based on full partition functions—that is, integrations over all configuration space. Such free energy changes are pertinent, for instance, to the specific heat measured in calorimetry (Section 7.8) or for relative-binding affinity calculations (Section 9.4.3).

We are thus interested in free energy differences resulting from changes in either the temperature or the potential energy function. Either kind of change can be described conveniently by the dimensionless energy

$$U^*(\mathbf{r}^N, T) \equiv \frac{U(\mathbf{r}^N)}{k_B T}. \tag{12.14}$$

In general, then, we are interested in calculating free energy differences resulting from changing the initial dimensionless energy function U_0^* into some other function U_1^*. If we use the old trick of writing $U_1^* = U_0^* + \Delta U^*$, then the resulting free energy difference is given by

$$e^{-(F_1-F_0)/k_B T} = \frac{\int d\mathbf{r}^N \exp\left(-U_1^*\right)}{\int d\mathbf{r}^N \exp\left(-U_0^*\right)} = \frac{\int d\mathbf{r}^N \exp\left[-\left(U_0^* + \Delta U^*\right)\right]}{\int d\mathbf{r}^N \exp\left(-U_0^*\right)}$$

$$= \langle \exp\left(-\Delta U^*\right)\rangle_0, \tag{12.15}$$

where the subscript "0" indicates an average calculated based on the U_0^* ensemble.

12.8.2.1 Advanced Tricks for Thermodynamic Free Energy Calculations

There are once again many different algorithms for calculating free energy differences defined by Equation 12.15. Almost all of thermodynamic methods involve some kind of "staging" approach in which intermediate energy functions U_λ^* are introduced at a set of λ values between 0 and 1—for example, $\lambda = 0.1, 0.2, \ldots, 0.9$. Most simply, using a "thermodynamic integration" strategy, one can write the desired free energy difference as a sum of incremental differences: $F_1 - F_0 = (F_1 - F_{0.9}) + (F_{0.9} - F_{0.8}) + \cdots + (F_{0.1} - F_0)$. Each incremental free energy difference is calculated independently using the appropriate pair of λ values substituted for 0 and 1 in Equation 12.15; hence, a series of independently simulated equilibrium ensembles at every λ value is required.

Many approaches attempt to improve on the basic thermodynamic-integration strategy. One can use a strategy based on "annealing" (Section 12.5.1), whereby one generates the series of intermediate ensembles based on the (reweighted) configurations from the preceding λ. Finally, we should also mention the now-famous nonequilibrium methods developed based on the work of Chris Jarzynksi (Section 11.6), which only require an equilibrium ensemble at one endpoint (e.g., $\lambda = 0$) but not at intermediate λ values. The paper by Neal explains the fascinating relation between Jarzynski-type methods and annealing.

For further information on a variety of free energy methods, see the books by Frenkel and Smit, and by Chipot and Pohorille.

12.8.2.2 On Binding Free Energy Calculations

The theory of the binding free energy (difference), ΔG_0^{bind}, was explained in detail in Chapter 9, as were basic elements of a thermodynamic-cycle strategy for calculating relative affinities—that is, differences in ΔG_0^{bind} values. As a reminder, although ΔG_0^{bind} is a free energy difference, it is defined with reference to a standard but unphysical state. Therefore, the relative affinities $\Delta\Delta G^{\text{bind}}$ have the most direct physical meaning.

As discussed in Chapter 9, the calculation of $\Delta\Delta G^{\text{bind}}$ is typically done using the "alchemical" thermodynamic cycle originally proposed by Tembe and McCammon. The key point is that the required alchemical free energy differences are precisely of the "thermodynamic" form (12.15). The necessary alchemical changes are described as changes to the forcefield (i.e., potential energy function) and therefore correspond to defining suitable U_0^* and U_1^* functions. As a simple example, one atom type can be changed to another simply by changing the van der Waals, electrostatic, and angle parameters. In a classical forcefield, it is these parameters that actually define the atom type.

We note that it is also possible to calculate ΔG_0^{bind} directly. This complex topic is beyond the scope of this chapter, and the reader is referred to articles by Gilson et al., and by Woo and Roux.

Finally, an overall warning. It's clear that the binding affinity depends sensitively on atomic details such as shape complementarity, electrostatic and H-bonding interactions, as well as solvation effects (recall Section 9.3). Hence, the use of an atomistically detailed model—for example, a standard MM forcefield—is almost a requirement. At the same time, the necessary free energy calculations depend intrinsically on having an equilibrium ensemble of the system (see Equation 12.15). We have already discussed the current near-impossibility of obtaining atomistic equilibrium ensembles of protein systems, except perhaps based on extraordinary resources. This would seem to rule out reliable binding affinity estimates and is hard to dispute. However, one hope underlying many relative affinity computations is that similar errors due to undersampling will cancel out when two ΔG_0^{bind} values are subtracted to yield $\Delta\Delta G^{\text{bind}}$.

12.8.3 APPROXIMATE METHODS FOR DRUG DESIGN

This book has focused on a careful treatment of statistical mechanics, and this chapter has similarly described rigorous statistical calculations. Nevertheless, numerous approximate calculations are of value—and commonly used—in the field of drug design. The reader is referred to the book by Leach for an introduction to some of these methods.

The rough estimation of binding affinities using "docking" methods is a key example. The process of docking typically involves an MC style algorithm for fitting a molecule into a protein-binding site. Random trial moves of the ligand are evaluated based on an energy function. Different ligand conformations may also be tested, and occasionally different receptor conformations are also used. If a reasonable conformation and energy value for the ligand–receptor complex can be reached, that same energy is often used as a docking "score"—that is, an approximate binding free energy. The advantage of such a relatively simple, nonstatistical calculation is that large numbers of molecules can rapidly be tested. In an industrial drug design setting, performing docking and related calculations are seen as a way of "screening" a large library of compounds for an initial list that are worth investigating experimentally.

It is also very interesting to note that "virtual screening" of ligands for a certain target can be performed without knowing the structure of the target receptor. The idea is simple. If you know the structures of a number of ligands that bind a certain target, you can then devise a quantitative description of the common features that seem to enable binding—that is, devise a "pharmacophore." For example, such an analysis might reveal an aromatic group and an H-bond acceptor separated by a certain distance, with a hydrophobic group at another location. A large library of chemical structures can then be rapidly tested to see if there are matches to the specified pharmacophore. This type of analysis is just one example from a broad class of methods employing "quantitative structure–activity relationships" (QSAR), which may include information about the receptor structure if it is available.

The reader should be aware that, beyond estimating binding affinities, there are numerous other aspects to the drug design process. For instance, a candidate molecule may bind strongly to the desired protein target, but this is of little value if the molecule is otherwise toxic to the cell or perhaps rapidly metabolized (chemically decomposed) by the cell. Another complex issue is the molecule's ability to reach and enter the target cell type in a specific tissue. The interested reader should consult the voluminous literature on drug design and medicinal chemistry.

12.9 PATH ENSEMBLES: SAMPLING TRAJECTORIES

In Chapter 11, we emphasized the ensemble view of dynamics. That is, any dynamical process can happen in myriad ways, as shown by Czerminski and Elber. Observed behavior will reflect an ensemble of dynamical trajectories. In analogy to the description of configurations in equilibrium, each trajectory can be weighted with a probability based on a Boltzmann-like factor (11.1). Here, we are interested in paths starting from state A and ending in state B after traversing a complex energy landscape.

It should not be surprising that if a trajectory can be assigned a weight, then we can perform a simulation that generates an ensemble of trajectories (or, informally, "paths"). While this is a very advanced topic, the basic ideas are quite interesting and can be expressed in the same language we have been employing until now.

Before discussing the statistical sampling of paths, it is worth mentioning that many methods have been developed for optimizing paths—that is, searching for the path with the highest probability. Such optimization methods typically start from an initial guess for the pathway, as in energy minimization. However, it would seem that making a first attempt to sample paths—perhaps in a reduced model—could greatly improve the chances of finding an optimal path in a complex landscape. There are also methods for generating approximate thermal paths between pre-specified states A and B, but we are not reviewing such nonstatistical approaches.

12.9.1 THREE STRATEGIES FOR SAMPLING PATHS

The first path-sampling method for molecules, suggested by Lawrence Pratt, is to run an MC simulation in the space of trajectories. Thus, given the current ("old") trajectory, one modifies it to obtain a trial ("new") trajectory, which is then accepted or rejected based on a Metropolis criterion derived from path weights. Numerous papers using this method have been published; see the review by Bolhuis et al.

A second method is a simple variant on the reweighting idea of Section 12.5. In this case, as shown by Zuckerman and Woolf, one can bias the dynamics of a system to go from A to B—that is, one generates trajectories from an alternative distribution to increase the likelihood of transitioning to B. The resulting trajectories must be reweighted using an analog of Equation 12.8 for paths. Because of the (extremely) high dimensionality of the path space, however, it is difficult to attain good overlap with the unbiased distribution except in special cases. Nevertheless, the reweighting approach is useful in generating an initial, highly diverse set of paths.

A third approach, developed by Huber and Kim, implicitly uses polymer growth ideas. This "weighted ensemble" strategy divides configuration space into several regions and initializes a number of trajectories, say m of them, from some starting configuration or distribution. Each trajectory has an equal weight in the ensemble initially, and is therefore assigned a probability $1/m$. All m trajectories are run for a time and then examined. If one reaches a new region, that trajectory is split into m identical copies ("daughters"), which share the probability of the "mother" trajectory—that is, each daughter has a weight $1/m^2$, preserving the total probability of 1. This process is repeated: the trajectories are all run further, then checked, and perhaps split. In this manner, weighted-ensemble simulations can yield unbiased transition trajectories, as well as the configuration-space probability distribution evolving with time; the rate can be calculated from the latter. See the article by Zhang et al. Mathematically, the trajectories can be viewed as polymers by equating a time step and a polymer segment (monomer). The idea to "split" growing polymers, computationally, into multiple copies is quite old. Related ideas were explored in the "forward flux method" developed by Allen and coworkers.

12.10   PROTEIN FOLDING: DYNAMICS AND STRUCTURE PREDICTION

We all know the "protein-folding problem" is very famous, but it is important to realize there are two folding problems. The first is better termed the structure prediction problem: Given a protein's sequence of amino acids, what is its three-dimensional structure? (Since we know statistical mechanics, we might rephrase this as, "What are the most important configurational states and their relative populations?") The most common approaches to structure prediction use "homology modeling"—that is, they start from experimental structures of a protein with a similar sequence. This is far outside the realm of statistical mechanics and will not be discussed here.

Structures are also predicted using physical principles and molecular simulation. Most commonly, CG models are used, since these permit the greatest exploration of configuration space. Some methods use a combination of CG modeling and information from existing structures. Either way, limitations in the CG model accuracy will necessarily affect the precision with which structures can be predicted.

The second protein-folding problem is the issue of how proteins fold. What types of pathways and intermediates tend to be exhibited in general? Given a specific system, what are the specific intermediates and rate constants? The reader is referred again to Chapters 10 and 11 as well as the article by Snow et al.

The "how" of protein folding is often addressed with CG models. One of the most famous approaches, due to Lau and Dill, treats proteins as consisting of just two types of amino-acid "beads": hydrophobic or hydrophilic. A model simplified to two bead-types clearly is aimed at characterizing general features of the folding process. Intermediate-resolution models can be used to probe the folding processes of specific proteins.

It should seem almost obvious to readers that path-sampling algorithms may also aid the investigation of folding. This was indeed shown by Juraszek and Bolhuis, who used transition path sampling to study detailed folding trajectories of a small protein.

Other path-sampling methods should also be applicable. Generally speaking, better algorithms will permit better sampling and hence better models.

12.11 SUMMARY

Many of the most important ideas in biomolecular simulation can be described concisely. First, simulations are planned based on a balance of computational resources and model detail. If statistical mechanical conclusions are desired for a large system, then a less detailed model must be chosen that permits full sampling. Simpler, "CG" models require dramatically less computation—and often are used for slow processes like protein folding and conformational transitions.

Sampling itself can be performed in myriad ways, the most basic of which mimic the motions a molecule would undergo naturally. Such dynamical simulation, equivalent to making a movie, must be run long enough for the movie to depict all the important motions several times over. In general, good sampling of atomistic protein models is impossible with resources typically available in 2009. Although hardware capabilities are rapidly advancing, even supercomputer simulations can be far from adequate, as described by Grossfield et al. It is therefore important to ascertain the statistical reliability of any calculated "observables," as well as of the overall sampling. The ideas of correlation times and sample size are central to assessing sampling quality.

A number of more advanced topics were broached to give the reader a sense for the research side of the field. Based on just two algorithmic "devices"—namely, the Metropolis algorithm and reweighting—a large variety of sampling procedures can be constructed. These advanced methodologies can employ a range of temperatures, models of differing resolutions, and assembly of molecular fragments. Path-sampling strategies were also sketched briefly, as were specialized applications to free energy estimation and protein folding.

No single chapter could encapsulate decades of progress in the field of biomolecular simulation, but hopefully the reader has been made to appreciate basic—often unspoken—ideas of computation. No attempt has been made to provide detailed recipes for implementing the approaches that have been sketched. Rather, the references below should prove a valuable resource for the practical aspects of computation.

FURTHER READING

Allen, R.J, Warren, P.B., and ten Wolde, P.R., *Physical Review Letters*, 94:018104, 2005.

Ayton, G.S., Noid, W.G., and Voth, G.A., *Current Opinion in Structural Biology*, 17:192–198, 2007.

Bolhuis, P.G., Chandler, D., Dellago, C., and Geissler, P., *Annual Review in Physical Chemistry*, 59:291–318, 2002.

Chipot, C. and Pohorille, A. (editors), *Free Energy Calculations*, Springer, Berlin, Germany, 2007.

Czerminski, R. and Elber, R., *Journal of Chemical Physics*, 92:5580–5601, 1990.

Frantz, D.D., Freeman, D.L., and Doll, J.D., *Journal of Chemical Physics*, 93:2769–2784, 1990.

Frenkel, D. and Smit, B., *Understanding Molecular Simulation*, 2nd edition, Academic Press, Amsterdam, the Netherlands, 2001.

Garel, T. and Orland, H., *Journal of Physics A*, 23:L621, 1990.

Gilson, M.K., Given, J.A., Bush, B.L., and McCammon, J.A., *Biophysical Journal*, 72: 1047–1069, 1997.

Grossfield, A. and Zuckerman, D.M., *Annual Reports in Computational Chemistry*, 5:23–48, 2009.

Grossfield, A., Feller, S.E., and Pitman, M.C., *Proteins*, 67:31–40, 2007.

Huber, G.A. and Kim, S. *Biophysical Journal*, 70:97–110, 1996.

Huber, G.A. and McCammon, J.A., *Physical Review E*, 55:4822–4825, 1997.

Juraszek, J. and Bolhuis, P.G., *Proceedings of the National Academy of Sciences, USA*, 103:15859–15864, 2006.

Kumar, S., Bouzida, D., Swendsen, R.H., Kollman, P.A., and Rosenberg, J.M., *Journal of Computational Chemistry*, 13:1011–1021, 1992.

Lau, K F. and Dill, K.A., *Macromolecules*, 22:3986–3997, 1989.

Leach, A., *Molecular Modelling*, Prentice Hall, Harlow, U.K., 2001.

Liu, J.S., *Monte Carlo Strategies in Scientific Computing*, Springer, Berlin, Germany, 2001.

Lyman, E. and Zuckerman, D.M., *Journal of Physical Chemistry B*, 111:12876–12882, 2007.

Lyman, E. and Zuckerman, D.M., *Journal of Chemical Physics*, 127:065101, 2007.

Lyman, E., Ytreberg, F.M., and Zuckerman, D.M., *Physical Review Letters*, 96:028105, 2006.

Neal, R.M., *Statistics and Computing*, 11:125–139, 2001.

Pratt, L.R., *Journal of Chemical Physics*, 85:5045–5048, 1986.

Senn, H.M. and Thiel, W., *Topics in Current Chemistry*, 268:173–290, 2006.

Snow, C.D., Sorin, E.J., Rhee, Y.M., and Pande, V.S., *Annual Reviews in Bio-Physics and Biomolecular Structure*, 34:43–69, 2005.

Swendsen, R.H. and Wang, J.-S., *Physical Review Letters*, 57:2607–2609, 1986.

Tembe, B.L. and McCammon, J.A., *Computers and Chemistry*, 8:281–283, 1984.

Tozzini, V., *Current Opinion in Structural Biology*, 15:144–150, 2005.

van Gunsteren, W.F. et al., *Angewandte Chemie International Edition*, 45:4064–4092, 2006.

Voth, G.A. (editor), *Coarse-Graining of Condensed Phase and Biomolecular Systems*, CRC Press, Boca Raton, FL, 2009.

Wales, D.J., *Energy Landscapes*, Cambridge University Press, Cambridge, U.K., 2003.

Woo, H.J. and Roux, B., *Proceedings of the National Academy of Sciences, USA*, 102:6825–6830, 2005.

Zhang, B.W., Jasnow, D., and Zuckerman, D.M., *Proceedings of the National Academy of Sciences, USA*, 104:18043–18048, 2007.

Zuckerman, D.M. and Woolf, T.B., *Journal of Chemical Physics*, 111:9475–9484, 1999.

Index

325

Boltzmann factor, 8, 28, 57–60, 62–64, 66–71, 73,
    76–78, 81, 85, 90, 101, 116, 117,
    124–127, 132, 137, 138, 140, 141, 145,
    146, 150, 151, 154–156, 164, 165, 169,
    171, 180, 184, 199, 204, 213, 219, 221,
    259, 260, 265, 267–270, 273, 274, 290,
    294, 300, 305, 306, 308, 309, 311,
    314, 318
Butane
    computer simulation trajectory, 7–8
    correlated sampling, 32
    distribution, 26
    energy landscape, 8
    transition path ensemble, 277–278

C

Calculus, 7, 24, 26, 28, 36, 117, 154,
    199, 274
Calorimetry, 182–184, 211–212, 318
Cartesian coordinates, 36, 111, 117, 119–120,
    123, 124, 127, 132, 139
Catalyst, 90, 91, 229
cdf, *see* Cumulative distribution functions
Chemical potential, 171–173, 184
Chevron plots, 263–264
Classical *vs.* quantum mechanics, 13, 106, 107,
    113, 119, 121, 292
Coarse-grained model, 298–300, 322, 323
Coarse graining (CG), 235–236, 299–300
Commitment probability, 280–281
Configuration, 5, 7, 11, 18, 58, 60, 65–66, 76, 77,
    80, 112, 114, 115, 123, 133–135
Conditional probability, 51–53, 55, 89, 93, 238,
    250, 271, 273, 307
Confidence intervals, 43
Configuration space, 5, 7, 18, 31, 43, 50, 78, 93,
    112, 114, 115, 117, 119, 133–135, 138,
    141, 199, 236, 237, 241, 251, 254,
    257–260, 264, 265, 267, 270, 275, 277,
    279–281, 284, 289, 291, 294, 302, 303,
    305, 310, 311, 316, 318, 322
Conformational change, 9, 211, 214, 223,
    235–266, 279, 282; *see also*
    Transitions
    kinetics, *see* Kinetic analysis
Constant-volume free energy (*F*),
    164–173
Convergence, *see* Sampling problems
Convolutions, 31, 99, 245, 246, 285, 286
    correlated sampling, 32
    diffusion, 39–40, 98, 285
    integral definition, 32–34
    variance of a sum, 39–40
Cooperativity, 222–229, 285

Coordinates and forcefields
    Cartesian coordinates, 36, 117, 119–120, 123,
        124, 127, 132, 139, 141, 284
    forcefield, 119–126, 129, 131, 132, 218, 219,
        299, 304, 305, 319, 320
    internal coordinates, 120–121
    Jacobian factors, 36, 65, 117, 119, 123–124,
        135, 145, 147, 203, 305
    modeling water, 122
    quantum mechanics, 122–123
Correlations, 144, 145
    complex, 48–50
    definition, 45
    linear, 46–48
    mutual information, 49–50
    physical origins, 50
    probability theory, 127–128
    time, 53, 315, 316, 323
    time and function, 103–107
    visual inspection, 48–49
Cumulative distribution functions (cdf), 26–28,
    31, 54
Current, 281–283, 288, 290

D

de Broglie wavelength (λ), 82
Debye–Hückel theory, 198, 200–202
Detailed balance, 94
Degeneracy number, 126–129, 131,
    133, 135
    vs. symmetry number, 129
Dielectric behavior, water
    constant, 193
    dipole review, 194
    PMF analogy, 194
    polarizability, 194–195
    protein charge, 196–197
    reorientation, 195–196
    water solvation, nonpolar
        environment, 196
Diffusion
    binding and, 100
    constant and square-root law, 98–100
    convolution, 39–40, 98, 285
    differential equation, 287–289
    diffusion-limited catalysis, 231
    ensemble dynamics, 265–294
    Fokker–Planck/Smoluchowski equation,
        289–291
    issues, 291–293
    linear landscape, 285–286
    probability distributions, 283
    protein, 1–3
    trajectory picture, 284–285

Printed in the United States
by Baker & Taylor Publisher Services